Lecture Notes in Energy

Volume 98

Lecture Notes in Energy (LNE) is a series that reports on new developments in the study of energy: from science and engineering to the analysis of energy policy. The series' scope includes but is not limited to, renewable and green energy, nuclear, fossil fuels and carbon capture, energy systems, energy storage and harvesting, batteries and fuel cells, power systems, energy efficiency, energy in buildings, energy policy, as well as energy-related topics in economics, management and transportation. Books published in LNE are original and timely and bridge between advanced textbooks and the forefront of research. Readers of LNE include postgraduate students and non-specialist researchers wishing to gain an accessible introduction to a field of research as well as professionals and researchers with a need for an up-to-date reference book on a well-defined topic. The series publishes single- and multi-authored volumes as well as advanced textbooks.

Indexed in Scopus and EI Compendex The Springer Energy board welcomes your book proposal. Please get in touch with the series via Anthony Doyle, Executive Editor, Springer (anthony.doyle@springer.com)

Soheil Saraji · Si Chen

Sustainable Oil and Gas Using Blockchain

With contributions from: Dayo Akindipe, Karisma Karisma, Fred J. McLaughlin, Bertrand W. Rioux, Katerina Serada, Pardis M. Tehrani, and Joseph Wyer

Springer

Soheil Saraji
Energy and Petroleum Engineering
University of Wyoming
Laramie, WY, USA

Si Chen
Open Source Strategies, Inc
Los Angeles, CA, USA

ISSN 2195-1284 ISSN 2195-1292 (electronic)
Lecture Notes in Energy
ISBN 978-3-031-30696-9 ISBN 978-3-031-30697-6 (eBook)
https://doi.org/10.1007/978-3-031-30697-6

This Springer imprint is published by the registered company Springer Nature Switzerland AG
The registered company address is: Gewerbestrasse 11, 6330 Cham, Switzerland

Foreword

A global energy transition is underway, focusing on clean energy and reducing carbon emissions from traditional fossil fuels. There are also fast technological innovations in renewable energies, reducing costs and making them economically competitive. In addition, the oil and gas industry is under growing pressure from investors, activists, and customers to make the industry comply with Environmental, Social, and Governance (ESG) standards and become more transparent about its emissions. These changes and pressures have already impacted the oil and gas industry and promoted sustainability initiatives from within. It is clear that the industry is going through its own transition to stay relevant in the future of the energy sector. This poses new challenges and opportunities for the industry. One area of challenge/opportunity for the industry is quantifying, monitoring, verifying, and reporting carbon emissions and managing its carbon footprints. This is a critical area for the industry if it wants to become transparent about its emissions and access the growing pool of EGS-compliant funding for further growth. However, the main challenges are the cost, the lack of existing measurement standards, and the negative public persecution of the industry.

A new emerging technology, blockchain, seems to be an ideal tool for monitoring, reporting, and verifying carbon emissions and other environmental attributes, especially between non-trusting parties. This technology can potentially solve some of the challenges the oil and gas industry faces and provide new opportunities and economic incentives for quantifying the emission data. The latter is sometimes called "measurement economy," where entities that invest in reliable monitoring emission equipment and procedures could benefit from detailed emission data attached to their fuel products. Furthermore, blockchain has shown to be an excellent tool for circular economy and supply chain management. However, technical, cultural, and regulatory issues around blockchain still need to be resolved.

This book started as a white paper on using blockchain technology for carbon capture and storage. We eventually decided to expand it to a whole book on "sustainable oil and gas" and invited more experts to contribute to this book. It took us more than a year to collaborate with authors from around the world on various topics, from software to oil and gas field technology to the law. We looked at how they could be applied to carbon capture and storage, carbon markets, methane reduction, sustainable aviation fuels, and even plastics.

We focused our efforts on two central questions. The first question was, "*Can the oil & gas industry help solve the climate problem?*" We believe this is possible. Getting there, though, would require significant capital investments and supportive policies. That, in turn, would require the trust of the investment community and the general public, including environmentalists and climate groups. But, if we manage it well, the potential payoff would be huge. The world still relies on fossil fuels and will continue to do so for decades. So, the oil and gas industry is truly "where the action is" for reducing carbon emissions. Just eliminating methane leakage from the oil and gas industry alone, for example, would have a significant climate impact in the world today.

For the oil and gas industry, we hope this book will give a fresh perspective on how to manage the energy transition successfully. A transition to a low-carbon economy is unavoidable. The question is, how will it happen? The key to a favorable outcome will be getting ahead of the events rather than being overtaken by them. It will require transition strategies that inspire the trust of the general public and the financial markets. Done correctly, the energy transition could help the industry become not just environmentally but also financially more sustainable.

For those outside the oil and gas industry, we hope this book will help you see the potential for working with the industry to solve the climate problem. Like it or not, oil and gas are the beating heart of the modern world. We can't just stop using them or divest from the industry. This industry must be part of any transition strategy, even one that eventually takes us out of fossil fuels. At the same time, this industry is taking steps to address the climate problem and has enormous resources that could aid us in the transition. So, can we help it play a positive role and guide, even challenge, it to do the most it can?

It won't be easy, but what's the alternative? A world split into green versus brown economies, where the oil and gas industry is shut out of the public financial markets and major advanced economies while it continues to release greenhouse gasses, protected by a few supportive governments and financed by private equity and sovereign wealth funds? Does anyone in the oil and gas industry really want this to happen? Will this solve the climate problem?

Ultimately, solving the climate problem means we all must work together, the oil and gas industry, utilities, airlines, investors, banks, and even environmentalists. This brings us to the second fundamental question that motivated writing this book, "*Can blockchain play a role in facilitating this transition in the oil and gas industry?*"

When we started working on the book in 2021, cryptocurrencies were all the rage. However, as we're wrapping up the book, many cryptocurrencies have crashed, the exchange FTX has filed for bankruptcy, and a cryptowinter is upon us. So, what is the place for the blockchain going forward?

It's not magic, but as a technology, blockchain is about a lot more than cryptocurrency. It's ultimately a collaboration technology that allows people to share data and transactions, and we believe that makes it an important tool for large-scale, global problems such as climate change. It is a tool that can bring a group of non-trusting stakeholders together and develop an economy around a shared area of concern, i.e., carbon emissions. This is the topic that we explore in this book.

The book is divided into three sections: *Foundations*, *Applications*, and *Interviews*. In *Foundations*, we cover a range of foundational topics, including the challenges and opportunities that the oil and gas industry is facing in transition (Chapter 1), climate and financial markets and their impacts on oil and gas operations (Chapter 2), an introduction to the blockchain technology (Chapter 3), and legal and regulatory issues of the blockchain (Chapter 4). In the second part, *Applications*, we explore different use cases of blockchain in developing a sustainable oil and gas industry. This part starts with an overview of blockchain and sustainability (Chapter 5). We then focus on specific blockchain applications in reducing methane emissions (Chapter 6), carbon capture and storage (Chapter 7), sustainable aviation-transportation fuels (Chapter 8), sustainable plastics (Chapter 9), and carbon credit markets (Chapter 10). The last section, *Interviews*, is the text of eight interviews we conducted with experts knowledgeable in the energy sector and blockchain. Some of the interview materials were used in Chapters 1 and 5. The complete text for these interviews is provided for interested readers to dive deeper into the perspective of each consulted expert.

This book covers a diverse range of topics from upstream to downstream operations. The applications discussed here, however, are not an exhaustive list, and other interesting applications did not make it to this book due to time constraints. Also, examples mentioned in the text and the experts interviewed were selected based on the authors' familiarity with the individuals and use cases. There are many more examples and knowledgeable people we could not explore or talk to as time became a limiting factor. We invite experts from within and outside the oil and gas industry to give us their feedback and help improve this book.

Acknowledgments

We sincerely thank our co-authors and contributors to this book, Dayo Akindipe, Karisma Karisma, Fred J. McLaughlin, Bertrand W. Rioux, Pardis M. Tehrani, and Joseph Wyer. This book would not be possible without their dedication and contribution. We also would like to thank all the experts that accepted to interview for this book: Kelly Bott, Dan Cearnau, Bryan Hassler, Kari Hassler, Aaron Lohmann, Karl Osterbuhr, Sriram Srinivasan, Steve Swanson, John Westerheide, and Jasmine Zhu. The discussion with them informed and impacted the direction of this book.We also appreciate Katerina Serada for the great discussions on the plastics circular economy, which inspired Chapter 9 of this book. Finally, we like to thank Benjamin Nweke for his help in plotting the graphs in this book and editing some of the interview texts.

Soheil Saraji would like to especially thank Dr. Glen Murrell and Mr. Bryan Hassler for their constant encouragement throughout writing this book and generous support of his early efforts in developing a research program around the Blockchain application in energy. I also thank Sherona Simpson for proofreading Chapters 3 and 5; Joana Olsen for collecting the energy use-cases of blockchain in Chapter 5; Autumn Bizon and Andrea Frosinini for great discussions and for helping me to expand my network of blockchain experts. Ultimately, I would like to dedicate this book to Kai and Sherona for being the greatest inspiration in anything I do.

Si Chen would like to thank the Hyperledger Foundation for its support of our development of an open-source carbon accounting platform using the blockchain; Shaun Frankson, Pedro Carvalho, Gabe Malek, Charles Ford, and Mike Matthews for sharing their time and perspectives; and Andrea Frosinini for connecting Soheil and me. I would like to dedicate my work to Carolina Con. Thank you, Carolina, for your input, which I have found so interesting, and for your support, without which none of this would have been possible.

Contents

About the Authors

Dr. Soheil Saraji is an associate professor of Energy and Petroleum Engineering, an adjust professor at the School of Energy Resources, and co-director of the Hydrocarbons Research Laboratory at the University of Wyoming. He has eighteen years of research experience and more than 35 peer-reviewed journal publications in subsurface energy extraction, storage, and carbon geo-sequestration. Furthermore, Dr. Saraji is a pioneer in applied blockchain research for the oil and gas industry. He has developed new courses and research initiatives on this topic at the University of Wyoming.

Mr. Si Chen is the president of Open Source Strategies, Inc. in Los Angeles, CA, which specializes in open-source software for climate finance and investing. He leads the development of open-source blockchain carbon accounting software at Hyperledger Labs. Previously, he has managed investment portfolios for institutional pension funds, central banks, and hedge funds and has been published in *The Journal of Portfolio Management*. He is also the co-founder and CTO of GraciousStyle.com, an online retailer.

Part I
Foundations

Chapter 1
Energy Transition: Challenges and Opportunities for the Oil & Gas Industry

Soheil Saraji and Dayo Akindipe

1.1 Introduction

The oil and gas industry (OGI) is among the world's largest, most complex, and crucial industries, involving upstream, midstream, and downstream sectors. The upstream sector is the first phase in the life cycle of oil and gas, and consists of the exploration and development operations, drilling and well completion, production and optimization, reservoir engineering, facilities management, decommissioning, and reclamation. The midstream segment involves transportation, storage, and trading of raw materials (oil and gas). The downstream operations include refining the raw materials into more valuable products, product distribution, and transportation. The OGI played a significant role in the previous energy transition from coal in the early twentieth century and provided the energy source used to create wealth over the past century (Murphy 2019).

However, the utilization of fossil fuels has come with attendant issues that affect both the natural and human environment. Oil spills, gas flaring, post-combustion emissions, seismic tremors, and the Dutch disease are notable environmental and social issues that have placed a negative tag on the OGI (O'Rourke and Connolly 2003; McGarr et al. 2015; Corden 1984). A significant contemporary challenge for the OGI is its lifecycle emissions, mostly greenhouse gases (CO_2, CH_4, and N_2O) and other toxic compounds (e.g., SO_2, NOx, and volatile organic compounds), that are respective precursors to the greenhouse effect and ground-level ozone (IPCC 2014; U.S. EPA 2016a). Climate scientists have unanimously opined that the exacerbating

S. Saraji (✉) · D. Akindipe
Energy and Petroleum Engineering, University of Wyoming, Laramie, WY, USA
e-mail: ssaraji@uwyo.edu

D. Akindipe
e-mail: Dayo.Akindipe@nrel.gov

D. Akindipe
National Renewable Energy Laboratory, Golden, CO, USA

S. Saraji and S. Chen, *Sustainable Oil and Gas Using Blockchain*,
Lecture Notes in Energy 98, https://doi.org/10.1007/978-3-031-30697-6_1

greenhouse effect, which leads to global surface temperature rise, is the primary cause of the imminent climate change (IPCC 2014) and have recommended a 1.5 °C surface temperature rise limit over the pre-industrial baseline to mitigate its adverse effects (IPCC 2018). The spotlight is now on the industry and its response to the climate challenge. Initial reluctance to accept the industry's contribution to the climate crisis has slowed progress in decarbonization efforts. However, as political and economic pressures intensify and as environmental, social, and governance (ESG) becomes a global investment standard, industry players are now exploring ways to decarbonize existing operations and expand their energy portfolio to renewable and low-carbon sources (Oxford Institute for Energy Studies 2020; Beck et al. 2020). The global drive for accelerated decarbonization in the energy industry has created the term *energy transition*. In its current use, energy transition connotes that the world's primary energy supply is shifting from carbon-intensive sources (fossil fuels) to renewable and low-carbon sources (e.g., wind, solar, geothermal, hydrogen, and biogenic sources) (S&P Global 2020). The future of fossil fuels lies in the ability of the industry to adapt to the market, environment, and social trends within the energy transition framework. Failure to do so could sidetrack global climate targets and render the industry obsolete.

1.2 Oil and Gas in Energy Transition

The traditional oil and gas industry has been experiencing challenges in the past decade as renewable energy has become competitive in price. Major investment firms, such as BlackRock, have announced their pledge to reduce their exposure to fossil fuel, and demands for reduced emissions from energy customers have been growing. However, these challenges have also brought new opportunities for sectors within the OGI to become more efficient, develop new products for low-carbon energy markets, and expand their operations to emerging sectors. In this section, we first look at the long-term impact of the energy transition on the OGI, and then explore the new potential opportunities. We have interviewed a few energy experts on this topic and will draw on their insights in this section.

Historically, one of the main consumers of fossil fuels has been the electric grid. Over the past decade, this sector has been increasing its wind and solar resources to replace carbon-based fuels, especially coal. Is it possible to operate the electric grid with only renewable energy in the near future? Kari Hassler (Senior Manager of Market Operations, Xcel Energy) has emphasized the importance of fuel diversity for the electric grid: "*I think the challenges go back to the complacency of the markets and the energy grid as a whole, especially in the Midwest. We've become complacent because there was always a large surplus of dispatchable capacity. Even when we first started bringing on the wind, there were still plenty of dispatchable resources to accommodate for [intermittency of] that wind. So, I think developing policies and the tools that are needed to ensure that incentives are there for retaining the dispatchable resources as they're needed, but also encouraging investment in resources that have the flexibility, the availability, and the capability to manage the ramps of the intermittent resources [is essential]. … one of my overarching theories is that fuel diversity and resource diversity is critical as we move through this new age of resources. That's great to have renewable resources, but until batteries become more economic, we still need to maintain the diversity of the fleet through the fuel that the resources use as well as the locational diversity.*"

It seems inevitable for the energy sector of the future to benefit from a diverse basket of fuels, including fossil fuels, depending on the availability and local economy of each geographical location. In our interview with John Westerheide (Senior Director of Customer Solutions, Project Canary), he lays out his view of energy future: "*I don't think that our energy future and the energy transition is a binary choice between what I'll call traditional fuels, and zero emissions, or what is called Green Energy Fuels. The truth is that energy transition is going to be predicated on different paces at different places. What I mean by that is that each location is on its own energy adoption curve. So, what energy looks like in parts of Africa is very different from what energy looks like in Wyoming or in Oklahoma where I'm sitting. Equally, those different places have different proprietary energy endowments. For example, Costa Rica has a lot of geothermal. Other places may have proprietary endowments for wind and solar and other places may not. So, when people talk about the energy transition, One, it's not bright binary. And two, it's not homogeneous across all geographies. You're always going to be balancing the societal need with the proprietary endowments and economics of energy resources locally, to figure out what that right balance and mix are…. Regardless of where you're at, we need to move to lower greenhouse gas intensity and lower carbon intensity sources of energy continuously. And realistically, we've done that, and going from a biomass world to a coal world, to liquid petroleum-driven energy infrastructure, to gas infrastructure, to hydrogen, and to nuclear fuel, etc.*"

There is growing competition in the energy sector, and the OGI's share of the market has been slowly declining. Against this background, what has been the response from oil and gas companies? Jasmine Zhu (VP of Market Development, Xpansiv) says: "*We've seen a lot of the big oil companies now shifting some of their assets and their portfolio into cleaner energy, whether it be developing hydrogen or coming up with low carbon fuel solutions, or just acquiring smaller companies who are on the verge of making better technology come into place. You've seen Chevron and BP all taking leads in generating renewable natural gas as part of their portfolio. They're acquiring farmland and they're acquiring landfill. So, all these new assets that historically was now part of what you think a traditional energy company would care about, is now a driving division within that organization.*"

The change is coming. The markets are shifting, and OGI needs to adapt itself to the new environment. What will the oil and gas industry look like 10–30 years from now? We asked this question of Bryan Hassler, Chairman of PureWest Energy, and he replied: "*30 years from now we're looking at many companies saying they want a net zero carbon footprint. And I ask a lot of these companies, what does a net zero carbon footprint look like? And essentially, I think they still will anticipate using hydrocarbon fuels in what resources they've got remaining. Most of those will be natural gas-fired generation units, there is a transition toward hydrogen. I think, from an oil and gas perspective, you're already seeing some of the very large majors, and some of the intermediate-sized companies try to evaluate what does that energy transition look like? It will still incorporate natural gas and crude oil as reigning feedstocks in the electric generation stack of fuels. I think you'll still see coal, but the environmentalists have pretty much put the nail in the coffin of the future of coal. Also, carbon capture and sequestration may factor into some of that, but as you know, it's very expensive, heavily subsidized with tax incentives, and very difficult to establish. But there is a developing carbon capture sequestration and utilization industry developing, and I think that'll factor into the extension of utilization of crude oil, natural gas, certainly in the feedstock for the future. And you'd get natural gas and crude oil factor into many of the everyday things we utilize in life; or we depend upon food, feedstock fertilizers, plastics, clothing, you name it. So, hydrocarbon molecule sits in our everyday lives, and we'll continue to do so.*"

Energy has been a key element of our modern society. We have relied on fossil fuels for about 200 years (from the pre-industrial age), and it has fueled our economic growth. We should be concerned with the rate of transitioning to other energy sources that may not be as reliable. This was certainly pressure tested with the recent developments in Europe. Bryan Hassler agrees: "*if you look at European nation's energy prices, $35 to $40 per MMBtu, here in the United States for the month of July, [it is] $6.5 per MMBtu. Countries are reliant upon our low-cost supplies to give them alternatives to issues like the Russia-Ukrainian War, where natural gas supplies to Europe have, in essence, gone to zero or close to zero. Russian crude oil kind of shifted, at least out of the European markets and in the North American markets. It has put a constraint on the system. What happens when you put a constraint on the system? Even if you've got every great intention of lowering your carbon intensity and carbon footprint as rapidly as possible, it puts a burden on the populace, so to speak, in terms of inflation in energy, fertilizer, food, and down the chain.*"

In order to get another perspective, we posed the same question to Jasmine Zhu (VP of Market Development, Xpansiv), and she said: "*If you were to ask me that last year [(2021)], I probably would have said the trend is that fossil fuels will be phased out sooner than we would anticipate because we're seeing a lot of new projects. That's purely green-driven, like when projects are out here in New Jersey, and New York, and some of the northeastern states are basically saying we're going to have very aggressive Net Zero plans in place. And now, with everything that's going on with Europe, Ukraine, and Russia. I think this really has put a stop to some of those realities, which have shifted a little bit, and you see a bit of the supply shortage towards Germany, France, and other European countries that may have triggered demand that we in the United States haven't considered or priced in from 10–15-year type of a forecast. Now we're saying we should drill more and produce more liquefied natural gas so that we could send and help our allies overseas. So, the existence of natural gas suddenly becomes a necessity again. I think as much as anyone wants to think 15 years out, we'd like to be in a much greener place. It's hard to say that for now, when there's such a shortage in this market, and people have to really build in certain hours where they cannot have electricity just so that they can manage the grid and the cost of that generated power.*"

Ultimately, energy security is just one of many important challenges in this technology-driven modern age that depends on abundant energy. The demand for energy is consistently growing year-on-year despite the contraction in 2020 due to the COVID-19 pandemic (IEA 2021a). However, supply and availability of energy sources to satisfy the growing demand of modern society are limited. This is even more heightened with the climate-change issues emanating from the combustion of fossil fuels since the industrial era. Energy supply diversification and transition to renewable and low-carbon resources is a sustainable pathway to energy security and climate change mitigation. However, current global utility-scale renewable energy (e.g., wind, solar, and geothermal) generation and supply pales in comparison to fossil-based (e.g., coal and natural gas) energy (IEA 2021a). In addition, variable renewable energy sources such as wind and solar are not sufficient to provide utility-scale baseload electricity without an extensive energy storage infrastructure (Matek and Gawell 2015).

Therefore, it seems that the OGI will be an important part of the energy sector for the foreseeable future. However, we expect a gradual reduction in the traditional fossil fuel market share and a significant tightening of investments in this sector. If companies in this sector continue their business as usual, the future looks bleak. However, as we mentioned previously, any challenge will bring opportunities for companies that are willing to innovate and adapt to the markets. We would like to go even further and claim that the OGI could play a key role in the future of the energy sector if it embraces the energy transition sooner than later. We have identified three main growth areas for the OGI that could help establish its role in the coming decades (Table 1.1): (a) modernization and digitization to make the industry more efficient, reduce carbon emissions, and improve the safety and human welfare of its workers, (b) developing premium and low-carbon fuels with low carbon intensity compared to traditional fossil fuels, (c) leveraging OGI's expertise and infrastructure in subsurface hydrocarbon extraction and offshore operations for clean energy development and storage. In the preceding sections, we dive deep into each of these growth areas and provide examples of undergoing investments and projects.

1.3 Modernization and Digitization

Facing challenges associated with the energy transition, the OGI will have to become technologically and commercially sophisticated in order to continue to be a part of the world's sustainable energy mix. OGI companies are looking for digital technology to drive efficiency and productivity, reduce carbon emissions, and transform their operations to meet these challenges. The industry generates huge amounts of data, which is often recorded manually across multiple, disconnected paper records or spreadsheets by siloed functional teams (How Do You Reshape When Today's Future May Not Be Tomorrow's Reality? - Oil and Gas Digital Transformation and the

Table 1.1 Summary of opportunities for OGI in the energy transition

Opportunities	Modernization & digitization	Low-carbon and differentiated fuels	Harnessing underground & offshore facilities & expertise
Goals	Improve efficiency Improve safety and human welfare Monitor and reduce emissions	Reduce operational emissions Develop competitive fuels for low-emission markets	Repurpose the infrastructure for clean energy Leverage the underground expertise for underground storage and clean energy Leverage the offshore operation expertise for clean energy
Enabling Technologies, Products, or Processes	Multi-well drilling pads Internet of Things Blockchain Data Analytics Artificial Intelligence and Machine Learning	Biofuels Sustainable Aviation Fuel Digital Fuels Certified/ Differentiated Gas Blue hydrogen	Converting OGI pipelines to CO_2 or H_2 pipelines Convert abandoned wells and mature fields to geothermal or carbon storage Carbon & Hydrogen Geo-storage in saline aquifers Geothermal Energy Development Offshore Wind & Solar

Workforce Survey 2020, n.d.). This data, if organized and processed in a timely manner, could be utilized towards achieving the digitization and modernization goals of the industry.

Some examples of utilizing advanced digital technologies to improve operations in the OGI are (Hanga and Kovalchuk 2019; Lu et al. 2019): (a) utilizing real-time data streams generated by sensors to ensure better control and optimization of crude production, (b) robotics in offshore fields for drilling, inspection and damage control to enhance efficiency and personal safety, (c) wireless sensor networks to monitor and enhance production, as well as to detect and prevent issues with regards to health and safety, (d) the radio-frequency identification (RFID) technology for asset management, oil rig site management, pipeline inspection, safety, and security, etc. In the following sections, we focus on three emerging digital technologies that have a massive potential to transform OGI, making it profitable in the energy transition era.

1.3.1 Internet of Things

The Internet of Things (IoT) is a network of items embedded in smart devices (sensors and actuators) connected to the internet. The term "IoT" was coined by Kevin Ashton, executive director of the Auto-ID Center, in 1999, when he presented a radio-frequency identification (RFID)-based supply chain optimization system to Procter & Gamble (P&G) (Wanasinghe et al. 2020). The IoT enables machine-to-machine communication over a network without requiring human-to-computer interaction. The physical layer consists of the IoT nodes used for data acquisition and control of the OGI equipment and facilities (Elijah et al. 2021).

As discussed earlier, data is the core concept that drives the digital transformation of the OGI industry. IoT-based devices are at the heart of this transformation as they provide reliable data streams for various applications in upstream, midstream, and downstream operations (Gai et al. 2020; Klein et al. 2018). A recent literature review demonstrated that service companies are the main drivers of innovation in this space, far ahead of academic institutions and oil companies (Wanasinghe et al. 2020). In addition, the available literature was mainly focused on conceptual demonstrations rather than actual case studies (Wanasinghe et al. 2020). Nevertheless, several use cases are being tested in the industry in six critical areas (Wanasinghe et al. 2020): (i) remote monitoring, operation, and asset optimization (e.g., pressure relief valve, sucker-pump working condition, production and well monitoring/optimization, etc.), (ii) predictive maintenance, (iii) automation and control (e.g., automation of drilling systems, production systems, well completion systems, etc.), (iv) occupational safety and health compliances, situational awareness, and personnel tracking, (v) supply chain and fleet management, (vi) security. For example, the use of wearable watches, smart helmets, and smart glasses by oil field engineers in offshore fields for real-time assistance, safety, and communication with the control room for navigation and enhanced collaboration was highlighted by Hanga and Kovalchuk (Hanga and Kovalchuk 2019) as a successful use case of IoTs in the OGI.

However, there are serious challenges to deploying this technology at scale in the OGI. A comprehensive list of challenges was provided by Wanasinghe et al. that includes (Wanasinghe et al. 2020): (a) cybersecurity concerns, (b) lack of technological readiness, (c) lack of interoperability, adaptability, and standardization, (d) challenges in data storage and analytics, (e) maintenance cost, and obsolescence, (f) mindset of employees. Among these challenges, cybersecurity concerns were the most cited constraint for the industry-wide adaptation of IoTs (Wanasinghe et al. 2020). Other concerns regarding a centralized IoT system are (Restuccia et al. 2019): (i) can make illegitimate use of IoT data, for example, mass-surveillance programs. (ii) can expose the system to hacking by malicious activities, and (iii) authentication of IoT entities that will be primarily deployed in the wild with little supervision. IoTs are a relatively new technology, and further innovation is required for widespread adoption in the industry. Nevertheless, the IoTs, in combination with other digital technologies such as Artificial Intelligence, Big Data Analytics, Blockchain, etc., provide a massive opportunity for the future of OGI.

1.3.2 Big Data Analytics and Artificial Intelligence

As discussed in the previous section, the emergence of various IoT devices in the OGI accelerated the accumulation of massive data. Therefore, data processing and management have become a challenge for the industry and, at the same time, provided opportunities for the application of new data processing technologies such as Big Data analytics (BD), Artificial Intelligence (AI), and Machine Learning (ML). BD deals with the huge amount of data being collected from a variety of sources (volume), the speed at which the data is being collected in real-time (velocity), and the formats in which the data are collected (variety) (Elijah et al. 2021). BD analytics refers to the process of researching these massive amounts of data in order to uncover hidden patterns and correlations. BD has been used to help with exploration, drilling, oil recovery, and production (Agwu et al. 2018; Bello et al. 2015; Bravo et al. 2013; Choubey and Karmakar 2021; Desai et al. 2021; Elijah et al. 2021; Gupta and Shah 2021; Li et al. 2021b; Nguyen et al. 2020; Patel et al. 2020; Perrons and Jensen 2015).

AI involves the use of computer algorithms in an attempt to mimic the operations of human brains or thoughts to understand and make decisions. AI can offer to handle the ever-increasing amount of data for prediction, prevention, and optimization (Hanga and Kovalchuk 2019). ML is a subset of AI focusing on human-like learning with great potential in OGI applications, especially when it comes to analyzing and interpreting data. The main goals of ML are making predictions, performing clustering, extracting association rules, and making decisions from a piece of given information or data. It offers better ways of developing drilling plans, as well as diagnosing, monitoring, predicting, and performing real-time optimization at a minimal cost (Hanga and Kovalchuk 2019). Evidence for the successful implementation of ML algorithms in the OGI are (Hanga and Kovalchuk 2019; Tariq et al. 2021): (i) maximizing the oil and gas production rate while minimizing the volume of produced water and sand in oil production, (ii) production pattern data recognition, (iii) minimizing the cost of lifting in the wellbore, (iv) enhancing reservoir modeling, (v) applications in anomaly detection and preventative maintenance. ML can help identify patterns and predict failures based on relationships between valid readings.

These digital tools are also identified as building blocks of a futuristic vision of oilfields. i.e., intelligent oilfields (Elijah et al. 2021). Some of the characteristics of intelligent oilfields are self-diagnostics, control and monitoring systems, autonomous operations, the use of advanced mathematical models for control of equipment, and the real-time exchange of data for controlled objects. Overall, BD, AI, and ML have a massive potential for transforming the OGI, making it safer and more efficient. This is manifested in the emergence and growth of many startup companies focused on bringing these technologies to various sectors in the OGI.

1.3.3 Blockchain Technology

Blockchain is a distributed ledger technology whose records are batched into timestamped blocks, and each block is identified by its cryptographic hash. Each block references the hash of the block that came before it, which establishes a link between the blocks, and thus creates a chain of blocks or a Blockchain. A blockchain network is a set of non-trusting writers sharing a digital database with no trusted intermediary. Properties of a Blockchain provide: (i) a robust, truly distributed peer-to-peer system that is tolerant of node failures and (ii) transparency, verifiability, and auditability of the network's activity. Blockchain was started merely as a decentralized platform to secure cryptocurrencies by solving digital ownership or double-spending problems (Nakamoto, n.d.). It then evolved into a world computer enabling smart contracts, decentralized autonomous organizations, etc. (Buterin, n.d.). Currently, it is accepted as the potential backbone of the future of the internet (i.e., web3) (Wood, n.d.).

In OGI, Blockchain has emerged as a promising innovation that could play an essential role in delivering the technological and commercial capabilities that the sector will need to achieve its goals of modernization and digitization. It could offer operational cost reductions, increased efficiency, fast and automated processes, and transparency (Saraji and Khalaf 2022). Blockchain has much to offer in the OGI, including blockchain-enabled transactional digital platforms for trading (Andoni et al. 2019; Mehta et al. 2021), carbon accounting and the management of carbon emissions (Saraji and Borowczak 2021), the design and construction of wells and facilities, the tracking of drilling equipment history and maintenance, the automation of drilling as well as the optimization of drilling operations, and supply-chain management (Elijah et al. 2021). Current examples of employed applications include guaranteeing the authenticity of wellbore rock and fluid samples and creating a shared consensus about the progress of drilling campaigns (Perrons and Cosby 2020). Another area of exploration is the integration of blockchain and IoT. This combination alleviates some of the challenges mentioned in the IoT section (e.g., decentralization, non-repudiation, etc.). It also opens up new opportunities, such as enabling autonomous and self-organized machine-to-machine (M2M) communications, automatic and safe firmware updates, and data security.

A wide range of challenges has contributed to the slow adaptation of this technology in the industry, ranging from technical challenges to unclear regulations and legal frameworks around this technology and the lack of a technical workforce with the required skills. Technical challenges include the lack of standardization and interoperability among technology platforms within the blockchain technology space and the inability of the blockchain (alone) to protect the integrity of the physical assets (Perrons and Cosby 2020). The latter is known as the oracle problem and emerges when trying to build physical-digital bridges (Caldarelli 2020). However, this is a fast-growing field, and the rate of innovation in the space is very promising. It is a matter of time before commercial use cases emerge in the industry.

1.4 Low-Carbon Fuels

Based on California's Low Carbon Fuels Standard (LCFS), low-carbon fuels are fuels that inherently have minimal to net-zero carbon intensity over their lifecycle or have been engineered to significantly decrease their lifecycle carbon intensity compared to conventional fossil fuels (California Air Resources Board 2020). Typical low-carbon fuels (LCF) are from biogenic sources (e.g., agricultural waste and wood biomass), while others are fossil-based sources such as certified gas and blue hydrogen. In the former class of LCF, the major energy source is from plants and trees, which, over their lifetime, have captured and stored significant amounts of atmospheric CO_2. Therefore, at the point of resource extraction, biogenic LCFs are carbon negative. When these fuels are combusted for electricity and heat generation, the ensuing CO_2 emissions are counteracted by pre-stored CO_2, making the process almost carbon neutral. The nearness to carbon neutrality for biogenic LCF will depend on other ancillary processes like transportation that contribute to the lifecycle emissions profile.

LCF from fossil sources include fuels derived from hydrocarbon extraction that have been strictly monitored for their emissions profile. Monitored emissions are mostly fugitive CH_4 that has 27–30 times the global warming potential (GWP) of CO_2 over a 100-year period (U.S. EPA 2016b). Other monitored emissions are CO_2 from gas flaring activities during source extraction. Although carbon neutrality is not achieved with emissions monitoring, these fuels are still categorized as LCF because of their lower lifecycle emissions compared to conventional fossil fuels (California Air Resources Board 2020).

The energy return on energy invested (EROI) of most biofuels is currently very low, and the commercial potential of these resources as stand-alone fuels in the long term is uncertain (Murphy 2019). "*Biogas is nothing more than methane, that's been emitted into the environment from municipalities, waste facilities, you name it. We've got to embrace that cleaner energy from the oil and gas sector can be much more cost-effective. And I think what we're all looking for is just standardization of measurement of our carbon intensity or methane intensity, and standardization of the monitoring techniques, and [development of] firm's that are going to independently verify and validate emissions. [Also,] I think we want to get a level playing field; a set of rules and regulations that everybody is familiar with and abides with. I think, at least as a methane producer and natural gas producer, we want to make the case that in the future of energy in whatever the energy transition is to produce responsibly and to assist in lowering carbon and methane emissions across North America and across the world for that matter.*" Said Bryan Hassler.

There are indications that markets are interested in LCF from fossil sources. "*You think commodity is just a physical movement of something, but now you add one more layer on top of that, which is the data for each one of these commodities. Then you create this whole new market. Like I said, you got the gas, you got the oil, you could do the same thing for hydrogen, sustainable aviation fuel. ... When people make these responsibly sourced claims, where's the data behind it? Well, now you got the access to those data, which is all being captured from production, from inception, and is being stored within a digital fuels registry. So, one really could say, today I care about the methane intensity of something, and some of these other environmental attributes like water, and* SO_X *and* NO_X *are less important. Fine, maybe I'll pay a little bit more for that methane intensity. Six months later, there's a water shortage, I want to buy the commodity that takes up the least water, and I'll pay your premium for that. So, you get to really build out a lot more markets, because now you've got 20 attributes to work with, rather than just this one MMBtu of gas which is traded for a few bucks at Henry Hub. Now, the data becomes a lot more richer and the technologies are being developed to kind of put further confidence around that data that's being captured.*" Said Jasmine Zhu.

In the following sections, we shall discuss some examples of LCF in more detail and provide a suitable case study that the OGI can adopt.

1.4.1 Biofuels

Biomass are primarily solid fuels sourced from plants and animals (usually animal waste). They include renewable resources such as standing forests, waste (agricultural, human, municipal), and energy crops (Saxena et al. 2009; Ram and Mondal 2022). The OGI is gradually increasing investments in biomass utilization for energy and the production of non-fuel products. Between 2010 and 2018, BP and Shell spent 2.3% and 1.33% of their capital expenditures (CAPEX) on clean energy—biomass energy accounted for more than one-half of the CAPEX (Li et al. 2022). Within the same period, ExxonMobil derived all of its clean energy (0.22% of CAPEX) from biomass (Li et al. 2022). These percentages are meager compared to enormous amounts still being spent on fossil fuel production (IEA 2020b). Therefore, the OGI needs to marry its annual pledges with active investments in bioenergy generation.

Bioethanol is a biofuel derived from the fermentation of crops and agricultural waste. It is majorly used as a renewable additive to gasoline or directly combusted as a biofuel (Mussatto et al. 2010). Feedstocks for bioethanol production are sugar-containing crops (e.g., sugarcane, sugar beet, and molasses), starch-containing grains (e.g., corn, barley, and wheat) and root crops (e.g., cassava and potato), and lignocellulosic biomass such as straw, agricultural waste, crop and wood residues (Mussatto et al. 2010; Bušić et al. 2018). The downstream OGI is already using bioethanol in gasoline blends. In the United States, most of the gasoline sold contains at least 10% bioethanol (E10 blend) derived from corn starch, and up to 15% bioethanol in gasoline (E15 blend) has been allowed for medium-duty vehicles manufactured since 2001 (EIA 2022). In other parts of the world, the bioethanol composition in gasoline is still very low or non-existent. For example, the European Union (EU) still limits ethanol content in gasoline to 10%—most EU countries use E5 (5%) gasoline (Gray 2020). It is apparent that drastic policy changes that include consultations with both OGI stakeholders and automobile manufacturers are required to enable higher bioethanol participation in the global low-carbon fuel mix.

Renewable natural gas (RNG) is another significant low-carbon biofuel that has gained traction. Also known as biomethane, this product is a high-purity gas derived from the anaerobic microbial degradation (or digestion) of food, agricultural, and municipal solid waste. The low-quality biogas (containing 45–60% CH_4) derived from digestion is post-treated and upgraded to attain the acceptable purity levels (greater than 90% CH_4) required in methane-powered engines (U.S. EPA 2020). A major player being actively engaged by OGI companies is Clean Energy Fuels Corporation, based in California. Since 2018, they have formed partnerships and joint ventures with BP, Chevron, and Total Energies. In December 2020, BP agreed to provide $50 million in investments to Clean Energy in developing and constructing RNG facilities (Clean Energy 2020b). The joint venture was initiated in March 2021 specifically for carbon-negative RNG from dairy farms and other agricultural sites to be used as a low-carbon fuel for road transportation (Clean Energy 2021b). Total Energies recently signed a 100 million dollar- joint venture with Clean Energy to develop carbon-negative RNG production facilities in the United States and expand Clean Energy's downstream RNG fueling infrastructure (Clean Energy 2021a; Reuters 2021a). Similarly, in July 2020, Chevron signed an $8 million agreement with Clean Energy to participate in their Adopt-a-Port program. Within this program, Chevron provides carbon-negative RNG to truck operators that haul goods from the Los Angeles and Long Beach ports and supplies RNG to stations near these ports. It also subsidizes the purchase of RNG-powered trucks and the cost of retrofitting existing trucks with RNG-capable engines (Clean Energy 2020a). Based on its initial success, Chevron has invested an additional $20 million for program expansion (Clean Energy 2021c). In September 2021, Shell launched its first RNG plant in the United States in Junction City, Oregon. This facility uses cow manure and excess agricultural residues as feedstocks and is estimated to produce 736,000 MMBtu a year of RNG. Shell is also collaborating with RNG producers from dairy farms in Kansas and Idaho, with a potential for an additional 900,000 MMBtu of RNG annually (Shell 2021).

1.4.2 Sustainable Aviation Fuels

The aviation industry is one of the most carbon-intensive sectors worldwide. Despite the COVID-19 pandemic-induced slowdown of emissions from air transport, it still contributes to between 2–3% of global CO_2 emissions (IEA 2021c). The industry heavily relies on expensive and scarce carbon offsets, especially in developing markets. One major solution for low-carbon air transport is sustainable aviation fuels (SAF). SAF is generally biogenic jet fuel derived from the conversion of biomass into a high-energy-density fuel. Conversion processes vary depending on the type of feedstock used. Common feedstocks include starch, sugar, animal fats, and vegetable oil (Michaga et al. 2021). These first-generation sources are gradually being replaced with second-generation oleo-chemical (e.g., used cooking oil) and lignocellulosic feedstocks (e.g., agricultural residues, nonedible biomass, and waste streams) that do not significantly affect food security (Michaga et al. 2021).

So far, SAF supply to aviation is still very low. In 2021, the proportion of SAF used in global air transport was a meager 0.05% of the 57 billion gallons of jet fuel consumed (IATA 2021a, b). Based on recent projections by the International Air Transport Association (IATA), about 119 billion gallons of SAF will be required to meet the 2050 net-zero goals of the international aviation industry (IATA 2021a). One challenge limiting the use of SAF is the high cost of procuring and converting feedstocks to jet fuel standards. Another is interoperability issues, especially because aviation is a global industry (BP 2021; Michaga et al. 2021). Therefore, SAF supply needs to be streamlined and accessible in multiple geographical locations worldwide. Currently, most of the supply is limited to North America and Europe; however, the Asian market is also coming on stream (Neste 2022). Another issue is the challenge of compatibility with existing turbines and jet engines. This has restricted the SAF-to-conventional jet fuel blending ratio to 50% (Michaga et al. 2021).

The downstream OGI is the main supplier of fossil-derived jet fuels for the aviation industry. Major players include BP, Shell, and Total Energies. Among these legacy OGI companies, BP is the most active in the SAF market and, in collaboration with SkyNRG (a leading SAF producer), was the first to make SAF commercially available to the aviation industry at the Oslo Airport in Norway (BP 2021). Other players in the SAF market are Neste, a Finnish downstream (mostly oil refining and product distribution) company that has now rebranded as a renewable fuels producer and supplier. Neste is now a major global player in the SAF market with a reach across North America, Europe, and Asia. It currently produces around 34 million gallons (100,000 tons) of SAF annually at its refinery in Porvoo, Finland (Neste 2022). New production streams are expected to be operational in 2023 from refineries in Rotterdam (The Netherlands) and Singapore to increase annual capacity to 515 million gallons (1.5 million tons) (Neste 2022). In addition, the company recently signed a joint venture agreement with Marathon Petroleum Corporation (a leading OGI company in the United States) to convert Marathon's Martinez refinery in California to produce sustainable biofuels, including SAF (Reuters 2022). This is another success story of converting conventional refineries to biorefineries that the OGI seriously needs to consider for SAF to become mainstream.

1.4.3 Certified Gas

Methan as a greenhouse gas has eight times the global warming potential of CO_2, at least in the short and medium-term. Therefore, CH_4 emissions across the value chain of natural gas production and supply are seriously concerning. If the OGI wants to continue supplying natural gas as a bridge fuel, there is a need for stricter regulation and enforcement on CH_4 emissions. Certified gas is one product that addresses this gap. Certified gas is natural gas that has been certified by third-party authorities to have minimal CH_4 pre-combustion emissions and meet global ESG best practices.

Before fossil-based natural gas can be certified, a lifecycle (mostly Scope 1 and 2) emissions analysis and an ESG assessment are implemented across the upstream (e.g., fracking, production, and separation), transportation, and storage value chain (Russo 2021).

Some OGI upstream and midstream companies in the United States and Canada are adopting this certification process due to investor and stakeholder (mainly gas utilities) intensification of the need to decarbonize and adopt ESG standards (DiChristopher 2021). Presently, certified gas is available through two leading independent companies—Equitable Origin and Project Canary. Equitable Origin uses its EO100™ certification standard to evaluate natural gas producers' compliance with certified gas standards across five main criteria/principles: (i) corporate governance, transparency, and business ethics, (ii) human rights, social impacts, and community development (iii) indigenous peoples' rights, (iv) fair labor & working conditions, and (v) climate change, biodiversity & environment (Equitable Origin 2017). In addition, participating producers are scored according to performance targets PT 1, PT 2, or PT 3 if they meet (PT 1), exceed (PT 2), or lead (PT 3) global best practices in ESG-based natural gas provision (Equitable Origin 2017). Presently, the company has certified up to 10 Bcf of natural gas production from 18 producers in North America (DiChristopher 2021). Similarly, Project Canary, through its TrustWell™ standard, monitors site-level CH_4 emissions at high resolutions and in real-time. This is in conjunction with other ESG-relevant data, such as air quality, water usage, and land restoration (Project Canary 2021).

Certified gas producers are exclusively medium-sized OGI companies aiming to differentiate their products from regular natural gas by demanding a premium over the market price. These companies include PureWest Energy, Southwestern Energy, Chesapeake Energy, Énergir, and Seven Generations Energy, all active in the North American gas market (Russo 2021; DiChristopher 2021). There is currently no indication on whether OGI majors will adopt certified gas. This may be because environmentalists could perceive certified gas adoption as greenwashing fossil-based natural gas instead of focusing on renewables and other low-carbon ventures (Feinstein 2022). Notwithstanding, the Norwegian-based major, Equinor ASA, has engaged both Equitable Origin and Project Canary to certify natural gas produced from the Utica Shale basin through partnerships with Southwestern Energy and Chesapeake Energy (DiChristopher 2021). The future of certified gas will lie with policy measures that make CH_4 monitoring and emissions mitigation mandatory (and not voluntary) for natural gas producers. An example is the recent United States Environmental Protection Agency's proposed rule that aims to cause a 41 million ton reduction in CH_4 emissions from oil and gas activities by 2035 (U.S. EPA 2021a).

1.4.4 Blue Hydrogen

Among emerging low-carbon fuels, hydrogen is attracting global attention as a key low-carbon energy carrier for decarbonizing hard-to-decarbonize sectors, such as aviation, shipping, trucking, cement production, and steelmaking (IEA 2019). In addition, hydrogen is expected to play a major role in mitigating the intermittency of wind and solar energy by balancing the energy supply from renewable energy plants and the energy demand of customers (i.e., peak shaving).

We currently do not have many large-scale hydrogen-producing facilities around the world. Though, many governments, including the US government, have pledged to fund the required infrastructures for a hydrogen economy. How will an energy sector with hydrogen look in the future? "*I think hydrogen will be located at hubs that can consume it directly. So, [for example] the Magnum project in Utah. They've got salt that can store hydrogen. They've got plenty of renewables and water to generate hydrogen, [and] fill those hydrogen caverns and, and co-run 800 megawatts of power, initially on 30% Hydrogen-70% natural gas, but the technology is there for the turbine to be retrofitted to burn 100% hydrogen in the future. … [In] Southwest Wyoming, they could build a hydrogen hub that co-locates with the facility in Utah, but also feeds the Pacific Northwest* [via] *Ruby pipeline, [which is a] fairly new vintage and internally coated [pipeline].*" Said Bryan Hassler.

Blue hydrogen refers to the process of generating hydrogen molecules from methane or natural gas when the generated carbon (by-product) from the process is captured and stored to generate a net-zero carbon fuel. In contrast, green hydrogen is directly generated from renewable energy via the electrolysis process and hence natively is carbon neutral.

> I've had some discussions with Bloomberg's hydrogen folks. … I said, how much incremental wind and solar do you need to install to make sure that you can generate the green hydrogen that meet 100% of your projections? The renewables folks said it's physically impossible to put that much renewable in place to generate hydrogen. So, we have to look at blue hydrogen. …It's [probably] more economic than green hydrogen, it's [also] more reliable. [Regarding] the carbon associated with the generation, the carbon dioxide molecule can be sequestered effectively.—Bryan Hassler

There are oil and gas companies that are interested in expanding their portfolio into blue hydrogen. This is an attractive growth area as OGI companies typically have access to feedstock natural gas needed, they have geologic and underground expertise for carbon geo-storage and access to a pipeline network for future distribution of hydrogen.

1.5 Harnessing Underground & Offshore Facilities & Expertise

Another way the OGI can navigate and thrive within the energy transition is by applying its expertise in subsurface resource exploration, extraction, and production, as well as offshore technology and operation in new fields. The industry in the last century has recovered hydrocarbon resources from shallow and deep reservoirs (greater than 10,000 feet) with temperatures ranging from near-surface to very high temperatures (up to 250 °C) (DeBruijn et al. 2008). These capabilities are useful in contemporary challenges encompassing energy security and sustainability, energy storage, and climate change mitigation. In the following sections, we will discuss how the industry can actively lead the development and sustainable deployment of technologies that address these challenges.

1.5.1 Underground Hydrogen Storage

Common ways to store energy for electricity are through pumped hydroelectric, battery, and thermal storage (U.S. EPA 2021b). Pumped hydroelectric storage is topography dependent and restricted to certain geographical locations (Rahman et al. 2015). Storing energy with lithium-ion, lead acid, and lithium-iron batteries puts pressure on existing mineral resources (i.e., rare earth metals) and leads to issues with toxic waste disposal, which in both cases is not sustainable (U.S. EPA 2021b). Thermal storage in surface tanks incurs significant energy losses during charging-discharging cycles (Sarbu and Sebarchievici 2018; Enescu et al. 2020). Other forms of non-fossil energy storage such as hydrogen and compressed air storage are not entirely suitable at surface conditions due to volume, pressure, temperature, and surface reactivity (in the case of hydrogen) restrictions. Therefore, we believe that underground energy storage is a necessary component in the global energy transition.

Surface hydrogen storage requires cryogenic tanks where liquid hydrogen is stored below −253 °C (hydrogen's boiling point) (Sundén 2019; Langmi et al. 2022). In cryogenic storage, about 35% of the energy stored is used to liquefy the gaseous hydrogen. Venting is occasionally required because the tanks can absorb heat from its surroundings, causing the liquid to boil (Sundén 2019). Hydrogen can also be stored in metallic tanks as a compressed gas. However, in the long term, this becomes unsafe because hydrogen can permeate the rigid structure of these metals due to its small atomic size leading to leakage and extensive hydrogen embrittlement (i.e., making metals significantly less ductile) (Murakami 2019; Langmi et al. 2022). Underground hydrogen storage (UHS) has been proposed to assuage these issues.

Hydrogen can be stored underground in aquifers, salt caverns, and depleted gas reservoirs (Caglayan et al. 2020; Bünger et al. 2016; Tarkowski 2019; Tartakovsky et al. 2020; Zivar et al. 2021). These subsurface reservoirs have the suitable capacity, porosity, and permeability, among other reservoir properties, to allow for both storage and production of hydrogen when needed. Although numerous underground gas storage sites worldwide are run by OGI producers, dedicated hydrogen storage sites are limited to four sites in operation. One is in Teesside in the United Kingdom, while the rest—Clemens Dome, Moss Bluff, and Spindletop—are located in Texas, United States (Tarkowski 2019; IEA 2021d; Zivar et al. 2021). Upscaling and commercializing this technology are necessary for hydrogen to become a mainstream carrier fuel. However, some existing challenges need to be addressed (Matos et al. 2019; Heinemann et al. 2021; IEA 2021d; Lysyy et al. 2021). There are open areas for research and development that need to be implemented via industry-led collaborations with academic and research institutions. The OGI has extensive experience with such partnerships that have accelerated onshore and offshore exploration and production of hydrocarbons over the years. The direct transfer of research outcomes from lab-scale research to field pilots has been facilitated by professional organizations like the Society of Petroleum Engineers (SPE), the American Association of Petroleum Geologists (AAPG), and the Society of Exploration Geophysicists (SEG). As OGI majors become integrated energy companies with significant participation in the electric utility market, efficient, safe, and sustainable underground hydrogen storage is necessary.

1.5.2 Geological Carbon Sequestration

Atmospheric carbon dioxide has been described as the dominant greenhouse gas contributing to existential climate change. One major way to mitigate CO_2 emissions and their effects is through carbon capture, utilization, and storage (CCUS). CCUS is a four-step process that ultimately leads to geological CO_2 sequestration or the repurposing of CO_2 for other industrial uses (Benson et al. 2005; IEA 2020a). Geological CO_2 sequestration is the only viable solution for large-scale carbon storage. The underground storage sites include deep saline aquifers, depleted hydrocarbon reservoirs, and unmineable coal seams (Benson et al. 2005; Aminu et al. 2017). An extensive monitoring. Reporting, and verification program would follow CO_2 injection to quantify and ensure storage integrity (Cuéllar-Franca and Azapagic 2015; Hepburn et al. 2019). Primary conditions for storage integrity in the aforementioned formations are a CO_2 impermeable caprock, good reservoir porosity, and permeability, sealing faults (if faults exist), and extensive reservoir connectivity. Other factors affecting storage integrity are rock (geomechanical and geochemical) properties, brine salinity, well placement and completion, and well abandonment procedures (Akindipe et al. 2021; Benson et al. 2005; Li et al. 2021a; Leung et al. 2014; Aminu et al. 2017; Saraji et al. 2013, 2014).

Since the 1950s, the OGI industry has been using CO_2 as a working fluid for oil recovery, first through the injection of carbonated water (i.e., CO_2-enriched brine) (Martin 1951a, b; Hickok et al. 1960; Christensen 1961) and then via direct CO_2 injection for enhanced oil recovery—starting in the 1970s with the SACROC project (Han et al. 2010; Merchant 2017). Through several pilot tests and injection optimization schemes, CO_2 utilization for enhanced oil recovery (EOR) has commercially produced several billion barrels of oil over the years. In both CO_2 injection approaches, significant amounts of the injected CO_2 are also stored as a distinct flowing phase or dissolved within the remaining oil and brine in the reservoir. CO_2-EOR projects are concentrated in North America (mostly the United States) (Global CCS Institute 2021). This is due to a working infrastructure (e.g., CO_2 pipelines) and a market for CO_2-EOR. For example, within the state of Wyoming, USA, ExxonMobil has been supplying CO_2 from its Shute Creek Treating Facility for EOR locally and around the Rocky Mountain region through existing pipeline networks since the 1990s (Parker et al. 2011).

Direct injection of CO_2 for storage is still emerging but rapidly accelerating. The first field-scale injection of CO_2 for dedicated storage was accomplished by Statoil (now Equinor) in their Sleipner Field, offshore Norway. This saline aquifer CO_2 storage project started in 1996, is still operational, and stores about 1 Mt of CO_2 annually. As of May 2022, there are six dedicated (non-CO_2 EOR) commercial CO_2 storage projects worldwide—in the United States (Illinois-Decatur), Canada (Quest), Norway (Sleipner and Snøhvit), Australia (Gorgon), and just recently Iceland (Orca) (Global CCS Institute 2021). The Orca project is the first commercial DAC project with an annual capture capacity of 4000 tons. CO_2 is stored in a mafic formation where it is mineralized—another industry first (Climeworks 2021; Carbfix 2021; Reuters 2021b). Other carbon-intensive countries like China and Brazil are intensifying their CCUS efforts through dedicated storage and CO_2-EOR projects at different stages of development (Global CCS Institute 2021).

Although the technology behind geological carbon storage has been applied for many years, the number of commercial projects and the total annual CO_2 storage considerably lag behind storage requirements for 2050 net-zero carbon goals. Specifically, 40 Mt of CO_2 (2020 data) is being stored annually, whereas a storage rate of 5000 Mt per year is needed to meet the 2050 goal of 220 Gt of total CO_2 stored as prescribed by the International Energy Agency (IEA 2021b). However, based on current storage potential estimates, existing underground storage sites have the capacity to store up to 55,000 Gt of CO_2 (IEA 2021b). The OGI's existing infrastructure and subsurface expertise could help to accelerate the rate of adaptation of CCUS as a critical mid-term solution in the energy transition.

1.5.3 Geothermal Energy

In geothermal energy systems, the natural heat energy from the earth's core is harnessed for electricity generation and heating. Geothermal energy is one of the oldest forms of energy that is manifest in surface systems like geysers, warm springs, and fumaroles within specific regions of the world with significant volcanic activity. The heat energy within the hot water and wet/dry steam has formed a primary source of energy for millennia (U.S. DOE 2019). These surface systems are directly linked to hydrothermal reservoirs in the subsurface. Analogous to hydrocarbon reservoirs, hydrothermal reservoirs are formed when hot water and steam migrate upwards from deeper hotter rocks near the earth's core through faults and fissures toward the earth's surface. Before reaching the surface, the migrating fluids could become trapped when they reach a region of impermeable rock (serving as a caprock), thereby forming a hydrothermal reservoir (DiPippo 2016; U.S. DOE 2019). As of 2020, the geothermal installed capacity stands at 14.6 GW, with the United States accounting for the majority (about 25% or 3.673 GW) of this capacity (IEA Geothermal 2020).

Most of the geothermal plants around the world used for electricity generation are driven by thermal energy from hydrothermal systems found in places like California in the United States, Indonesia, Japan, and Iceland (i.e., regions with high volcanic activity) (Richter 2020). In such plants, wells are drilled to produce hot water and steam used to drive turbines for electricity generation. These plants can run as dry steam, flash, or binary cycles. In dry steam plants, the dry steam from the reservoir is directly used to drive the turbine (U.S. DOE 2019). In flashed systems, the produced wet steam from the reservoir is flashed in a high-pressure column to separate the steam from the hot liquid water. Flashing could be single or multiple staged (Harvey and Wallace 2016; Moya et al. 2021). The resulting dry steam is used to drive the turbine while the hot water can be reused for on-site or industrial heating and/or re-injected into the reservoir. In binary cycle plants, the produced fluid from the reservoir is not directly used to drive the turbine. Instead, it is charged into a heat exchanger as a working fluid for vaporizing a secondary working fluid that is then used to drive a power-generating turbine (Mines 2016). Dry steam and flash systems are generally applicable for high temperature reservoirs (more than 180 °C). Binary power cycles, like the Organic Rankine Cycle, have the advantage of being able to produce power from lower temperature (not less than 100 °C) reservoirs (U.S. DOE 2019).

Recently, enhanced geothermal systems (EGS) have been developed (Gong et al. 2020). These are man-made reservoirs formed by injecting water and other fluids to induce fractures in hot (>150 °C) low-permeability rocks. EGS is usually first implemented on existing hydrothermal wells (in-field EGS) and can be gradually extended to other adjacent reservoirs (near-field EGS). EGS is still an evolving technology with most projects still at the demonstration stage (Doughty et al. 2018). Deep EGS, involving drilling to and stimulating deeper (up to 7 km and above) hot rocks, is also emerging (U.S. DOE 2019). One major issue limiting EGS commercialization is the high levelized cost of electricity (LCOE)—the cost per unit of energy generated—from EGS compared to conventional systems and other renewable energy sources (e.g., wind and solar). From recent 2021 estimates, the LCOE for EGS in the United States is between $156–$416 per MWh of electricity produced (NREL 2021a). Meanwhile, conventional hydrothermal and utility-scale solar photovoltaic (PV) are about $60–$85 and $30–$50 per MWh, respectively (NREL 2021a, b). Some might argue that the difference in technology performance is the opportunity cost incurred to provide a more reliable, dispatchable, and flexible baseload electricity—a major advantage of EGS (and geothermal energy as a whole). There are, however, many ongoing studies such as those being implemented by the U.S. department of energy (DOE) and other laboratories to optimize EGS and reduce its LCOE (U.S. DOE 2019; Robins et al. 2021; Hamm et al. 2021).

From the preceding discussion it is obvious that in the future, geothermal systems are not just going to increase in number and scale but in depth. This poses a major challenge for the global energy industry because of the very high temperatures that have never been encountered in other subsurface exploration and production ventures like oil and gas production. Drill bits, strings, tubing, casings, drilling fluids, cements, and sensors would need to be redesigned and fabricated with materials enabled to withstand such super-elevated temperature and pressure conditions. Well planning and design must be optimized to avoid issues around over-pressurization, lost circulation, and blowouts that could lead to higher drilling costs. Extensive rock and fluid characterizations at such conditions would need to be carried out. Since EGS is fracturing-intensive, the geomechanical properties of these reservoir systems need to be determined. EGS is heavily reliant on hydraulic fracturing (or fracking), which is banned in some states in the U.S. (e.g., New York, Washington, New Jersey, Pennsylvania, and Delaware) (DRBC 2021; Herrera 2020) and in other countries (e.g., France, Bulgaria, and Denmark) (Herrera 2020). This, alongside fracking's notorious reputation and seismic risks, may pose another challenge for global geothermal energy expansion.

The OGI is well primed for geothermal energy production due to its vast experience with directional drilling of vertical and lateral wells and hydraulic fracturing in challenging environments. However, geothermal energy production creates bigger challenges, which can spur similar industry innovation as was seen with deep-water offshore hydrocarbon production. Presently, the budding geothermal energy market is becoming attractive to investors, and many startup companies founded by former OGI experts (such as Fervo, Eavor, and Geothermix) are springing up (Roberts 2020). These companies are bridging the gap between demonstration and commercialization of innovative technology within geothermal energy. Oil majors are not particularly active in this market. However, it is very probable that they will join in as the market and regulatory barriers lessen just as in the broader renewables (wind, solar, and hydro) markets.

1.5.4 Offshore Wind and Solar

Wind and solar energy together comprise the majority (54% as of 2021) in the renewable energy portfolio (IRENA 2022). Overall costs of electricity generation from these sources have substantially decreased in the past two decades—with solar PV even becoming cheaper than non-renewable electricity from natural gas and coal in some parts of the world (IRENA 2021). One major challenge with these resources is the need for space, especially for utility-scale electricity generation with solar PV, solar thermal (or concentrating solar power [CSP]) arrays, and wind turbines. About 28% of the world's solar energy generation comes from land-based installations on the rooftops of residential and commercial buildings, as separate installations within an estate, and as CSP arrays for thermal energy generation (IRENA 2022). Similarly, 25.1% of global wind energy generation comes from onshore wind turbine installations (IRENA 2022). The effect of wind and solar energy installations on land use and the resulting land use change has been extensively studied in the literature (Poggi et al. 2018; Wu et al. 2020; Ioannidis and Koutsoyiannis 2020; McKenna et al. 2022). These studies have revealed that if the ongoing development trend continues, especially with the rising demand for low-carbon and renewable energy, land availability for other uses (such as for agriculture) would significantly dwindle (Merrill 2021).

One way to increase wind and solar energy generation without necessarily changing land use is by accelerated offshore development. Majority of the current offshore developments are wind farms where generated electricity is transferred from offshore substations through subsea cables to onshore substations for onward transmission and distribution. This technology is being spearheaded by countries in Western Europe (e.g., United Kingdom, The Netherlands, and Denmark) that do not have sufficient land area for onshore installations (IEA 2019). Offshore solar energy generation, on the other hand, is relatively new. By offshore solar we mean those that are installed out at sea and not floating solar arrays that can be installed on surface reservoirs and lakes. The first offshore solar array developed by Oceans of Energy—the Zon-op-Zee (Solar-at-Sea) project—was installed in 2019 in the Dutch North Sea with an initial generating capacity of 17 kW, expandable to 1 MW (Snieckus 2020). Nowadays, these modular arrays are being developed as a hybrid renewable energy system installed between offshore wind turbine masts (Snieckus 2020).

The development of offshore wind energy projects started in 1991 with the Vindeby installation developed by Ørsted (a Danish company) comprising 11 turbines with an installed capacity of 5 MW. As of 2021, the global installed generation capacity has risen to 56 GW, accounting for 1.8% of renewable energy generation worldwide (IRENA 2022). In these wind farms, the turbines are installed on rigid platforms piled into the seabed. Floating platforms are gradually being deployed but still account for just 0.2% of installed capacity (Bureau Veritas 2021). Over the years, the United Kingdom has been the major player in the offshore wind business. However, in recent years, China has experienced the largest growth with a current installed capacity of 26 MW (IRENA 2022). Outside China, the top active developers in the offshore wind market are Ørsted, SSE Renewables, ENGIE, and CGN—all renewable energy companies (Woodworth 2021). Legacy OGI companies are not yet active in this market. However, some oil servicing companies, such as Saipem, and Hyundai Heavy Industries (HHI), have been actively engaged in the construction and installation of offshore wind platforms (Reuters 2020; Bureau Veritas 2021).

In the past two decades, wind turbines and solar PV panels have provided low-carbon energy on offshore oil production platforms. These have not been utility-scale installations as they have only served power redundancy functions for topside equipment. Although late in the game, oil majors are beginning to diversify into utility-scale offshore wind development. They are rapidly buying up offshore wind leases around the world. As of 2021, four of the 18 offshore wind leases in the United States are controlled by oil majors—Equinor, BP, and Shell—in joint ventures (Wasser and Storrow 2021). In the United Kingdom, BP and TotalEnergies recently harnessed their financial advantage to outbid Ørsted and other renewable energy developers for major project development rights of an 8-GW offshore wind project (Reed 2021; Brooks 2021). Apart from China, which is accelerating development through state-owned companies, most OGI firms involved in large-scale offshore wind are European entities. This could be due to more stringent EU directives and targets on energy decarbonization. However, in April 2021, Chevron, a U.S. major, announced a partnership with Moreld AS, a Norwegian industrial firm, to invest in the floating offshore wind turbine technology developed by Ocergy Inc., a California tech firm (Saul and Crowley 2021). This is not a direct project installation investment as Chevron does not currently control any offshore wind lease.

Offshore renewable energy development comes with significant costs and risks that have limited development relative to land-based technologies. Based on current estimates, the LCOE for offshore wind ($67–160 per MWh) is more than double that of land-based wind ($27–75 per MWh) (NREL 2022a, b). Innovative methods to reduce associated costs in offshore projects are imperative. One way might be retrofitting decommissioned oil production platforms with wind turbines and solar PVs. Research on this has gained traction; however, demonstration projects are needed to confirm commercial feasibility (Alessi et al. 2019; Leporini et al. 2019; Mendes et al. 2021; Solomin et al. 2019). In addition to cost reduction, inherent risks need to be managed or mitigated. Major risks to development are severe weather patterns and natural disasters such as hurricanes, typhoons, tsunamis, and earthquakes that can disrupt both installation and production operations. Other risks are technical (ease of maintenance) and ecological (impact on marine life and bird migration). Although the OGI is vastly experienced in managing offshore-related risks, pivoting to wind and solar installations may require partnerships with existing players to reduce their risk exposure.

1.6 Summary and Conclusion

The OGI has been subject to transformation to find a place in the new trend of the global energy transition. There are several internal and external forces that are fueling this transformation. First, with much of the world's "easy oil" already produced, the companies will have to use increasingly sophisticated technologies to find and deliver these energy sources to the market (Perrons and Cosby 2020). Second, the expectations placed upon the sector by many of its stakeholders have grown considerably with regards to environmental stewardship, safety, and human welfare (Perrons and Cosby 2020). Third, the market volatility in the OGI, the dwindling demand for oil due to the impact of COVID-19, predictions of a long-term decline in demand, and the significant reduction in capital investment in the industry (Elijah et al. 2021).

Facing these challenges, the industry will require an increasing degree of technological and commercial sophistication to continue to be a part of the world's sustainable energy mix. To meet these challenges, the OGI has opportunities to employ digital technology to drive efficiency and productivity into operations, help develop standards and technologies for low-carbon fuels, and provide its subsurface and offshore expertise in various areas related to energy/carbon storage, geothermal energy, and offshore wind and solar industry.

References

Agwu OE, Akpabio JU, Alabi SB, Dosunmu A (2018) Artificial intelligence techniques and their applications in drilling fluid engineering: a review. J Petrol Sci Eng 167:300–315. https://doi.org/10.1016/j.petrol.2018.04.019

Akindipe D, Saraji S, Piri M (2021) Salt precipitation during geological sequestration of supercritical CO_2 in saline aquifers: a pore-scale experimental investigation. Adv Water Resour 155:104011. https://doi.org/10.1016/j.advwatres.2021.104011

Alessi L, Correia JAFO, Fantuzzi N (2019) Initial design phase and tender designs of a jacket structure converted into a retrofitted offshore wind turbine. Energies 12(4): 659. https://doi.org/10.3390/en12040659

Aminu MD, Nabavi SA, Rochelle CA, Manovic V (2017) A review of developments in carbon dioxide storage. Appl Energy 208:1389–1419. https://doi.org/10.1016/j.apenergy.2017.09.015

Andoni M, Robu V, Flynn D, Abram S, Geach D, Jenkins D, McCallum P, Peacock A (2019) Blockchain technology in the energy sector: a systematic review of challenges and opportunities. Renew Sustain Energy Rev 100:143–174. https://doi.org/10.1016/j.rser.2018.10.014

Beck C, Rashidbeigi S, Roelofsen O, Speelman E (2020) The future is now: how oil and gas companies can decarbonize (Energy Transition and Decarbonization). McKinsey & Company. https://www.mckinsey.com/industries/oil-and-gas/our-insights/the-future-is-now-how-oil-and-gas-companies-can-decarbonize

Bello O, Holzmann J, Yaqoob T, Teodoriu C (2015) Application of artificial intelligence methods in drilling system design and operations: a review of the state of the art. J Artif Intell Soft Comput Res 5(2):121–139. https://doi.org/10.1515/jaiscr-2015-0024

Benson S, Anderson J, Bradshaw J, Deguchi G, Gale J, Heidug W, Holloway S, Keith D, Lloyd P, Thomson J, Zarlenga F, Zhou D, Celia M, Lombardi S, Oldenburg C, Stevens S, Borm G, Hawkins D (2005) Underground geological storage. In: Metz B, Davidson O, de Coninck H, Loos M, Meyer L (eds) IPCC special report on carbon dioxide capture and storage. Cambridge University Press, pp 195–276. https://www.ipcc.ch/site/assets/uploads/2018/03/srccs_chapter5-1.pdf

BP (2021) What is sustainable aviation fuel (SAF) and why is it important? | News and views. Air Bp, July. https://www.bp.com/en/global/air-bp/news-and-views/views/what-is-sustainable-aviation-fuel-saf-and-why-is-it-important.html

Bravo C, Saputelli L, Rivas F, Pérez AG, Nikolaou M, Zangl G, de Guzmán N, Mohaghegh S, Nunez G (2013) State of the art of artificial intelligence and predictive analytics in the E&P industry: a technology survey. SPE J 19(04):547–563. https://doi.org/10.2118/150314-PA

Brooks C (2021, February 9) BP, Total swoop in pricey UK offshore wind auction. IHS Markit. https://cleanenergynews.ihsmarkit.com/research-analysis/bp-total-swoop-on-pricey-uk-offshore-wind-auction-.html

Bünger U, Michalski J, Crotogino F, Kruck O (2016) 7—large-scale underground storage of hydrogen for the grid integration of renewable energy and other applications. In: Ball M, Basile A, Veziroğlu TN (eds) Compendium of hydrogen energy. Woodhead Publishing, pp 133–163. https://doi.org/10.1016/B978-1-78242-364-5.00007-5

Bureau Veritas (2021, December 1) HHI receives an Approval in Principle for Hi-Float Floating Offshore Wind Turbine Foundation. Marine & Offshore. https://marine-offshore.bureauveritas.com/newsroom/hhi-receives-approval-principle-hi-float-floating-offshore-wind-turbine-foundation

Bušić A, Marđetko N, Kundas S, Morzak G, Belskaya H, Ivančić Šantek M, Komes D, Novak S, Šantek B (2018) Bioethanol production from renewable raw materials and its separation and purification: a review. Food Technol Biotechnol 56(3):289–311. https://doi.org/10.17113/ftb.56.03.18.5546

Buterin V (n.d.) Ethereum: a next-generation smart contract and decentralized application platform, 36

Caglayan DG, Weber N, Heinrichs HU, Linßen J, Robinius M, Kukla PA, Stolten D (2020) Technical potential of salt caverns for hydrogen storage in Europe. Int J Hydrogen Energy 45(11):6793–6805. https://doi.org/10.1016/j.ijhydene.2019.12.161

Caldarelli G (2020) Understanding the blockchain oracle problem: a call for action. Information 11(11):509. https://doi.org/10.3390/info11110509

California Air Resources Board (2020) Low carbon fuel standard. California Air Resources Board. https://ww2.arb.ca.gov/sites/default/files/2020-09/basics-notes.pdf

Carbfix (2021) The world's largest direct air capture and CO_2 storage plant is ON. Carbfix Website, September 7. http://carbfix.com/the-worlds-largest-direct-air-capture-and-co2-storage-plant-is-on

Choubey S, Karmakar GP (2021) Artificial intelligence techniques and their application in oil and gas industry. Artif Intell Rev 54(5):3665–3683. https://doi.org/10.1007/s10462-020-09935-1

Christensen RJ (1961) Carbonated waterflood results—Texas and Oklahoma, May 26. https://doi.org/10.2118/66-MS

Clean Energy (2020a) Chevron partners with clean energy on adopt-a-port initiative to reduce emissions. Clean Energy Fuels, July 7. https://www.cleanenergyfuels.com/press-room/chevron-partners-with-clean-energy-on-adopt-a-port-initiative-to-reduce-emissions

Clean Energy (2020b) Clean energy to make more carbon-negative fuel available for transportation with BP. Clean Energy Fuels, December 21. https://www.cleanenergyfuels.com/press-room/clean-energy-to-make-more-carbon-negative-fuel-available-for-transportation-with-bp

Clean Energy (2021a) Clean energy and total sign joint venture to develop carbon-negative fuel and infrastructure. Clean Energy Fuels, March 4. https://www.cleanenergyfuels.com/press-room/clean-energy-and-total-sign-joint-venture-to-develop-carbon-negative-fuel-and-infrastructure

Clean Energy (2021b) Clean energy and BP create JV to invest in RNG fuel production. Clean Energy Fuels, March 10. https://www.cleanenergyfuels.com/press-room/clean-energy-and-bp-create-jv-to-invest-in-rng-fuel-production

Clean Energy (2021c) Chevron, clean energy fuels extend adopt-a-port initiative to reduce emissions. Clean Energy Fuels, May 11. https://www.cleanenergyfuels.com/press-room/chevron-clean-energy-fuels-extend-adopt-a-port-initiative-to-reduce-emissions

Climeworks (2021) Orca is Climeworks' new large-scale carbon dioxide removal plant. Climeworks Website. https://climeworks.com/roadmap/orca

Corden WM (1984) Booming sector and Dutch disease economics: survey and consolidation. Oxf Econ Pap 36(3):359–380. https://www.jstor.org/stable/2662669

Cuéllar-Franca RM, Azapagic A (2015) Carbon capture, storage and utilisation technologies: a critical analysis and comparison of their life cycle environmental impacts. J CO_2 Utilization 9:82–102. https://doi.org/10.1016/j.jcou.2014.12.001

DRBC (2021, February 25) New DRBC Regulation Prohibits High Volume Hydraulic Fracturing in the Delaware River Basin. Delaware River Basin Commission. https://www.state.nj.us/drbc/home/newsroom/news/approved/20210225_newsrel_HVHF-rulemaking.html

DeBruijn G, Skeates C, Greenaway R, Harrison D, Parris M, James S, Mueller F, Ray S, Riding M, Temple L, Wutherich K (2008) High-pressure, high-temperature technologies. Oilfield Rev 20(3):46–60. https://www.slb.com/-/media/files/oilfield-review/high-pressure-high-temperature

Desai JN, Pandian S, Vij RK (2021) Big data analytics in upstream oil and gas industries for sustainable exploration and development: a review. Environ Technol Innov 21:101186. https://doi.org/10.1016/j.eti.2020.101186

DiChristopher T (2021) Market for responsibly sourced gas begins to take root, stakeholders say. S&P Global - Market Intelligence, September 14. https://www.spglobal.com/marketintelligence/en/news-insights/latest-news-headlines/market-for-responsibly-sourced-gas-begins-to-take-root-stakeholders-say-66578408

DiPippo R (2016) 8 - Overview of geothermal energy conversion systems: Reservoir-wells-piping-plantreinjection. In: DiPippo R (ed) Geothermal power generation. Woodhead Publishing, pp 203–215. https://doi.org/10.1016/B978-0-08-100337-4.00008-5

Doughty C, Dobson PF, Wall A, McLing T, Weiss C (2018) GeoVision analysis supporting task force report: Exploration. https://escholarship.org/uc/item/4v7054cw

EIA (2022) Biofuels explained: ethanol. U.S. Energy Information Administration (EIA), April 7. https://www.eia.gov/energyexplained/biofuels/ethanol.php

Elijah O, Ling PA, Abdul Rahim SK, Geok TK, Arsad A, Kadir EA, Abdurrahman M, Junin R, Agi A, Abdulfatah MY (2021) A survey on industry 4.0 for the oil and gas industry: upstream sector. IEEE Access 9:144438–144468. https://doi.org/10.1109/ACCESS.2021.3121302

Enescu D, Chicco G, Porumb R, Seritan G (2020) Thermal energy storage for grid applications: Current status and emerging trends. Energies 13(2):340. https://doi.org/10.3390/en13020340

Equitable Origin (2017) EO100™ standard for responsible energy development. Equitable Origin. https://energystandards.org/wp-content/uploads/2020/10/EO100-Standard-for-Responsible-Energy-Development_2017_PT.pdf

Feinstein L (2022). Responsibly sourced gas: boon or boondoggle? Sightline Institute, February 10. https://www.sightline.org/2022/02/10/responsibly-sourced-gas-boon-or-boondoggle/

Gai H, Beath J, Fang J, Lou HH (2020) Alternative emission monitoring technologies and industrial internet of things–based process monitoring technologies for achieving operational excellence. Curr Opin Green Sustain Chem 23:31–37. https://doi.org/10.1016/j.cogsc.2020.04.009

Global CCS Institute (2021) The global status of CCS: 2021. https://www.globalccsinstitute.com/wp-content/uploads/2021/11/Global-Status-of-CCS-2021-Global-CCS-Institute-1121.pdf

Gong F, Guo T, Sun W, Li Z, Yang B, Chen Y, Qu Z (2020) Evaluation of geothermal energy extraction in Enhanced Geothermal System (EGS) with multiple fracturing horizontal wells (MFHW). Renewable Energy 151:1339–1351. https://doi.org/10.1016/j.renene.2019.11.134

Gray R (2020) Why raising the alcohol content of Europe's fuels could reduce carbon emissions. The EU Research and Innovation Magazine, Horizon, March 9. https://ec.europa.eu/research-and-innovation/en/horizon-magazine/why-raising-alcohol-content-europes-fuels-could-reduce-carbon-emissions

Gupta D, Shah M (2021) A comprehensive study on artificial intelligence in oil and gas sector. Environ Sci Pollut Res. https://doi.org/10.1007/s11356-021-15379-z

Hamm SG, Anderson A, Blankenship D, Boyd LW, Brown EA, Frone Z, Hamos I, Hughes HJ, Kalmuk M, Marble A, McKittrick AMW, Metcalfe E, Morse LJ, Nieto A, Payne J, Porse SL, Prisjatschew A, Stutz GR, Tasca C, Vandermeer W, Watson G, Weathers M, Winick, J (2021) Geothermal energy R&D: an overview of the U.S. Department of Energy's geothermal technologies office. J Energy Res Technol 143(10). https://doi.org/10.1115/1.4049581

Han WS, McPherson BJ, Lichtner PC, Wang FP (2010) Evaluation of trapping mechanisms in geologic CO_2 sequestration: case study of SACROC Northern Platform, a 35-year CO_2 injection site. Am J Sci 310(4):282–324. https://doi.org/10.2475/04.2010.03

Hanga KM, Kovalchuk Y (2019) Machine learning and multi-agent systems in oil and gas industry applications: a survey. Comput Sci Rev 34:100191. https://doi.org/10.1016/j.cosrev.2019.08.002

Harvey W, Wallace K (2016) 10 - Flash steam geothermal energy conversion systems: Single-, double-, and triple-flash and combined-cycle plants. In: DiPippo R (ed) Geothermal Power Generation. Woodhead Publishing, pp 249–290. https://doi.org/10.1016/B978-0-08-100337-4.00010-3

Heinemann N, Alcalde J, Miocic JM, Hangx SJT, Kallmeyer J, Ostertag-Henning C, Hassanpouryouzband A, Thaysen EM, Strobel GJ, Schmidt-Hattenberger C, Edlmann K, Wilkinson M, Bentham M, Haszeldine RS, Carbonell R, Rudloff A (2021) Enabling large-scale hydrogen storage in porous media – the scientific challenges. Energy Environ Sci 14(2):853–864. https://doi.org/10.1039/D0EE03536J

Hepburn C, Adlen E, Beddington J, Carter EA, Fuss S, Mac Dowell N, Minx JC, Smith P, Williams CK (2019) The technological and economic prospects for CO_2 utilization and removal. Nature 575(7781):87–97. https://doi.org/10.1038/s41586-019-1681-6

Herrera H (2020, January 6) The legal status of fracking worldwide: An environmental law and human rights perspective. The Global Network for Human Rights and the Environment (GNHRE). https://gnhre.org/human-rights/the-legal-status-of-fracking-worldwide-an-environmental-law-and-human-rights-perspective/

Hickok CW, Christensen RJ, Ramsay HJJ (1960) Progress review of the K&S carbonated waterflood project. J Petrol Technol 12(12):20–24. https://doi.org/10.2118/1474-G

How do you reshape when today's future may not be tomorrow's reality? - Oil and Gas Digital Transformation and the Workforce Survey 2020 (n.d.) TRUE Global Intelligence. ey.com/oilandgas/digitalskills

IATA (2021a) Net zero 2050: sustainable aviation fuels. IATA. https://www.iata.org/contentassets/b3783d24c5834634af59148c718472bb/factsheet_saf-.pdf

IATA (2021b) Industry statistics: fact sheet. IATA. https://www.iata.org/en/iata-repository/publications/economic-reports/airline-industry-economic-performance---october-2021b---data-tables/

IEA (2019) Offshore Wind Outlook 2019. International Energy Agency. https://iea.blob.core.windows.net/assets/495ab264-4ddf-4b68-b9c0-514295ff40a7/Offshore_Wind_Outlook_2019.pdf

IEA (2020a) CCUS in clean energy transitions. IEA. https://www.iea.org/reports/ccus-in-clean-energy-transitions

IEA (2020b) The oil and gas industry in energy transitions. International Energy Agency. https://iea.blob.core.windows.net/assets/4315f4ed-5cb2-4264-b0ee-2054fd34c118/The_Oil_and_Gas_Industry_in_Energy_Transitions.pdf

IEA (2021a) Global energy review 2021. International Energy Agency. https://iea.blob.core.windows.net/assets/d0031107-401d-4a2f-a48b-9eed19457335/GlobalEnergyReview2021.pdf

IEA (2021b) The world has vast capacity to store CO_2: net zero means we'll need it – analysis. IEA, April. https://www.iea.org/commentaries/the-world-has-vast-capacity-to-store-co2-net-zero-means-we-ll-need-it

IEA (2021c) Aviation—fuels & technologies. IEA, November. https://www.iea.org/fuels-and-technologies/aviation

IEA (2021d) Global hydrogen review 2021. IEA. https://iea.blob.core.windows.net/assets/5bd46d7b-906a-4429-abda-e9c507a62341/GlobalHydrogenReview2021.pdf

IEA Geothermal (2020) 2020 Annual Report. IEA Geothermal.

Ioannidis R, Koutsoyiannis D (2020) A review of land use, visibility and public perception of renewable energy in the context of landscape impact. Applied Energy 276:115367. https://doi.org/10.1016/j.apenergy.2020.115367

IPCC (2014) AR5 climate change 2014: mitigation of climate change. IPCC. https://www.ipcc.ch/report/ar5/wg3/

IPCC (2018) Summary for policymakers—global warming of 1.5 °C. https://www.ipcc.ch/sr15/chapter/spm/

IRENA (2021) Renewable Power Generation Costs in 2020. International Renewable Energy Agency (IRENA). https://www.irena.org/-/media/Files/IRENA/Agency/Publication/2021/Jun/IRENA_Power_Generation_Costs_2020.pdf

IRENA (2022) Technologies: Renewable Energy Technologies. International Renewable Energy Agency. https://www.irena.org/Statistics/View-Data-by-Topic/Capacity-and-Generation/Technologies

Klein L, Ramachandran M, van Kessel T, Nair D, Hinds N, Hamann H, Sosa N (2018) Wireless sensor networks for fugitive methane emissions monitoring in oil and gas industry. IEEE Inter Congr Internet Things (ICIOT) 2018:41–48. https://doi.org/10.1109/ICIOT.2018.00013

Langmi HW, Engelbrecht N, Modisha PM, Bessarabov D (2022) Chapter 13—hydrogen storage. In: Smolinka T, Garche J (eds) Electrochemical power sources: fundamentals, systems, and applications. Elsevier, pp 455–486. https://doi.org/10.1016/B978-0-12-819424-9.00006-9

Leporini M, Marchetti B, Corvaro F, Polonara F (2019) Reconversion of offshore oil and gas platforms into renewable energy sites production: Assessment of different scenarios. Renewable Energy 135:1121–1132. https://doi.org/10.1016/j.renene.2018.12.073

Leung DYC, Caramanna G, Maroto-Valer MM (2014) An overview of current status of carbon dioxide capture and storage technologies. Renew Sustain Energy Rev 39:426–443. https://doi.org/10.1016/j.rser.2014.07.093

Li D, Saraji S, Jiao Z, Zhang Y (2021a) CO_2 injection strategies for enhanced oil recovery and geological sequestration in a tight reservoir: an experimental study. Fuel 284:119013. https://doi.org/10.1016/j.fuel.2020.119013

Li H, Yu H, Cao N, Tian H, Cheng S (2021b) Applications of artificial intelligence in oil and gas development. Arch Computat Methods Eng 28(3):937–949. https://doi.org/10.1007/s11831-020-09402-8

Li M, Trencher G, Asuka J (2022) The clean energy claims of BP, Chevron, ExxonMobil and Shell: a mismatch between discourse, actions and investments. PLoS One 17(2):e0263596. https://doi.org/10.1371/journal.pone.0263596

Lu H, Guo L, Azimi M, Huang K (2019) Oil and Gas 4.0 era: a systematic review and outlook. Comput Ind 111:68–90. https://doi.org/10.1016/j.compind.2019.06.007

Lysyy M, Fernø M, Ersland G (2021) Seasonal hydrogen storage in a depleted oil and gas field. Int J Hydrogen Energy 46(49):25160–25174. https://doi.org/10.1016/j.ijhydene.2021.05.030

Martin JW (1951a) The use of carbon dioxide for increasing the recovery of oil. Q Bull Interstate Oil Compact 10(1):50

Martin JW (1951b) Additional oil production through flooding with carbonated water. Prod Mon 15(9):18

Matek B, Gawell K (2015) The benefits of baseload renewables: a misunderstood energy technology. Electr J 28(2):101–112. https://doi.org/10.1016/j.tej.2015.02.001

Matos CR, Carneiro JF, Silva PP (2019) Overview of large-scale underground energy storage technologies for integration of renewable energies and criteria for reservoir identification. J Energy Storage 21:241–258. https://doi.org/10.1016/j.est.2018.11.023

McGarr A, Bekins B, Burkardt N, Dewey J, Earle P, Ellsworth W, Ge S, Hickman S, Holland A, Majer E, Rubinstein J, Sheehan A (2015) Coping with earthquakes induced by fluid injection. Science 347(6224):830–831. https://doi.org/10.1126/science.aaa0494

McKenna R, Mulalic I, Soutar I, Weinand JM, Price J, Petrović S, Mainzer K (2022) Exploring tradeoffs between landscape impact, land use and resource quality for onshore variable renewable energy: An application to Great Britain. Energy 250:123754. https://doi.org/10.1016/j.energy.2022.123754

Mehta D, Tanwar S, Bodkhe U, Shukla A, Kumar N (2021) Blockchain-based royalty contract transactions scheme for Industry 4.0 supply-chain management. Inf Process Manag 58(4):102586. https://doi.org/10.1016/j.ipm.2021.102586

Mendes P, Correia JAFO, Castro JM, Fantuzzi N, Aidibi A, Manuel L (2021) Horizontal and vertical axis wind turbines on existing jacket platforms: Part 1 – A comparative study. Structures 32:1069–1080. https://doi.org/10.1016/j.istruc.2021.01.069

Merchant D (2017) Enhanced oil recovery – the history of CO_2 conventional wag injection techniques developed from lab in the 1950's to 2017, July 17. https://doi.org/10.7122/502866-MS

Merrill D (2021, April 29) U.S. Needs A Lot More Land to Go Green by 2050. Bloomberg.Com. https://www.bloomberg.com/graphics/2021-energy-land-use-economy/

Michaga MFR, Michailos S, Hughes KJ, Ingham D, Pourkashanian M (2021) 10—techno-economic and life cycle assessment review of sustainable aviation fuel produced via biomass gasification. In: Ray RC (ed) Sustainable biofuels. Academic Press, pp 269–303. https://doi.org/10.1016/B978-0-12-820297-5.00012-8

Mines G (2016) 13 - Binary geothermal energy conversion systems: Basic Rankine, dual–pressure, and dual–fluid cycles. In: DiPippo R (ed) Geothermal power generation. Woodhead Publishing, pp 353–389. https://doi.org/10.1016/B978-0-08-100337-4.00013-9

Moya D, Akinsipe OC, Kaparaju P (2021) Chapter 1—Various cycle configurations for geothermal power plants. In: Colpan CO, Ezan MA, Kizilkan O (eds) Thermodynamic analysis and optimization of geothermal power plants. Elsevier, pp 3–15. https://doi.org/10.1016/B978-0-12-821037-6.00005-6

Murakami Y (2019) 21—hydrogen embrittlement. In: Murakami Y (ed) Metal Fatigue, 2nd edn. Academic Press, pp 567–607. https://doi.org/10.1016/B978-0-12-813876-2.00021-2

Murphy D (2019) Renewable energy in the 21st century. Trubooks.

Mussatto SI, Dragone G, Guimarães PMR, Silva JPA, Carneiro LM, Roberto IC, Vicente A, Domingues L, Teixeira JA (2010) Technological trends, global market, and challenges of bio-ethanol production. Biotechnol Adv 28(6):817–830. https://doi.org/10.1016/j.biotechadv.2010.07.001

Nakamoto S (n.d.) Bitcoin: a peer-to-peer electronic cash system, 9.

Neste (2022) Neste annual report 2021. https://www.neste.com/sites/neste.com/files/attachments/corporate/investors/corporate_governance/neste_annual_report_2021.pdf

Nguyen T, Gosine RG, Warrian P (2020) A systematic review of big data analytics for oil and gas industry 4.0. IEEE Access 8:61183–61201. https://doi.org/10.1109/ACCESS.2020.2979678

NREL (2021a) Annual Technology Baseline 2021—Geothermal. National Renewable Energy Laboratory (NREL). https://atb.nrel.gov/electricity/2021/geothermal

NREL (2021b) Annual Technology Baseline 2021—Utility-Scale PV. National Renewable Energy Laboratory (NREL). https://atb.nrel.gov/electricity/2021/geothermal

NREL (2022a) Annual Technology Baseline 2022—Land-based Wind. National Renewable Energy Laboratory (NREL). https://atb.nrel.gov/electricity/2022/land-based_wind

NREL (2022b) Annual Technology Baseline 2022—Offshore Wind. National Renewable Energy Laboratory (NREL). https://atb.nrel.gov/electricity/2022/offshore_wind

O'Rourke D, Connolly S (2003) Just oil? The distribution of environmental and social impacts of oil production and consumption. Annu Rev Environ Resour 28(1):587–617. https://doi.org/10.1146/annurev.energy.28.050302.105617

Oxford Institute for Energy Studies (2020) Decarbonization pathways for oil and gas. Oxf Energy Forum 121:1–40. https://www.oxfordenergy.org/wpcms/wp-content/uploads/2020/03/OEF121.pdf

Parker ME, Northrop S, Valencia JA, Foglesong RE, Duncan WT (2011) CO_2 management at ExxonMobil's LaBarge field, Wyoming, USA. Energy Procedia 4:5455–5470. https://doi.org/10.1016/j.egypro.2011.02.531

Patel H, Prajapati D, Mahida D, Shah M (2020) Transforming petroleum downstream sector through big data: a holistic review. J Petrol Explor Prod Technol 10(6):2601–2611. https://doi.org/10.1007/s13202-020-00889-2

Perrons RK, Cosby T (2020) Applying blockchain in the geoenergy domain: the road to interoperability and standards. Appl Energy 262:114545. https://doi.org/10.1016/j.apenergy.2020.114545

Perrons RK, Jensen JW (2015) Data as an asset: what the oil and gas sector can learn from other industries about "Big Data." Energy Policy 81:117–121. https://doi.org/10.1016/j.enpol.2015.02.020

Poggi F, Firmino A, Amado M (2018) Planning renewable energy in rural areas: Impacts on occupation and land use. Energy 155:630–640. https://doi.org/10.1016/j.energy.2018.05.009

Project Canary (2021) How to tap certified responsibly sourced gas as a quick win in the transition to a clean energy future. Project Canary, September 10. https://www.projectcanary.com/white-paper/how-to-tap-certified-responsibly-sourced-gas-as-a-quick-win-in-the-transition-to-a-clean-energy-future/

Rahman, F, Baseer, MA, Rehman S (2015) Chapter 4—Assessment of electricity storage systems. In: Sørensen B (ed) Solar energy storage. Academic Press, pp 63–114. https://doi.org/10.1016/B978-0-12-409540-3.00004-9.

Ram M, Mondal MK (2022) Chapter 13 - Biomass gasification: a step toward cleaner fuel and chemicals. In: Gurunathan B, Sahadevan R, Zakaria ZA (eds) Biofuels and bioenergy. Elsevier, pp 253–276. https://doi.org/10.1016/B978-0-323-85269-2.00008-3

Reed S (2021, February 8) Oil Giants Win Offshore Wind Leases in Britain. The New York Times. https://www.nytimes.com/2021/02/08/business/oil-companies-offshore-wind-britain.html

Restuccia F, Kanhere SD, Salil S, Melodia T, Das SK (2019). Blockchain for the internet of things: present and future. http://arxiv.org/abs/1903.07448

Reuters (2020, August 25) Saipem to build 450 MW offshore wind farm in Italy. Reuters. https://www.reuters.com/article/saipem-wind-farm-idAFL8N2FR2Y1

Reuters (2021a) Clean energy, total sign JV for renewable natural gas production. Reuters, March 4. https://www.reuters.com/article/us-clean-energy-fuels-total-rng-idUSKBN2AW1AB

Reuters (2021b) World's largest plant capturing carbon from air starts in Iceland. Reuters, September 13. https://www.reuters.com/business/environment/worlds-largest-plant-capturing-carbon-air-starts-iceland-2021b-09-08/

Reuters (2022) Neste says to become first global renewable fuels producer. Reuters, March 2. https://www.reuters.com/business/energy/neste-team-up-with-us-oil-company-marathon-produce-renewable-fuels-2022-03-02/

Richter A (2020, January 27). The Top 10 Geothermal Countries 2019 – based on installed generation capacity (MWe). Think GeoEnergy. https://www.thinkgeoenergy.com/the-top-10-geothermal-countries-2019-based-on-installed-generation-capacity-mwe/

Roberts D (2020, October 21) Geothermal energy is poised for a big breakout. Vox. https://www.vox.com/energy-and-environment/2020/10/21/21515461/renewable-energy-geothermal-egs-ags-supercritical

Robins JC, Kolker A, Flores-Espino F, Pettitt W, Schmidt B, Beckers K, Pauling H, Anderson B (2021) 2021 U.S. Geothermal Power Production and District Heating Market Report (NREL/TP-5700-78291). National Renewable Energy Laboratory. https://www.nrel.gov/docs/fy21osti/78291.pdf

Russo TN (2021) Responsibly sourced gas: time to change the natural gas industry's narrative. Climate and Energy 37(7):22–27. https://doi.org/10.1002/gas.22215

Saraji S, Borowczak M (2021) A blockchain-based carbon credit ecosystem [White Paper]. http://arxiv.org/abs/2107.00185

Saraji S, Khalaf C (2022) Blockchain applications in the energy industry. In: Tehrani PM (ed) Regulatory aspects of artificial intelligence on blockchain. IGI Global, pp 159–180. https://doi.org/10.4018/978-1-7998-7927-5.ch008

Saraji S, Goual L, Piri M, Plancher H (2013) Wettability of supercritical carbon dioxide/water/quartz systems: simultaneous measurement of contact angle and interfacial tension at reservoir conditions. Langmuir 29(23):6856–6866. https://doi.org/10.1021/la3050863

Saraji S, Piri M, Goual L (2014) The effects of SO2 contamination, brine salinity, pressure, and temperature on dynamic contact angles and interfacial tension of supercritical CO_2/brine/quartz systems. Int J Greenhouse Gas Control 28:147–155. https://doi.org/10.1016/j.ijggc.2014.06.024

Sarbu, I., & Sebarchievici, C. (2018). A Comprehensive Review of Thermal Energy Storage. Sustainability, 10(1), 191. https://doi.org/10.3390/su10010191

Saul J, Crowley K (2021, April 13) Chevron Invests in Offshore Wind Energy for First Time. Bloomberg.Com. https://www.bloomberg.com/news/articles/2021-04-13/chevron-s-venture-arm-makes-first-investment-in-offshore-wind

Saxena RC, Adhikari DK, Goyal HB (2009) Biomass-based energy fuel through biochemical routes: a review. Renew Sustain Energy Rev 13(1):167–178. https://doi.org/10.1016/j.rser.2007.07.011

Shell (2021) Shell starts production at Shell New Energies Junction City, its first US renewable natural gas facility. Shell: 2021 Media Releases, September 28. https://www.shell.us/media/2021-media-releases/shell-starts-production-at-shell-new-energies-junction-city-its-first-us-renewable-natural-gas-facility.html

Snieckus D (2020, February 16) World's first offshore solar array rides out storm Ciara off Netherlands. Recharge. https://www.rechargenews.com/transition/worlds-first-offshore-solar-array-rides-out-storm-ciara-off-netherlands/2-1-757022

Solomin E, Kaixuan G, Yanming J, Minghan D (2019) Offshore PV and wind based energy generation. In: 2019 International Conference on Industrial Engineering, Applications and Manufacturing (ICIEAM), pp 1–6. https://doi.org/10.1109/ICIEAM.2019.8742965

S&P Global (2020) What is energy transition? S&P Global - Articles, February 24. https://www.spglobal.com/en/research-insights/articles/what-is-energy-transition

Sundén B (2019) Chapter 3—hydrogen. In: Sundén B (ed) Hydrogen, batteries and fuel cells. Academic Press, pp 37–55. https://doi.org/10.1016/B978-0-12-816950-6.00003-8

Tankimovich MR (2018) Big data in the oil and gas industry: a promising courtship [Thesis]. https://doi.org/10.15781/T2804Z300
Tariq Z, Aljawad MS, Hasan A, Murtaza M, Mohammed E, El-Husseiny A, Alarifi SA, Mahmoud M, Abdulraheem A (2021) A systematic review of data science and machine learning applications to the oil and gas industry. J Petrol Explor Prod Technol 11(12):4339–4374. https://doi.org/10.1007/s13202-021-01302-2
Tarkowski R (2019) Underground hydrogen storage: characteristics and prospects. Renew Sustain Energy Rev 105:86–94. https://doi.org/10.1016/j.rser.2019.01.051
Tartakovsky AM, Marrero CO, Perdikaris P, Tartakovsky GD, Barajas-Solano D (2020) Physics-informed deep neural networks for learning parameters and constitutive relationships in subsurface flow problems. Water Resour Res 56(5):e2019WR026731. https://doi.org/10.1029/2019WR026731
U.S. DOE (2019) GeoVision: Harnessing the heat beneath our feet (DOE/EE–1306). U.S. Department of Energy. https://www.energy.gov/sites/default/files/2019/06/f63/GeoVision-full-report-opt.pdf
U.S. EPA (2016a) Global greenhouse gas emissions data [Overviews and Factsheets], January 12. https://www.epa.gov/ghgemissions/global-greenhouse-gas-emissions-data
U.S. EPA (2016b) Understanding global warming potentials [Overviews and Factsheets], January 12. https://www.epa.gov/ghgemissions/understanding-global-warming-potentials
U.S. EPA (2020) An overview of renewable natural gas from biogas (EPA 456-R-20-001). United States Environmental Protection Agency. https://www.epa.gov/sites/default/files/2020-07/documents/lmop_rng_document.pdf
U.S. EPA (2021a) Reducing methane from the oil and natural gas source category. United States Environmetal Protection Agency. https://www.epa.gov/system/files/documents/2021-11/epas-proposed-oil-and-gasrules.presentation-11.2.2021.pdf
U.S. EPA (2021b, November 22) Electricity storage [Overviews and Factsheets]. U.S. Environmental Protectiono Agency. https://www.epa.gov/energy/electricity-storage
Wanasinghe TR, Gosine RG, James LA, Mann GKI, de Silva O, Warrian PJ (2020) The Internet of Things in the oil and gas industry: a systematic review. IEEE Internet Things J 7(9):8654–8673. https://doi.org/10.1109/JIOT.2020.2995617
Wasser M, Storrow B (2021, October 15) Big Oil wants to be Big Wind. Can fossil fuel companies be trusted? WBUR. https://www.wbur.org/news/2021/10/15/offshore-wind-shell-bp-equinor-concerns
Wood DG (n.d.) Polkadot: vision for a heterogeneous multi-chain framework, 21
Woodworth SB (2021, May 3) 2020's top wind developers are targeting significant wind capacity development, eyeing an equal split between onshore and offshore projects. IHS Markit. https://ihsmarkit.com/research-analysis/2020s-top-wind-developers-are-targeting-significant-wind-capac.html
Wu GC, Leslie E, Sawyerr O, Cameron DR, Brand, E, Cohen B, Allen D, Ochoa M, Olson A (2020) Low-impact land use pathways to deep decarbonization of electricity. Environ Res Lett 15(7): 074044. https://doi.org/10.1088/1748-9326/ab87d1
Zivar D, Kumar S, Foroozesh J (2021) Underground hydrogen storage: a comprehensive review. Int J Hydrogen Energy 46(45):23436–23462. https://doi.org/10.1016/j.ijhydene.2020.08.138

Dr. Soheil Saraji is an Associate Professor of Energy and Petroleum Engineering, an Adjust Professor at the School of Energy Resources, and co-director of the Hydrocarbons Research Laboratory at the University of Wyoming. He has eighteen years of research experience and more than 35 peer-reviewed journal publications in subsurface energy extraction, storage, and carbon geo-sequestration. Furthermore, Dr. Saraji is a pioneer in applied blockchain research for the oil and gas industry. He has developed new courses and research initiatives on this topic at the University of Wyoming.

Dr. Dayo Akindipe is an energy transition specialist with technical, research, and consulting experience that cuts across CCUS, geothermal energy, and oil and gas. He currently leads projects on geothermal technoeconomic analysis and repurposing of oil and gas infrastructure for energy storage. He has a BSc in Chemical Engineering from the University of Lagos, Nigeria, an MSc in Petroleum (Reservoir) Engineering from the Norwegian University of Science and Technology (NTNU), Norway, and a PhD in Petroleum Engineering (minor in Environment and Natural Resources) from the University of Wyoming, USA.

Chapter 2
Climate and Financial Markets

Si Chen

2.1 Introduction

Is there still a future for oil and gas companies?

In the past decade, major banks and asset managers have increasingly focused on climate change and shunned the oil and gas industry in the process. Coalitions such as Climate Action 100+, the Net Zero Asset Managers Alliance, and the Net Zero Banking Alliance signed up some of the world's biggest investors. Individual and institutional investors flocked to the Environmental, Social, and Governance (ESG) investing bandwagon. The oil and gas industry looked as if it might permanently lose the support of public financial markets.

Then, in March 2022, two major shocks hit the energy and financial markets: The war in the Ukraine and the SEC's proposal for climate disclosures. Together, they may bring investors and the industry together for one more try to move forward together in the energy transition.

2.2 Climate Investing and the ESG Boom

The climate is changing. Investors are taking note. In the early 2000s, there was a "clean tech boom," when venture capitalists funded cutting edge startups only to see many of them fail. In the following decade, major banks, pension funds, and mutual funds are taking action on climate change.

They see climate change pose real risks to assets and returns, in the form of two new types of financial risks: physical and transition risks. Physical risks refer to damage to physical assets due to extreme weather, such as storms and droughts.

S. Chen (✉)
Open Source Strategies, Inc, Los Angeles, CA, USA
e-mail: sichen@opensourcestrategies.com

S. Saraji and S. Chen, *Sustainable Oil and Gas Using Blockchain*,
Lecture Notes in Energy 98, https://doi.org/10.1007/978-3-031-30697-6_2

Perhaps the best examples of physical risk are the fires that led to $30 billion in liabilities and the bankruptcy of PG&E in 2019 and the winter Texas freeze of 2021, which led to insured losses of $10 to $20 billion and total economic loss of perhaps $80 to $130 billion (Golding et al. 2021).

Transition risks refer to a decline in revenue and profitability as governments and consumers respond to climate change, for example, through Greenhouse Gas (GHG) emissions regulations and taxes, as well as shifts in demand from high to low emission products. Transition risks due to regulations and taxes include the European Union's Emissions Trading Scheme (ETS), China's new emissions cap and trade program, and EU's proposed Cross Border Adjustment Mechanism (CBAM), which imposes carbon pricing on imported products. Consumer shifts due to climate change include calls to ban single-use plastics, switching to electric vehicles, vegan diets, and even grassroots movements to avoid air travel (Szegedi 2021).

As an example, Moody's has listed $4.5 trillion of rated debt securities as having high transition risk and $7.2 trillion as having high physical risks. Companies in the coal mining sector are classified as having the highest risks, while oil and gas, power utilities, steel, materials, and auto manufacturing companies are classified as high risks (Fig. 2.1). While they provide this as a warning, they have not acted to downgrade, because "*There is also typically greater scope for issuers to adjust to or manage environmental risks; for example, by adapting business models, changing policies (in the case of governments), or by passing on expected cost increases to customers or taxpayers*" (Moody's Investor Service 2020).

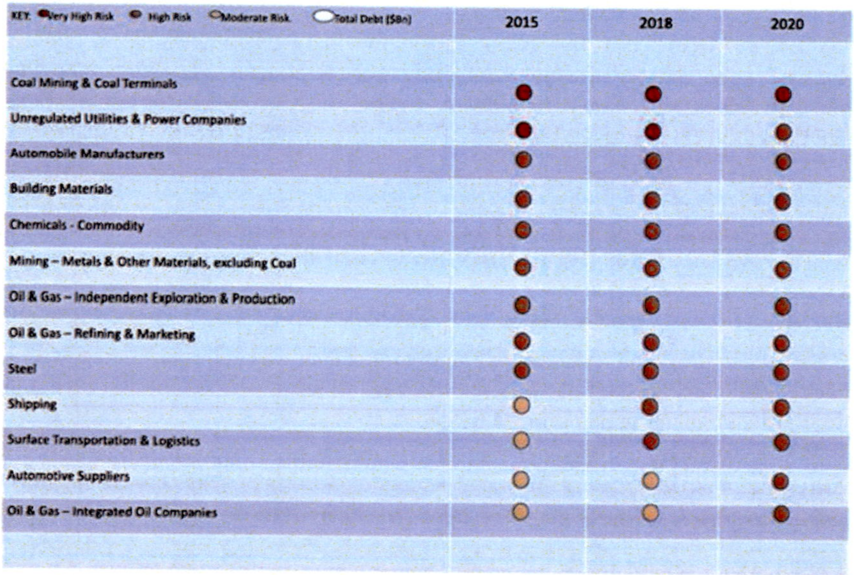

Fig. 2.1 Sectors with high environmental risks. *Source* Environmental Heat Map 2020 (Moody's 2020)

This focus on the financial risks of climate change has also drawn the attention of central banks and regulators. In 2017, eight central banks formed the Network of Central Banks and Supervisors for Greening the Financial System, which has grown to 108 members and 17 observers. They have conducted stress tests of banks and insurers for the effects of climate change on their portfolios, finding significant risks for loans to coal-fired power plants, equity investments in carbon-intensive industries, and real estate property values due to storms. The Bank of England and the European Central Bank are planning to increase capital requirements for climate risk, thus raising the cost of capital for industries with high GHG emissions. Other central banks are looking to support greater investment in green industries through bond purchases.

In response, banks and investors have started to focus on "financed emissions," or the amount of emissions attributable to a portfolio's holdings, as the new climate metric. The sheer amounts involved—hundreds of times greater than the banks' operational emissions—have made it a hot topic for the media (Hodgson 2021). This, in turn, is drawing out responses from the big banks to "do something about climate." Morgan Stanley, for example, became the first major U.S. bank to commit to net zero financed emissions by 2050 (Morgan Stanley 2021). The Net Zero Banking Alliance, sponsored by the UN Environment Programme, claims 114 member banks with $68 trillion in assets, or 38% of the global total (United Nations Environment Programme Finance Initiative, n.d.)

At the same time, new instruments have been introduced to finance the energy transition. Issuance of "green bonds," which began in 2007 to finance climate and environmental projects, have risen to over $500 billion by 2021 and is projected to reach $5 trillion by 2025 (Mutua). Sustainability-linked bonds, whose interest rates are tied to specific climate-related targets, are also growing rapidly, reaching over $135 billion in 2021 (Woellert 2022). Unlike green bonds, which are pledged for specific types of projects such as renewable energy, sustainability-linked bonds could be issued by oil and gas companies for targets such as reducing methane emissions. Both are in high demand and trade to "greemiums" over similar bonds without the green or sustainable labels.

Much more could be expected in these directions. Calculating financed emissions is complex and data-intensive, and many banks do not have the capability for it yet. How these bonds fit into financed emissions and net zero banking commitments will need to be worked out. Green and sustainability-linked bonds will probably need third-party standards and verification, as opposed to just being self-declared "green" today. Finally, as banks try to meet their targets, they will focus on high emissions sectors such as the oil and gas industry first. For example, HSBC is planning to reach its long-term net zero financed emissions commitment by setting targets to reduce financed emissions of its oil and gas and power generation utilities portfolio first (HSBC Group 2022).

The asset management industry has also embraced climate investing. Larry Fink, CEO of BlackRock with $9 trillion under management, made the headlines in his 2020 Letter to CEO's, when he wrote that

> Climate Risk Is Investment Risk. ... Our investment conviction is that sustainability- and climate-integrated portfolios can provide better risk-adjusted returns to investors. ... Even if only a fraction of the projected impacts is realized, this is a much more structural, long-term crisis. Companies, investors, and governments must prepare for a significant reallocation of capital. (Fink 2020)

In his 2022 Letter to CEO's, he challenged his fellow CEO's:

> Every company and every industry will be transformed by the transition to a net zero world. The question is, will you lead, or will you be led?

And the good news, for fund managers at least, is that climate change sells! Investors are pouring money into Environmental, Social, and Governance ("ESG") funds primarily out of concern about climate change and its impact. ESG funds began in the mid 1990s and were a niche category of the investment universe until about a decade ago. Today, they represent a staggering ½ of total US assets under management (Carlson 2020) and are expected to reach over $50 *trillion* by 2025, as shown in Fig. 2.2 (Henze and Boyd 2021).

This shift to climate investing has given fuel to the movement to divest from fossil fuels companies. In 2012, environmentalist Bill McKibben and student organizers launched the "Do the Math" tour, based partly on research from the CarbonTracker about a "carbon bubble" (Carbon Tracker Initiative 2011). Their argument was that most of the hydrocarbon reserves could not be burned if we were to reach the climate goals of the Paris Agreement and therefore are worthless. A decade later, the movement counts "*1,485 institutions publicly committed to at least some form of fossil fuel divestment, representing an enormous $39.2 trillion of assets under management.*" Most importantly, according to Divest Invest 2021 (2021):

> One of the most important victories for the movement has been the financial elite's gradual acceptance of the movement's core financial arguments. Fossil fuels are a bad bet financially.

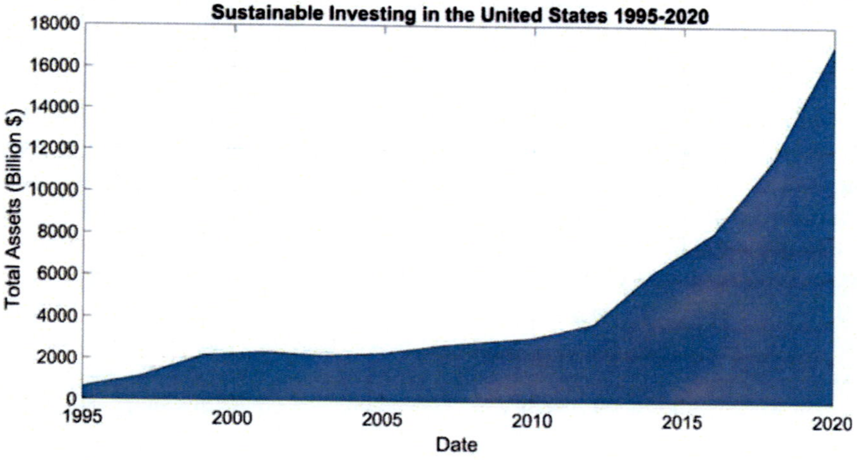

Fig. 2.2 Growth of ESG investing. *Source* Carlson (2020)

> It was once considered a fringe position to argue that the fossil fuel industry's value is dependent on "stranded assets"–fuel reserves still buried in the ground that will become worthless in a clean, renewable energy future. Now the concept is cited by BlackRock, the largest investment house in the world, as a reason to divest.
>
> The movement can now also offer solid proof that divestment is a sound financial strategy. Early adopters of divestment strategies are reporting positive financial results. Surveys and analyses by Wall Street firms support it. Divestment now moves in a positive feedback loop. As more and more institutions announce their plans to divest, many cite the financial reality that climate change will make fossil fuels obsolete and a renewable energy future inevitable. And by doing so, they are both hastening that change and convincing others of its inevitability.

Indeed, a BlackRock study for The Teachers' Retirement System of the City of New York on potential fossil fuel divestment found that "*Amongst surveyed peer institutions, materiality and risk-avoidance have been the driving consideration behind divestment research and implementation*" (Blackrock Sustainable Investing 2021). Also, the President of Harvard announced that it "*has no direct investments in companies that explore for or develop further reserves of fossil fuels. Moreover, HMC [Harvard Management Company] does not intend to make such investments in the future,*" because "*we do not believe such investments are prudent*" (Bacow 2021).

Many ESG funds overweigh industries with low perceived climate risk, such as tech and healthcare, at the expense of energy. Within the energy-related industry, they are shifting from coal and oil and gas stocks to renewable energy and technology stocks. This has given an added push to industries that were already in transition. For example, coal was already in decline due to competition from natural gas. It thus became an easy target for commitments to "phase out coal." While the majority of the institutions in the Divestment Database have pledged a full divestment of fossil fuels, most of the for-profit corporations who made divestment pledges targeted coal or coal and tar sands (Stand.earth, n.d.). This has led to a string of coal industry bankruptcies, including Peabody, Murray, and Westmoreland. In a sign of the industry's shift, Peabody, the world's largest private-sector coal company, staved off a second bankruptcy with a debt restructuring, then announced its new business model: It was getting into renewable energy.

Meanwhile, the financial markets have also pushed the automotive industry further in its transformation from internal combustion engines to electric vehicles, thanks to the incredible success of EV makers. Tesla, for example, now has a market cap that exceeds that of Toyota, Mercedes-Benz, BMW, Volkswagen, Ford, and GM *combined.* Even Lucid Motors, which so far has delivered only 300 vehicles, has a market cap of $38 billion.[1] The major auto manufacturers do not want to be left behind in this race. GM has set a goal of all EV's by 2035, Honda and Jaguar by 2040, Volvo and Bentley by 2030. Even if not explicitly pledging 100% EV's, the others are greatly increasing EV sales while phasing out combustion vehicles. VW, for example, states that 2026 will be its last combustion vehicle platform. Aiding this transition is the long list of governments, including California, New York, Washington, Canada,

[1] As of March 7, 2022.

Kenya, Mexico, Turkey, and the UK, which have pledged that all new car and van sales will be electric by 2040 or earlier (Lambert 2021).

The shifting economics of renewables versus coal and natural gas is also helping some major utilities announce aggressive targets for reducing GHG emissions (St. John 2020). American Electric Power, for example, has committed to net zero carbon dioxide emissions by 2050, with 2030 emissions 80% below 2000 levels. It intends to do so by investing $9.9 billion in renewable energy through 2026, bringing renewables to 50% of its total capacity by 2030 (American Electric Power). Southern Company has "*established an intermediate goal of a 50% reduction in GHG emissions from 2007 levels by 2030 and a long-term goal of net zero GHG emissions by 2050*" (Southern Company). Dominion Energy has promised "*Net Zero carbon and methane emissions for both our electric and gas businesses by 2050*" and "*the total share of zero-carbon generation should rise to around 70 percent by 2035*" (Dominion Energy 2021).

2.3 Effect on the Oil and Gas Industry

What does this mean for the oil and gas industry? Already suffering from overproduction, the industry was further caught between calls for divestment and customer shifts to new technologies. The combination has been devastating for oil and gas investors. From 2010 to February 2022, the MSCI World index outperformed the MSCI World Energy index by 373% versus 132%. Even more dramatically, within the energy sector alone, the MVIS Low Carbon Energy index has returned 214% since 2010, while the S&P Oil & Gas Exploration index has *lost* −18% of its value during the same time (Fig. 2.3).[2]

While many events, from an oversupply of oil shale production to COVID-19, led to these results, a hidden undercurrent is that the economy has become less reliant on fossil fuels. The fall in oil and gas prices during 2014 to 2016 did not lead to additional demand (Stocker et al. 2018), and the overall percentage of energy from fossil fuels has started to decline (DeSilver 2020). This, coupled with the booming demand for ESG investments, has led to the poor performance of oil and gas stocks. More importantly, the major oil companies are shifting away from additional exploration towards renewable energy, causing the oil and gas exploration stocks to underperform.

Naturally, this led to lower valuations for the oil and gas companies. While most people are familiar with the Price/Earnings ratio for stocks, the Price/Book ratio is in fact a better measure of relative valuations for companies when earnings are very volatile, as is the case for the energy industry. The Price/Book ratio is defined as the market price of the stock divided by the accounting value per share, based on the balance sheet of the company. Of course, the book of value of a company's assets is not the real market value of those assets, just like the cost of a building is not its

[2] As of month end February 28, 2022, based on ETF returns.

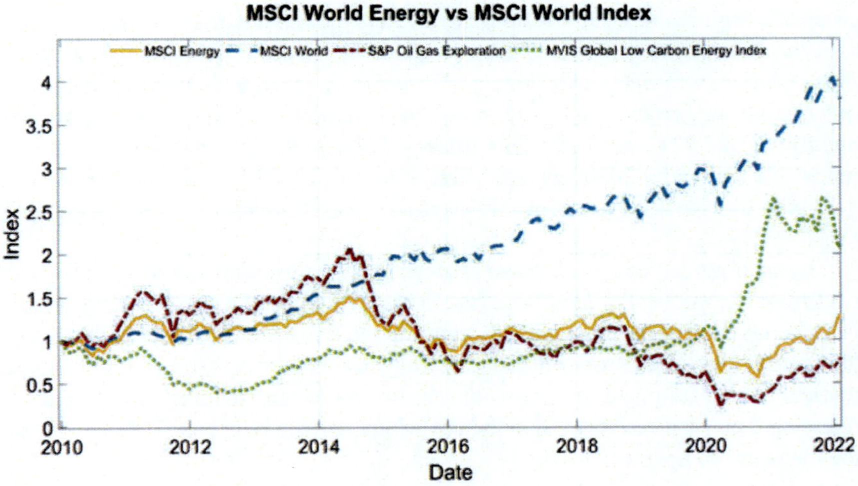

Fig. 2.3 Relative performance of energy stocks versus general market, 2010 to February 2022. *Source* Yahoo! Finance

current market value. Nevertheless, the Price/Book ratio is a good measure of the market's assessment of the value of a company's assets versus the historical cost basis of all the investments made in the company. It could therefore be thought of as the market's vote of confidence in the company's ability to produce future earnings, as well as a measure of how much value the company's management is creating with the investors' money.

In early March 2022, for example, the S&P 500 index had a Price/Book ratio of 4.01, reflecting near-record low interest rates and a rally going back to 2009. The S&P 500 Information Technology index has a Price/Book ratio of 9.48, reflecting the market's belief that the major tech companies will continue to grow earnings with minimal investment in physical assets long into the future. In contrast, even with a runup in energy stocks, the S&P Energy index sports a Price/Book ratio of only 2.44, nearly 40% less than the S&P 500 index and 75% less than 9.48 for the S&P 500 Information Technology index.[3] Another way to think about this is that $1 invested in the energy companies has become $2.44, versus $4.01 in the average large-cap company and $9.48 in a tech company.

Within the energy sector, oil majors such as Exxon, Chevron, and ConocoPhillips have Price/Book ratios of 1.6 to 1.7, whereas renewable energy companies such as NextEra and Vesta have Price/Book ratios of 4.77 to 6.55. In other words, the stock market is willing to pay $4 to $7 for each $1 invested in renewable energy, versus only about $1.6 in traditional oil and gas. This is yet another indication that the market believes that the future is with renewable energy.

Many oil and gas companies got the message. BP, for example, has "*set out a new strategy that will see us transform from being an International Oil Company focused*

[3] March 8, 2022.

on producing resources, to an Integrated Energy Company focused on delivering solutions for customers," where "*the cash flows from our oil, gas and refining activities enable our strategy, allowing us to invest in the energy transition and support our two growth areas – low carbon electricity and energy, and convenience and mobility*" (BP 2021). As part of this strategy, BP is promising to develop 50 GW of renewable energy by 2030, enough to meet the needs of 36 million people. Shell's goal is to provide enough renewable electricity for 50 million households by 2030 (Shell 2021).

These efforts are part of a broader strategy to align the oil majors with the realities of climate change and the demands of the financial markets. They are seeking to move from being oil and gas producers to capturing the downstream customer businesses that are moving away from oil and gas. They are also trying to meet the financial markets' overall demand for better climate disclosures and reduced climate risk, by reducing their own emissions and the emissions of their products. Thus far, we see the major oil companies:

1. Reporting Scopes 1, 2, and 3 emissions from their own operations and their customers' use of their products.
2. Committing to "net zero" targets of their own operations and their owned subsidiaries.
3. Committing to reducing the emissions of their customers. In BP's case, that means a 50% reduction in the carbon intensity of their business. Shell goes even further to target a full 100% reduction in carbon intensity with the use of carbon offsets.
4. Reducing the methane releases and flarings that add to the carbon intensity of their existing oil and gas output.
5. Moving downstream to develop renewable energy projects, electric vehicle charging, and biofuels for transportation aviation.

All this will be supported by the enormous cash flows from existing oil and gas assets. And enormous they are: Shell generated $30.2 billion in operating cash flows in 2021, BP $17.2 billion.

2.4 Who Will Own the Energy Transition?

These enormous amounts of cash, however, might not end up supporting strategies envisioned by the companies' management. This much cash looks very inviting to another group of players: Private Equity.

Alternatingly worshiped as the gods of finance and demonized as the worst excesses of extreme wealth, private equity funds are in fact the yin to the yang of the stock markets. Whereas public market investors value companies based on earnings and growth potential, private equity investors look for companies which have assets or cash flows but are in some ways neglected by the public market, either due to cyclical downturns, size, or limited growth potential. The best-known private equity strategy is a Leveraged Buyout or LBO, which uses a combination of bank debt, high yield bond financing, and long-term committed capital from their partners to buy a mature company. Other private equity strategies include looking for companies with hidden value, such as real estate or other saleable assets, or rolling up multiple midsized companies to sell as a larger entity at a premium. Because of their reliance on debt financing, they find companies with large bankable assets and high cash flows particularly attractive.

Oil and gas companies, with their vast assets in proven reserves, enormous cash flows, and low relative valuations, make very tempting targets to private equity funds hungry for deals. Today's private equity industry has over $7 trillion in assets under management (Wigglesworth 2021). Combined with additional financing from banks and high-yield investors, the industry has a total buying power of at least twice as much. All this cash that needs to be invested has created a fierce competition in deals, propelling the prices paid for assets even higher.

Private equity investors typically look at potential acquisition candidates using an "enterprise multiple," defined as the enterprise value of the company divided by its EBITDA. The enterprise value of the company is the total value of the company, including both equity, debt, and minus any cash. EBITDA, or Earnings Before Interest, Taxes, Depreciation, and Amortization, is an approximation of the cash flow available to the owner of the company. The ratio of enterprise value divided by EBITDA is thus a multiple of raw cash flow and helps gauge potential returns available after financing a purchase of the company with debt. A list of enterprise value multiplier for Oil & Gas Industry is summarized in Table 2.1.

In 2021, the average multiple for private equity acquisitions reached 15x (Farman 2022; Berkery Noyes, n.d.). This ratio reflects a general preference of private equity investors for companies with more stable cash flows. Some may object to the volatility of oil and gas businesses, but ultimately, the low valuation of oil and gas companies may prove too tempting for them to pass up.

Meanwhile, two other groups could be looking at buying assets from publicly traded oil and gas companies for reasons other than cash flows. In parallel and sometimes overlapping with private equity funds are the sovereign wealth funds, which are large pools of assets controlled by governments. They have risen in importance over the years and now also control over $7 trillion in assets. These funds invest both directly and through third-party managers such as private equity funds. Notably, the largest sovereign wealth funds include those of oil producing economies, such as Norway, Saudi Arabia, Abu Dhabi, Kuwait, Qatar, and Russia, and exporting economies, such as China and South Korea (Statista 2022). Their investment objectives often include both financial returns and strategic objectives of their governments.

Finally, there are the national oil companies, which are far larger than the publicly traded oil and gas companies. According to the IEA, national oil companies account for nearly two-thirds of the world's oil reserves and 57% of the world's oil production (IEA 2020). These huge entities operate with different agendas, with some looking for profitability and investment results, others looking to create jobs and economic development, and still some looking for strategic assets and energy security. As a

Table 2.1 EBITDA enterprise value multiplier for Oil & Gas Industry, March 2022

Ticker	Name	Sector	Sub-sector	Enterprise-to-ebitda
ENI.MI	Eni S.p.A	Energy	Oil & Gas Integrated	3.41
1605.T	Inpex Corporation	Energy	Oil & Gas E&P	3.78
5020.T	ENEOS Holdings, Inc	Energy	Oil & Gas Refining & Marketing	3.98
TTE.PA	TotalEnergies SE	Energy	Oil & Gas Integrated	4.16
APA	APA Corporation	Energy	Oil & Gas E&P	4.53
5019.T	Idemitsu Kosan Co., Ltd	Energy	Oil & Gas Refining & Marketing	4.60
REP.MC	Repsol, S.A	Energy	Oil & Gas Integrated	4.85
OMV.VI	OMV Aktiengesellschaft	Energy	Oil & Gas Integrated	4.93
GALP.LS	Galp Energia, SGPS, S.A	Energy	Oil & Gas Integrated	5.70
RDSB.L	Royal Dutch Shell plc	Energy	Oil & Gas Integrated	6.28
COP	ConocoPhillips	Energy	Oil & Gas E&P	6.40
IMO.TO	Imperial Oil Limited	Energy	Oil & Gas Integrated	6.45
MRO	Marathon Oil Corporation	Energy	Oil & Gas E&P	6.98
SU.TO	Suncor Energy Inc	Energy	Oil & Gas Integrated	7.00
OXY	Occidental Petroleum Corporation	Energy	Oil & Gas E&P	7.01
EOG	EOG Resources, Inc	Energy	Oil & Gas E&P	7.16
BP.L	BP p.l.c	Energy	Oil & Gas Integrated	7.42
CNQ.TO	Canadian Natural Resources Limited	Energy	Oil & Gas E&P	7.48
WPL.AX	Woodside Petroleum Ltd	Energy	Oil & Gas E&P	7.60
XOM	Exxon Mobil Corporation	Energy	Oil & Gas Integrated	8.10
ALD.AX	Ampol Limited	Energy	Oil & Gas Refining & Marketing	8.22
CVE.TO	Cenovus Energy Inc	Energy	Oil & Gas Integrated	8.53
CVX	Chevron Corporation	Energy	Oil & Gas Integrated	8.75
PKI.TO	Parkland Corporation	Energy	Oil & Gas Refining & Marketing	9.00
PXD	Pioneer Natural Resources Company	Energy	Oil & Gas E&P	9.12
MPC	Marathon Petroleum Corporation	Energy	Oil & Gas Refining & Marketing	9.18

(continued)

Table 2.1 (continued)

Ticker	Name	Sector	Sub-sector	Enterprise-to-ebitda
BKR	Baker Hughes Company	Energy	Oil & Gas Equipment & Services	9.49
FANG	Diamondback Energy, Inc	Energy	Oil & Gas E&P	9.93
VLO	Valero Energy Corporation	Energy	Oil & Gas Refining & Marketing	10.36
HES	Hess Corporation	Energy	Oil & Gas E&P	10.58
DVN	Devon Energy Corporation	Energy	Oil & Gas E&P	10.80
KMI	Kinder Morgan, Inc	Energy	Oil & Gas Midstream	10.87
KEY.TO	Keyera Corp	Energy	Oil & Gas Midstream	11.03
STO.AX	Santos Limited	Energy	Oil & Gas E&P	11.75
VPK.AS	Koninklijke Vopak N.V	Energy	Oil & Gas Equipment & Services	12.34
HAL	Halliburton Company	Energy	Oil & Gas Equipment & Services	12.93
SOL.AX	Washington H. Soul Pattinson and Company Limited	Energy	Thermal Coal	13.09
WMB	The Williams Companies, Inc	Energy	Oil & Gas Midstream	13.59
OKE	ONEOK, Inc	Energy	Oil & Gas Midstream	13.67
TRP.TO	TC Energy Corporation	Energy	Oil & Gas Midstream	14.24
SLB	Schlumberger Limited	Energy	Oil & Gas Equipment & Services	15.04
ENB.TO	Enbridge Inc	Energy	Oil & Gas Midstream	15.62
PPL.TO	Pembina Pipeline Corporation	Energy	Oil & Gas Midstream	15.72
NESTE.HE	Neste Oyj	Energy	Oil & Gas Refining & Marketing	16.99
PSX	Phillips 66	Energy	Oil & Gas Refining & Marketing	23.89
ORG.AX	Origin Energy Limited	Energy	Oil & Gas Integrated	23.94
LUNE.ST	Lundin Energy AB (publ)	Energy	Oil & Gas E&P	24.58

Source Yahoo! Finance

group, they have little transparency to the outside world (Manley et al. 2019; Natural Resource Governance Institute 2022).

All three groups are looking to acquire assets at a time when publicly traded companies are shedding assets to meet their climate goals. BP, for example, states that

> While these divestments may not directly lead to a reduction in absolute global emissions, they accelerate the pace bp can grow low-carbon businesses and underpin our aim to reduce our oil and gas production by around 40% by 2030.

In 2020, BP's divestment of its petrochemical and Alaska business units helped it reduce Scope 1 emissions by 5.4 million tons CO_2e, air emissions by 23%, methane emissions from upstream operations by 22%, as well as its water consumption (BP 2021). Shell, prodded by The Hague District Court's ruling that it must reduce emissions more aggressively, sold $11 billion of assets in 2021, including a $9.5 billion sale of its Permian basin assets to ConocoPhillips. Even ExxonMobil has been selling—$3.5 billion in 2021 (Rystad Energy 2022).

Altogether, the major oil and gas companies are looking to shed a staggering $140 billion in assets. Buyers include privately held companies such as Ineos, which was founded originally in 1998 by private equity investors and has subsequently purchased numerous petrochemical and energy assets. It recently added the veteran CFO of BP, Brian Gilvary, to its management team. After Ineos bought Hess's oil and gas assets in Denmark despite Denmark's announcement that it would halt oil production by 2050, Gilvary stated "*We have an appetite to acquire. We have the ability to do multibillion dollar deals. This deal was attractive to us. We know the fiscal regime under which we will be operating. We know what the endgame looks like*" (Raval 2021).

At an enterprise multiple of 6 financed with 50% debt financing, the cash-on-cash returns to the acquirer could be north of 30% per year. Buyers of oil and gas assets at these valuations don't have to worry much about the long-term energy transition. Just a few years of business-as-usual is enough to make the returns and earn the high performance fees of private equity funds. Thus, by pushing down the valuations of publicly traded companies, the divestment movement is perversely shifting assets to a group with less transparency and a shorter-term focus.

Finally, the underperformance of oil and gas companies is garnering interest from activist shareholders. The most famous example is Engine No. 1, a relatively small fund which won a proxy to elect three members to the board of Exxon. The fund appealed to the growing view that positive climate and ESG performance would lead to better long-term investment results, while Exxon's focus on short-term profits has led to "*diminished returns, high debt levels, and questions about its ability to maintain its dividend.*" Engine No. 1 found backing from a range of institutional investors, including BlackRock, Vanguard, and State Street (Henisz 2021).

A storming of the Bastille, though, it was not. A year after the proxy vote, Exxon released its "Advancing Climate Solutions 2022 Progress Report." The company now "*aims to achieve net-zero Scope 1 and 2 greenhouse gas emissions from its operated assets by 2050.*" This includes reducing methane emissions, investing in carbon capture technologies, hydrogen, and low-carbon biofuels. Still, it has no plans for partner assets, no targets for reducing Scope 3 emissions of its customers, and continues to rely on "Stated Energy Policy" scenarios in its long-term planning, even as governments around the world seem to enact new climate regulations daily.

Part of the problem is that shareholder proxy proposals have limited powers. SEC rules, for example, state that a shareholder proposal is excludable if it "*deals with a matter relating to the company's ordinary business operations*" (AutoZone, Inc.; Rule 14a-8 no-action letter 2021). It must also not exceed 500 words in length (Securities and Exchange Commission, n.d.). As such, only proposals which deal with top level governance, such as election of board members, or broad policies could be included on the proxy. It is a long way from the boardroom to actual implementation, especially at a large oil company. Time will tell what effects Board-level proxy activism really has.

An even bolder move is the activist investor Third Point calling for a breakup of Shell. Third Point argues that the company is undervalued because "*... you can't be all things to all people. In trying to do so, Shell has ended up with unhappy shareholders who have been starved of returns and an unhappy society that wants to see Shell do more to decarbonize.*" As a result, Shell's upstream, refining, and chemicals businesses which contribute 60% of the company's EBITDA are worthless at current market prices By breaking up the company, the legacy oil and gas businesses could reduce capital expenditures while generating high current returns to shareholders, while the rest of the company—renewables, Liquid Natural Gas, marketing—could grow more aggressively (Dunn 2021; Third Point Investors Limited 2021).

Such a move may make sense financially, but it also leaves important questions unanswered. For example, if major oil and gas companies are worth more split up than pursuing their current strategies, then:

- Do stock market investors ultimately not believe in Shell's bold pledge to be carbon neutral by 2050?
- Is spinning off legacy oil and gas assets the best way for major oil and gas companies to realize the value of their energy transition efforts?
- Is there special expertise in an integrated energy company, which would help re-direct cash flows from oil and gas assets to new energy investments?
- If integrated energy companies are split up like this, who will ultimately own their oil and gas assets? What will they do with the cash flows from those assets? What climate policies will they pursue? Will they invest in decarbonization or maximize near-term cash flows?

Like our understanding of climate itself, the financial markets' response to climate change is changing. The Divestment movement and ESG investing wave of the past decade are really just the first act of a long drama. So far, the underperformance of oil and gas companies has given them a "Have your cake and eat it too" moment. According to Divest-Invest 2021, divestment is simply a good investment decision:

> With ten years of data and studies from some of the most respected financial entities in the world, the verdict is now clear — there is no rationale for investors to keep fossil fuels in

> their portfolio. Divestment both increases financial returns and reduces long-term financial risk….
>
> Financial arguments against divest-invest no longer hold water.

But what if the last ten years were just a cycle that is ending? The underperformance of oil and gas companies has left them significantly undervalued, to a point where they are attractive targets for private equity funds, sovereign wealth funds, and national oil companies. At the same time, the shift towards ESG and asset divestitures of the oil majors have led to reduced investment in the industry. While climate activists continue to warn of a carbon bubble of unextractable oil reserves, the industry has been warning that the reduction in upstream oil and gas supply will lead to price volatility and loss of energy security (International Energy Forum, n.d.).

2.5 War and the SEC: A Reset

Then the world changed.

The Russia-Ukraine War that began on February 24, 2022, sent oil and natural gas prices skyrocketing and global economies scrambling to secure alternative energy supplies. Energy stocks rallied while broader stock market indices sank, leading to a reversal of the decade-long underperformance. The energy industry's calls for more exploration and reserves all of a sudden seem prescient, see Fig. 2.4 (Ugal 2021).

Less visible than the war is another event that could have significant long-term implications for the oil and gas industry. On March 21, 2022, the SEC put forward its long-awaited proposal to require companies to make climate risk disclosures.

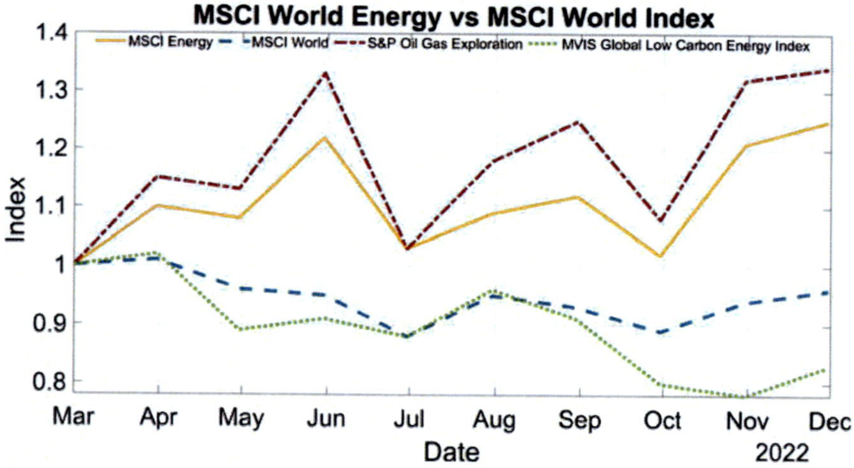

Fig. 2.4 Relative performance of energy stocks versus general market, March to November 2022. *Source* Yahoo! Finance

Included in the proposal are sweeping requirements for public companies to disclose material climate risk, their management strategies, and their direct (Scope 1), indirect (Scope 2), and supply chain (Scope 3) emissions (U.S. Securities and Exchange Commission 2022).

So far, the SEC's public input period for climate disclosures received 75% positive responses, and rules for mandatory climate disclosures are expected for later 2022 (Benson and Simon 2022). If adopted, the SEC will be joining the UK, which led the way by being the first to require climate disclosures on April 6, 2022, and the EU regulators are increasing their climate disclosure requirements. Regulators in other important financial markets are likely to follow in step, all of them based on the Taskforce for Climate Finance Disclosures (TCFD) framework as well as related standards such as the Carbon Disclosure Project (DCP), Sustainable Accounting Standards Board (SASB), and Global Reporting Initiative (GRI).

An even more aggressive bill was introduced in California. SB 260 would require all U.S.-registered entities doing business in California, whether public or private, to disclose their Scopes 1, 2, and 3 emissions without the phase-ins and exemptions of the SEC rules (Smith and Lan 2022). After passing in the Assembly, it failed to gain the votes necessary to pass in the Senate, with business groups opposing it on the grounds that "*Because there is no objective criteria for assessing Scope 3 emissions data ... it will be nearly impossible for ARB to 'verify' emissions data that is, by its very nature, subjective, inaccurate, and often incomplete.*" However, since industry groups argued that California's disclosure requirements "*should be compatible with the standards proposed by the SEC and International Sustainability Standards Board,*" this failure may well just be a de facto acceptance of SEC disclosure requirements (Cooley LLP 2022).

Does the oil and gas industry view these new disclosure requirements as added regulatory burdens? A recent Columbia study found that while all 15 major public oil companies in the study are already making climate disclosures, altogether, they fell short of what the investment community has been demanding (Wong et al. 2022). Only 6 of 15 were disclosing their Scope 3 emissions, while only 4 of the 15 companies used third-party verified emissions disclosures.

Beyond the additional work, though, these climate disclosure requirements also offer some interesting opportunities for the industry.

First, the regulators can now step in to mediate the industry's climate disclosures and climate initiatives and the investment community's needs for climate action. This will probably be a welcome change to bankers, who are loath to give up long-term relationships and genuinely believe they could do more by engaging with their energy industry clients (Chen 2022). It will also help better define the role of fund managers, who are ultimately responsible for investment results. Speaking about the disclosures on Bloomberg, Larry Fink of BlackRock sounded downright conciliatory when he said he "does not want to be the climate police." He then proceeded to laud all the energy industry CEO's he's met who are serious about climate change and about transitioning to low-carbon businesses (Brush and Massa 2022).

Second, the additional climate disclosure requirements may benefit the oil and gas industry more than it hurts. While some in the industry still do not want to make Scope

3 emissions disclosures, these emissions figures are in fact not difficult to calculate, because they are mostly from customers' combustion of fuels. Anyone could estimate an oil and gas company's Scope 3 emissions by multiplying its output by published emissions factors, which are readily available for free from government and research sources (see for example U.K. Department for Business, Energy & Industrial Strategy 2021). The implications of those Scope 3 emissions are also clear: Emissions from the combustion of fuels are exactly what regulators, investors, and the general public are targeting with everything from carbon taxes to product switching to outright bans. Thus, the industry's Scope 3 emissions are the source of its greatest risk. Refusing to disclose the Scope 3 emissions neither obscures them from observers nor makes them less of a risk.

On the other hand, if Scope 3 emissions disclosures are generally required across all industries, then the oil and gas industry's low-carbon products, including those from carbon capture and storage, alternative fuels, hydrogen, and methane reductions, could be counted as reduced emissions by its customers. For example, any company using steel made from natural gas with carbon capture or green hydrogen would be able to reduce their emissions accounts. Similarly, any company which flies with an airline buying sustainable aviation fuel could reduce their emissions accounts. Scope 3 emissions disclosures would thus create marketable benefits for the low carbon output of the oil and gas industry.

Finally, these climate disclosure requirements will mean that climate investing will just be a part of investing, instead of a vote for or against the oil and gas industry. An example of this is Sustainability-Linked Bonds discussed earlier and related Sustainability-Linked Loans. While the oil and gas industry has been locked out of the "Green Bonds" market due to their specific use of proceeds requirements, it has successfully tapped the Sustainability-Linked Bonds and Loans market. The interest rates of these bonds and loans are tied to specific metrics such as reductions of methane emissions. When the issuer meets targets, it could receive a reduction in interest rates. Conversely, in some cases, failure to meet targets could trigger higher interest rates.

Through instruments like these, climate-focused investors could invest in the oil and gas industry by funding its decarbonization. The industry could create more investment vehicles for such investors, such as securities linked to methane reduction, carbon capture and storage, and sustainable fuels. Demand for climate-aligned investments in the public markets could help the industry secure large amounts of capital at favorable rates. Thus, the oil and gas industry could follow the banking industry by offering a variety of investment vehicles suited to different investor interests. Some would offer higher current yields for legacy fossil fuel assets with limited lifespans, while others would offer long-term investments in decarbonization with greater growth potential. This would solve the valuation problem raised by Third Point, without breaking up companies.

All this would hinge on climate disclosures, which over time may drive everything from borrowing costs to investment capital decisions and mergers and acquisitions in the industry. Meanwhile, as the SEC and EU regulators crack down on ESG funds which do not live up to their marketing promises, fund managers' ESG claims and banks' financed emissions targets will also become serious commitments that their investors will hold them accountable for. Missed targets could lead to serious financial liabilities. The fund managers and banks would, in turn, hold their investee companies liable for their climate targets, if only to protect themselves. Thus, climate disclosures would take on the importance and seriousness of financial disclosures. Companies that revise their climate disclosures or miss analysts' estimates could face falling stock prices and shareholder lawsuits. All this will make climate disclosures and the data supporting them central to the industry's access to the financial markets, instead of another nice-to-have for appeasing "woke" investors. If the ESG investing boom was a cause-driven movement, then climate disclosures will turn climate and emissions into hard-nosed financial metrics.

2.6 Summary and Conclusion

Thus begins the next act.

Geopolitical turmoil and supply fears have pointed to a continued need for the oil and gas industry during the transition to a low-carbon future. Spikes in oil and gas prices and the outperformance of energy stocks show that it is dangerous for fund managers to ignore the industry completely. Disclosure requirements will provide an opening for the oil and gas industry and the financial markets to work together while addressing climate change.

Both sides must take the opportunity before it is lost again. For the industry, it's a chance to prove that its strategies of carbon capture and sequestration, hydrogen, sustainable low-carbon fuels, and even carbon credits and offsets produce real climate benefits. If it's successful, it could continue to enjoy access to the public markets, which could award it with higher valuations that, in turn fuel the growth of companies with attractive business models For investors, it's a chance to influence this critical industry on the right path, lest it falls out of public ownership and accountability, while diversifying their portfolios and improving risk-adjusted returns.

In the end, the oil and gas industry is going through the energy transition like every other industry. Its role is both pivotal and simple. Its product is the primary cause of the GHG emissions that are causing climate change. Yet as businesses, it also simply responds to the market. If there is demand for oil, natural gas, and even coal, then there will be suppliers who will produce it. Divesting from them won't make them go away. Divesting at current low valuations would simply make it profitable for someone else to buy those assets and take them out of the reach of public-facing regulators and investors.

Thus, unless the industry and investors could work together to make that transition successful, there is no stopping climate change. The industry is already committed to a number of strategies. In the rest of this book, we'll take a look at how we could make them work.

References

American Electric Power (n.d.) AEP's clean energy future. AEP.com. https://www.aep.com/about/ourstory/cleanenergy. Accessed 1 Aug 2022

AutoZone, Inc.; Rule 14a-8 no-action letter (2021) SEC.gov. https://www.sec.gov/divisions/corpfin/cf-noaction/14a-8/2021/asyousowautozone080621-14a8-incoming.pdf

Bacow L (2021) Climate change: update on Harvard action. Harvard University. https://www.harvard.edu/president/news/2021/climate-change-update-on-harvard-action/

Benson E, Simon Z (2022) Getting (dis)closure: U.S. regulatory changes leading to climate-related financial disclosures. CSIS. https://www.csis.org/analysis/getting-disclosure-us-regulatory-changes-leading-climate-related-financial-disclosures

Berkery Noyes (n.d.) Q3 2021 Private Equity Industry Trends - Berkery, Noyes & Co., LLC. Berkery Noyes. https://berkerynoyes.com/q3-2021-private-equity-industry-trends/. Accessed 1 Aug 2022

Blackrock Sustainable Investing (2021) Investment and fiduciary analysis for potential fossil fuel divestment. https://ieefa.org/wp-content/uploads/2021/03/BlackRock-Phase-One.pdf

BP (2021) Sustainability report 2020. https://www.bp.com/content/dam/bp/business-sites/en/global/corporate/pdfs/sustainability/group-reports/bp-sustainability-report-2020.pdf

Brush S, Massa A (2022) Fink says BlackRock doesn't want to be 'environmental police'. Bloomberg.com, June 2, 2022. https://www.bloomberg.com/news/articles/2022-06-02/fink-says-blackrock-doesn-t-want-to-be-environmental-police

Carbon Tracker Initiative (2011) Unburnable Carbon – are the world's financial markets carrying a carbon bubble? https://www.banktrack.org/download/unburnable_carbon/unburnablecarbonfullrev2.pdf

Carlson D (2020) ESG investing now accounts for one-third of total U.S. assets under management. MarketWatch, November 17, 2020. https://www.marketwatch.com/story/esg-investing-now-accounts-for-one-third-of-total-u-s-assets-under-management-11605626611

Chen S (2022) Interview - CDP and climate disclosures: a corporate perspective. In: The open climate investing book. Open Source Strategies, Inc, N.p. https://climate-investing-book.opensourcestrategies.com/v/main/book/analyzing-investments/interview-climate-disclosure-corporate-perspective

Cooley LLP (2022) California's proposed Climate Corporate Accountability Act goes belly up—for now at least. JD Supra. https://www.jdsupra.com/legalnews/california-s-proposed-climate-corporate-5849795/

Department for Business, Energy & Industrial Strategy (2021) Greenhouse gas reporting: conversion factors 2021. GOV.UK. https://www.gov.uk/government/publications/greenhouse-gas-reporting-conversion-factors-2021

DeSilver D (2020) Renewable energy is growing fast in the U.S., but fossil fuels still dominate. Pew Research Center. https://www.pewresearch.org/fact-tank/2020/01/15/renewable-energy-is-growing-fast-in-the-u-s-but-fossil-fuels-still-dominate/

Divest Invest 2021 (2021) https://www.divestinvest.org/wp-content/uploads/2021/10/Divest-Invest-Program-FINAL10-26_B.pdf

Dominion Energy (2021) Dominion Energy Climate Report 2021. https://www.dominionenergy.com/-/media/pdfs/global/company/esg/2021-climate-report.pdf

Dunn K (2021) As break-up calls grow louder, Shell delivers an earnings dud. Fortune, October 28, 2021. https://fortune.com/2021/10/28/shell-q3-third-point-investors-climate-activist-daniel-loeb/

Farman M (2022) European PE buy-and-build deals surge as valuations heat up. S&P Global. https://www.spglobal.com/marketintelligence/en/news-insights/latest-news-headlines/european-pe-buy-and-build-deals-surge-as-valuations-heat-up-68721456

Fink L (2020) Larry Fink's letter to CEOs: a fundamental reshaping of finance. BlackRock. https://www.blackrock.com/corporate/investor-relations/2020-larry-fink-ceo-letter

Golding G, Kumar A, Mertens K (2021) Cost of Texas' 2021 deep freeze justifies weatherization. Federal Reserve Bank of Dallas. https://www.dallasfed.org/research/economics/2021/0415.aspx

Henisz W (2021) Why engine No. 1's victory is a wake-up call for ExxonMobil and others. Knowledge at Wharton. https://knowledge.wharton.upenn.edu/article/engine-no-1s-victory-wake-up-call-for-exxonmobil-and-others/

Henze V, Boyd S (2021) ESG assets rising to $50 trillion will reshape $140.5 trillion of global AUM by 2025, Finds Bloomberg Intelligence | Press. Bloomberg.com. https://www.bloomberg.com/company/press/esg-assets-rising-to-50-trillion-will-reshape-140-5-trillion-of-global-aum-by-2025-finds-bloomberg-intelligence/

Hodgson C (2021) Banks feel the heat on financed emissions. Financial Times. https://www.ft.com/content/1a7f573c-9a5a-4e41-a113-a96ce7fa80a8

HSBC Group (2022) Announcement - HSBC sets financed emissions targets for oil and gas, and power and utilities. HSBC. https://www.hsbc.com/news-and-media/media-releases/2022/hsbc-sets-financed-emissions-targets-for-oil-and-gas-power-and-utilities

IEA (2020) The oil and gas industry in energy transitions – analysis. https://www.iea.org/reports/the-oil-and-gas-industry-in-energy-transitions

International Energy Forum (n.d.) IEF - IHS Markit investment report: investment crisis threatens energy security. International Energy Forum - IEF. https://www.ief.org/investment-report-2021. Accessed 29 July 2022

Lambert F (2021) Countries and automakers agree to go all-electric by 2040 in weak new goal set at COP26. Electrek, November 10, 2021. https://electrek.co/2021/11/10/countries-automakers-agree-go-all-electric-by-2040-weak-new-goal-cop26/

Manley D, Mihalyi D, Heller PR (2019) Hidden giants. International Monetary Fund. https://www.imf.org/Publications/fandd/issues/2019/12/national-oil-companies-need-more-transparency-manley

Moody's (2020) Environmental heat map. Moody's ESG Solutions. https://esg.moodys.io/reports

Moody's Investor Service (2020) Environmental risks global heatmap overview. https://www.moodys.com/sites/products/ProductAttachments/Infographics/Environmental-Risks-Global-Heatmap-Overview.pdf. Accessed 2 Aug 2022

Morgan Stanley (2021) Morgan Stanley announces 2030 targets for net-zero financed emissions commitment. Morgan Stanley. https://www.morganstanley.com/press-releases/2030-targets-for-net-zero-financed-emissions-commitment-#

Mutua DC (2022) Green bonds still have long way to go to dent climate crisis. Bloomberg.com, February 1, 2022. https://www.bloomberg.com/news/articles/2022-02-01/green-bonds-still-have-a-long-way-to-go-to-dent-climate-crisis

Natural Resource Governance Institute (2022) Oil & gas production - barrels of oil equiv./day. National Oil Company Database. https://www.nationaloilcompanydata.org/indicator

Raval A (2021) Big Oil eyes US$140B in asset divestments to meet net zero targets, leaving field to those that operate in the shadows. Financial Post, July 14, 2021. https://financialpost.com/commodities/energy/oil-gas/big-oil-eyes-us140b-in-asset-divestments-to-meet-net-zero-targets-leaving-the-field-to-those-that-operate-in-the-shadows

Rystad Energy (2022) Upstream M&A deals reached a three-year high of $181 billion in 2021, returning to pre-Covid levels. Rystad Energy. https://www.rystadenergy.com/newsevents/news/press-releases/Upstream-MA-deals-reached-a-three-year-high-of-181-billion-in-2021-returning-to-pre-Covid-levels/

Securities and Exchange Commission (n.d.) 17 CFR § 240.14a-8 - Shareholder proposals. Cornell Law School Legal Information Institute, N.p. https://www.law.cornell.edu/cfr/text/17/240.14a-8

Shell (2021) Shell's climate target. https://www.shell.com/energy-and-innovation/the-energy-future/our-climate-target/_jcr_content/par/relatedtopics.stream/1650460792140/c5833d4644bd8acd6dfc6c5a70b5b42a483639a8/our-climate-target-v1-ax.pdf

Smith DC, Lan D (2022) Analysis of SEC's and California Legislature's proposed corporate greenhouse gas reporting mandates. Manatt, Phelps & Phillips, LLP. https://www.manatt.com/insights/newsletters/client-alert/analysis-of-secs-and-california-legislatures-pro

Southern Company (n.d.) Additional environmental priorities. Southern Company. https://www.southerncompany.com/clean-energy/environment/climate.html. Accessed 1 Aug 2022

Stand.earth (n.d.) The database of fossil fuel divestment commitments made by institutions worldwide. Global Fossil Fuel Commitments Database. https://divestmentdatabase.org/. Accessed 5 Aug 2022

Statista (2022) Largest sovereign wealth funds worldwide 2022. Statista. https://www.statista.com/statistics/276617/sovereign-wealth-funds-worldwide-based-on-assets-under-management/

St. John J (2020) The 5 biggest US utilities committing to zero carbon emissions by 2050. Greentech Media. https://www.greentechmedia.com/articles/read/the-5-biggest-u.s-utilities-committing-to-zero-carbon-emissions-by-mid-century

Stocker M, Baffes J, Vorisek D (2018) What triggered the oil price plunge of 2014–2016 and why it failed to deliver an economic impetus in eight charts. World Bank Blogs. https://blogs.worldbank.org/developmenttalk/what-triggered-oil-price-plunge-2014-2016-and-why-it-failed-deliver-economic-impetus-eight-charts

Szegedi K (2021) Shifting sands: how consumer behaviour is embracing sustainability. Deloitte. https://www2.deloitte.com/ch/en/pages/consumer-business/articles/shifting-sands-sustainable-consumer.html

Third Point Investors Limited (2021) Third quarter 2021 investor letter. https://thirdpointlimited.com/wp-content/uploads/2021/10/Third-Point-Q3-2021-Investor-Letter-TPIL.pdf

Ugal N (2021) Oil and gas underinvestment raise fears of continued price shocks and volatility, says report. Upstream Online. https://www.upstreamonline.com/focus/oil-and-gas-underinvestment-raise-fears-of-continued-price-shocks-and-volatility-says-report/2-1-1117457

United Nations Environment Programme Finance Initiative (n.d.) Members – United Nations environment – finance initiative. United Nations Environment Programme Finance Initiative. https://www.unepfi.org/net-zero-banking/members/. Accessed 1 Aug 2022

U.S. Securities and Exchange Commission (2022) SEC proposes rules to enhance and standardize climate-related disclosures for investors. SEC.gov. https://www.sec.gov/news/press-release/2022-46

Wigglesworth R (2021) Private capital industry soars beyond $7tn. Financial Times, June 11, 2021. https://www.ft.com/content/4d0e6f18-2d56-4175-98c5-e13559bdbc25

Woellert L (2022) Sustainable debt steps up-POLITICO. Politico, April 28, 2022. https://www.politico.com/newsletters/the-long-game/2022/04/28/sustainable-debt-steps-up-00028520

Wong HX, Zimmermann N, Blanton E, Boersma T (2022) ESG investing and the US oil and gas industry: an analysis of climate disclosures. SIPA Center on Global Energy Policy. https://www.energypolicy.columbia.edu/sites/default/files/pictures/Upstream_ESG_Final%20(1).pdf

Si Chen is the President of Open Source Strategies, Inc. in Los Angeles, CA, which specializes in open-source software for climate finance and investing. He leads the development of open-source blockchain carbon accounting software at Hyperledger Labs. Previously, he has managed investment portfolios for institutional pension funds, central banks, and hedge funds and has been published in The Journal of Portfolio Management. He is also the co-founder and CTO of GraciousStyle.com, an online retailer.

Chapter 3
Introduction to Blockchain

Soheil Saraji

3.1 Introduction

Blockchain is a digitally-native multidisciplinary technology that owes its existence to advances in cryptography and other computer science fields, such as distributed systems. It also benefits from a game-theoretic economic design for its security. In technical terms, Blockchain is a distributed ledger technology whose records are batched into timestamped blocks, and each block is identified by its cryptographic hash (this concept is discussed later in this chapter). Each block references the hash of the block that came before it. This establishes a link between the blocks, thus creating a chain of blocks or a blockchain. A blockchain network is a set of non-trusting writers sharing a digital database with no trusted middleman. These properties of Blockchain provide: (i) a robust, truly distributed peer-to-peer system that is tolerant of node failures, (ii) transparency, verifiability, and auditability on the network's activity.

We can also consider this innovation an outcome of a techno-social movement among scholars that started in the 1980s and bloomed in the late 2000s. Blockchain was first created as a decentralized platform to secure cryptocurrencies by solving digital ownership or double-spending problems (Nakamoto, n.d.). It then evolved into a world computer enabling smart contracts, decentralized autonomous organizations, etc. (Buterin, n.d.). Currently, it is accepted as the potential backbone of the future of the internet (i.e., web3) (Wood, n.d.). However, this technology is quickly evolving, and new paradigms on what blockchain could become may come to light in the coming years.

S. Saraji (✉)
Energy and Petroleum Engineering, University of Wyoming, Laramie, WY, USA
e-mail: ssaraji@uwyo.edu

S. Saraji and S. Chen, *Sustainable Oil and Gas Using Blockchain*,
Lecture Notes in Energy 98, https://doi.org/10.1007/978-3-031-30697-6_3

To understand the origin of blockchain technology, we need to explore the intersection of two relatively older technologies: ledger and cryptography. The following sections briefly explore the history of money, ledger, and blockchain technology. Next, an introduction to cryptography and cryptographic algorithms used in Blockchain is provided, followed by a brief overview of distributed and decentralized systems. After that, a technical description of how blockchain and consensus mechanisms work is laid out. Finally, two practical concepts in modern blockchains, smart contracts and governance, are discussed.

3.2 A Brief History of Money and Ledger

Blockchain is sometimes called a distributed ledger technology. Therefore, the first step in understanding blockchain is to learn about ledgers. This section provides a brief historical overview of the ledger and another closely related concept: money. Money is defined by modern economists using the three roles it plays in the economy (Surowiecki 2012): (1) store of value, meaning that money allows you to defer consumption until a later date, (2) unit of account, meaning that it allows you to assign a value to different goods without having to compare them, and (3) medium of exchange, an easy and efficient way to trade goods and services with one another.

The true origin of money is subject to controversy and disagreement (Surowiecki 2012; The Origins of Money Are Murky and Mysterious | Science News 2018). However, there are two prominent theories on its origin. The first theory, shared mostly among economists, considers the bartering of goods and services as the origin of money. On the other hand, archaeologists and anthropologists believe that debt payment was the main reason for the creation of money in society. Several archeological pieces of evidence support the latter theory.

It is believed that money-friendly debt started about 5500 years ago in the form of silver shekels in Mesopotamia and Egypt (Surowiecki 2012). Humans have used different types of money throughout history. The first type adopted by human society is perhaps non-metal money in the form of beads, stones, and other collectibles, such as Wampum Beads, Salt Bars (Ethiopia), and Cowrie Shells (Nigeria). Another example is Snail shells used by Chumash Indians, who lived 2000 km north of Mesoamerican societies (Southern California) about 800 years ago (The Origins of Money Are Murky and Mysterious | Science News 2018). Another fascinating piece of evidence is from a small island in Micronesia called Yap, which dates back to at least 1000 years ago (A Brief History of Money 2019). They used huge donut-shaped discs of limestone, known as Rai stone, for money. These large pieces of stone were quarried from a nearby island and transported to Yap via primitive boats. This was an arduous process, and hence these stones held value in society until a foreign merchant with a more advanced boat learned about Yap stones and started to query and transport a large number of stones to the island.

Money in the form of precious metal pieces is another type of money that has been used in human society. The early examples are Bronze Aes Rude (Rome), silver shekels (Mesopotamia), Bronze Spade (China), and Cooper Plate (Sweden). The first minted coin was found in the Kingdom of Lydia, today's Turkey, around 2600 years ago (Surowiecki 2012). The other examples of certified (minted) metal money are Silver Dekadrachm (Greece), Bronze Yuan (China), and Gold Auers (Rome). The next generation of money was paper-based money, first found in China during the Song Dynasty (Jiaozi Promissory Note). This type of money was famously exported to Europe by Marco Polo, who visited China in the same era (A Brief History of Money 2019). The first European paper money was Swedish bank notes around 1661 AC (A Brief History of Money 2019). This was followed by the widespread use of Private Bank Notes in Europe and the United States. These early paper money were typically pegged to valuable assets like precious metals and were merely a receipt that could be exchanged for a precious metal of equivalent value. Today we use fiat currencies, which are paper money backed by central governments (Surowiecki 2012).

As mentioned earlier, a close-related concept to money is the ledger. A ledger is a principal recording of accounts that enables economic activity and financial relationships. It is practically an accounting system that tracks assets, values, and money for a group of people. Historical evidence for ledgers started over 5000 years ago in ancient Mesopotamia (A Brief History of Ledgers. Before Starting My Investigation Into… | by LLFOURN | Unraveling the Ouroboros | Medium, n.d.). The early ledgers were pictographic clay tablets with rows and columns. Within each cell, they drew a picture of the type of item (e.g., a bushel of barley, a sheep, etc.) and made holes indicating the quantity of it. The early types of ledgers were found in Uruk Period, excavated from Susa, Iran.

Around 700 years ago, a new accounting method emerged amongst northern Italy's merchants and moneylenders (A Brief History of Ledgers. Before Starting My Investigation Into… | by LLFOURN | Unraveling the Ouroboros | Medium, n.d.). In this method, every item must be entered twice, once as a credit and once as a debit. This is called the double-entry accounting system. According to scholars, the wide adaptation of the double-entry accounting system may have eventually led to what we now call "capitalism" (A Brief History of Ledgers. Before Starting My Investigation Into… | by LLFOURN | Unraveling the Ouroboros | Medium, n.d.). Prior to the invention of fiat money, ledger entries always measured a quantity of something tangible owned or owed. Nowadays, humanity's primal motivation to gather wealth is redirected away from real goods to ledger numbers on paper and electronic databases. As a result, the boundary between money (paper money) and the ledger (e.g., bank account) has become ambiguous.

About 40 years ago, a new ledger technology was introduced, allowing for the emergence of digitally native money, sometimes called internet money or, more famously, cryptocurrency. Blockchain, a cryptographically secured decentralized network, is the backbone of this ledger technology. In the next section, we explore the origin of this technology from a historical perspective.

3.3 A Brief History of Blockchain

The first application of Blockchain was to secure digital ledgers in the world of the internet. However, this innovation was a product of a social movement that started about 30–40 years ago. In 1992, three computer scientists in California launched a new mailing list for discussing cryptography, mathematics, politics, and philosophy. They called the members of this mailing list the *cypherpunks* (The Cypherpunks 2019). The word "Cypherpunk," coined by Jude Milhon, is composed of two independent words: "cypher," meaning encryption, and "punk," meaning rebel. At that time, computers were becoming widespread, and the internet had become available to the public. There was a growing concern regarding digital censorship and surveillance on the open internet by the authorities. "*Cypherpunks write code. We know that someone has to write software to defend privacy, and we're going to write it*" (A Cypherpunk's Manifesto, n.d.). This is a famous statement from a document named "A Cypherpunk's Manifesto" written by Eric Huges. Emphasizing the importance of privacy in the digital world, Huges continues, "*... privacy in an open society requires anonymous transaction systems...*" and "*... privacy in an open society also requires cryptography*" (A Cypherpunk's Manifesto, n.d.). Therefore, one of the main concerns of this movement was the need for anonymous transaction systems that were not controlled by central authorities.

This movement was inspired by previous academic works on digital privacy and, most prominently, the work of David Chaum. Chaum was a computer scientist and cryptographer who published his dissertation titled "Computer Systems Established, Maintained, and Trusted by Mutually Suspicious Groups" in 1982. This is the first known proposal for the blockchain protocol (How DigiCash Blew Everything, n.d.). Later, recognizing the growing need for digitally-native money to allow computer-based transactions, Chaum invented *DigiCash*, an anonymous electronic cash. *eCash*, the underlying algorithm of DigiCash, used novel cryptography to ensure user privacy while solving the *double spending problem*. The double spending problem is when a user can spend the same money more than once because the algorithm cannot effectively track all transactions and prevent unauthorized transactions promptly. To give a simplistic example, think of an image in the *jpeg* format on the internet. Anyone can download and send multiple copies of this image to other people. Therefore, a digital ledger is needed to secure electronic money. Unfortunately, despite being technically a perfect product, the company founded on eCash (DigiCash) went bankrupt in 1998 (How DigiCash Blew Everything, n.d.). Nevertheless, this early effort in developing anonymous digital cash earned David Chaum the title *the godfather of cryptocurrency*.

Following Chaum's footsteps, several attempts were made to develop electronic cash and privacy-preserving technologies. In 1997, *Hashcash* was invented by Adam Back as a way to prevent email spam (Hashcash, n.d.). Although this innovation was not directly addressing the need for electronic cash, it developed a mechanism known as proof of work, which was a precursor for the upcoming innovations. Founded in 1996 by Douglas Jackson, *e-gold* was one of the first dot-com companies to create a digital currency two years before PayPal (Zetter, n.d.). E-gold was a private, international currency that would circulate independently of government controls and stand impervious to the market's highs and lows. It issued a digital currency backed by gold reserves that anyone could hold and transfer. At its height, e-gold processed more than $2B in transfers a year. Eventually, a US court ruled e-gold guilty of money laundering and retroactively violating money transmitter laws (Zetter, n.d.). The founder was found criminally liable, and in 2008, all e-gold balances were frozen. This experience showed the shortcoming of having a trusted central company securing digital money. The other noteworthy innovations were *b-money* by Wei Dai (1998) (B-Money | Satoshi Nakamoto Institute, n.d.) and *BitGold* by Nick Szabo (2005) (Bit Gold | Satoshi Nakamoto Institute, n.d.). They both use public key cryptography for identity and proof-of-work (invented by Adam Back) to mint new coins. However, both algorithms were ultimately vulnerable to Sybil attacks. A Sybil attack is when a malicious user cheaply spins up many new "Sybils" or identities.

In late 2008, a white paper was published online titled "Bitcoin: A Peer-to-Peer Electronic Cash System" by an anonymous author who called himself/herself/themselves "Satoshi Nakamoto" (Nakamoto, n.d.). In an email, Nakamoto explained this system: "*I've been working on a new electronic cash system that's fully peer-to-peer, with no trusted third party*" (Nakamoto 2008). He continued by explaining the main properties of this system: "*Double-spending is prevented with a peer-to-peer network. No mint or other trusted parties. Participants can be anonymous. New coins are made from Hashcash style proof-of-work. The proof-of-work for new coin generation also powers the network to prevent double-spending*" (Nakamoto 2008). This innovation, by no means, was an isolated attempt by Nakamoto but rather an improvement on the previous work. Nakamoto explicitly credited *Hashcash* and *b-money* in his white paper (Nakamoto, n.d.). Also, there is strong evidence that he was a part of or at least a sympathizer to the Cypherpunk movement. At the time of publishing the Bitcoin whitepaper in 2008, there was a global financial crisis caused by predatory lending and excessive risk-taking by financial institutions. The US government deployed massive bailouts to save the economy and those institutions. This was against values held by Cypherpunks, who believed that money should be independent of political powers. In fact, the genesis block of Bitcoin (the first block in the Bitcoin network), mined in early 2009, contains this message: "*The Times 03/Jan/2009 Chancellor on brink of second bailout for banks*" (Genesis Block - Bitcoin Wiki, n.d.). Bitcoin has become the first successful cryptocurrency in the world with $440B market cap at the time of this writing.[1] The faith of Bitcoin, though, is still unclear as new regulations are expected around the world to restrict its use by the public based on fears of money laundering, other illicit financial activities, and tax evasion. In the context of this book, we can think of Bitcoin as a global experiment demonstrating how cryptography and decentralized networks could be used to secure data among non-trusting parties. In the coming sections, we tackle the technical aspect of blockchain by introducing its two essential components: cryptography and distributed/decentralized networks.

---

[1] January 31, 2023.

3.4 Introduction to Cryptography

Merriam-Webster dictionary defines cryptography as "*secret writing.*" A more technical definition is "*the practice and study of techniques for secure communication in the presence of adversarial behavior*" (Rivest 1990). A few definitions in the science of cryptography are important to know. *Plaintext* is the message that will be put into secret form; *ciphers* are methods of converting *plaintext* into a secret message. Most ciphers employ a key, which specifies such things as the arrangement of letters within a cipher alphabet (Fig. 3.1a), the pattern of shuffling in a transposition, or the settings on a cipher machine. The act of converting the *plaintext* to a secret message is called encryption, and the reverse (converting the secret message back to *plaintext*) is decryption (Fig. 3.1a).

Different types of ciphers have been used throughout history; simple physical items such as Scytale Cipher (a rode and a strap of leather with letters written on it) were used by Spartans in the 7th Century BC and Caesar Cipher (two co-centric disks with letters on each) used by the Roman Army in 100 BC (11 Cryptographic Methods That Marked History: From the Caesar Cipher to Enigma Code and Beyond, n.d.; The Story of Cryptography: Historical Cryptography, n.d.). Perhaps the most famous was the Enigma Machine (1918) used during World War II by the German army. This was an electro-mechanical machine in which the key was embedded in a set of rotating discs (3–5). In 1932, Marian Rejewski, a Polish cryptographer, discovered how Enigma works. Following her work, Alan Turing (1939) broke the cipher of

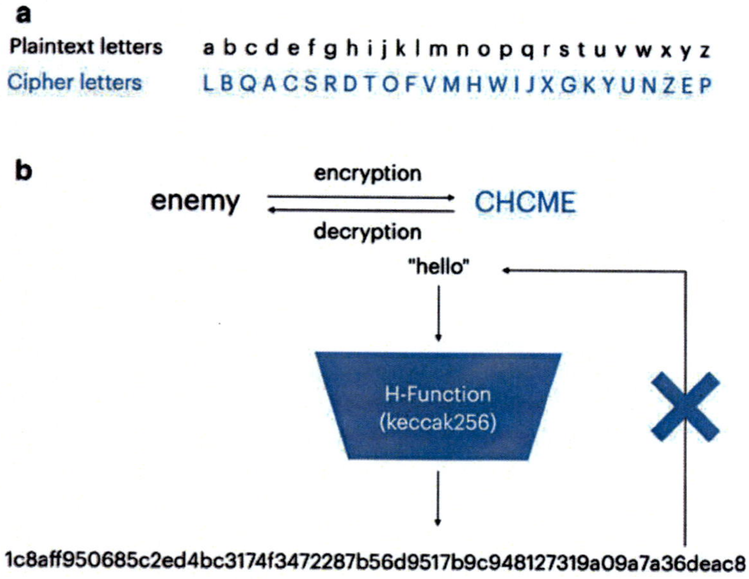

Fig. 3.1 **a** Cipher letters (keys) are used to convert plaintext to a secret message (encryption) or vice versa (decryption). **b** A one-way hash function to encrypt a message. *Source* Author

the Enigma machine (The Story of Cryptography: 20th Century Cryptography, n.d.). This gave a considerable advantage to the British army through the last few years of World War II and eventually helped them to win the war. The Hollywood movie "Imitation Game" is based on this story.

In the 1970s, with the emergence of computers and the digital realm, there was a need for more robust ciphers. In the early 1970s, Horst Feistel designed Lucifer Cipher at IBM (The Story of Cryptography: Modern Cryptography, n.d.). A few years later, in 1973, the United States National Bureau of Standards (NBS) asked for block cipher proposals, and IBM submitted a variant of Lucifer. In 1976, NBS eventually adopted Data Encryption Standard (DES) as the federal standard. Several digital ciphers were proposed, adapted, and ultimately broken and replaced in the following years. One can mark the start of modern cryptography with the publication of a peer-reviewed scientific paper, "New Directions in Cryptography" (Whitfield and Martin 1976).

One of the modern and secure cipher keys is a hash function. It is a function that accepts some arbitrary input and returns a fixed-length digest (summary value), typically in hexadecimal, sometimes called the hash value. As shown in Fig. 3.1b, an important property of this function is that it only works in one-way, meaning no one can use the hash value (i.e., secret message) to find the original value (*plaintext*). A variation of this function has been used to secure Blockchain networks, including SHA256 (Bitcoin) and Keccak-256 (Ethereum).

The cryptographic algorithms could be divided into two groups: *symmetric cryptography* and *asymmetric cryptography*. To explain, let us assume Alice wants to transmit a message m to Bob over a network while maintaining the secrecy of m in the presence of an eavesdropping adversary (Eve), see Fig. 3.2. If Alice and Bob can share a secret key, then Alice will encrypt the message (*plaintext*) using the key, send it to Bob, and he can decrypt the message using the same key. This process is called symmetric or *secret key cryptography*, where only one key can be used to encrypt or decrypt the message. Unfortunately, this is also a vulnerability for this algorithm, as Alice and Bob have to meet in person or find a secret channel to communicate the secret key beforehand.

In asymmetric cryptography, also known as *public key cryptography*, a key pair (public and private keys) is used instead. The public key could be shared with the public, including adversaries (e.g., Eve). However, the private key only belongs to the owner (e.g., Alice), and no one, including friends (e.g., Bob), should have access to it. It is important to note that public/private key pairs are not two random values but rather are related through elliptic-curve cryptography. In other words, knowing a private key, one can derive the paired public key (using elliptic-curve algorithms), but the reserve is not possible. Therefore, knowing the public key of a person does not jeopardize their private key, as it is secured by cryptography. Figure 3.2b shows how this scheme could be used for two different applications: encryption and digital signature. For encryption, Alice could encrypt the message using Bob's publicly available key. Only the private key holder (Bob) can decrypt a message encrypted with the corresponding public key. Therefore, only Bob can decrypt and read the original message. This is an improvement compared to secret key cryptography

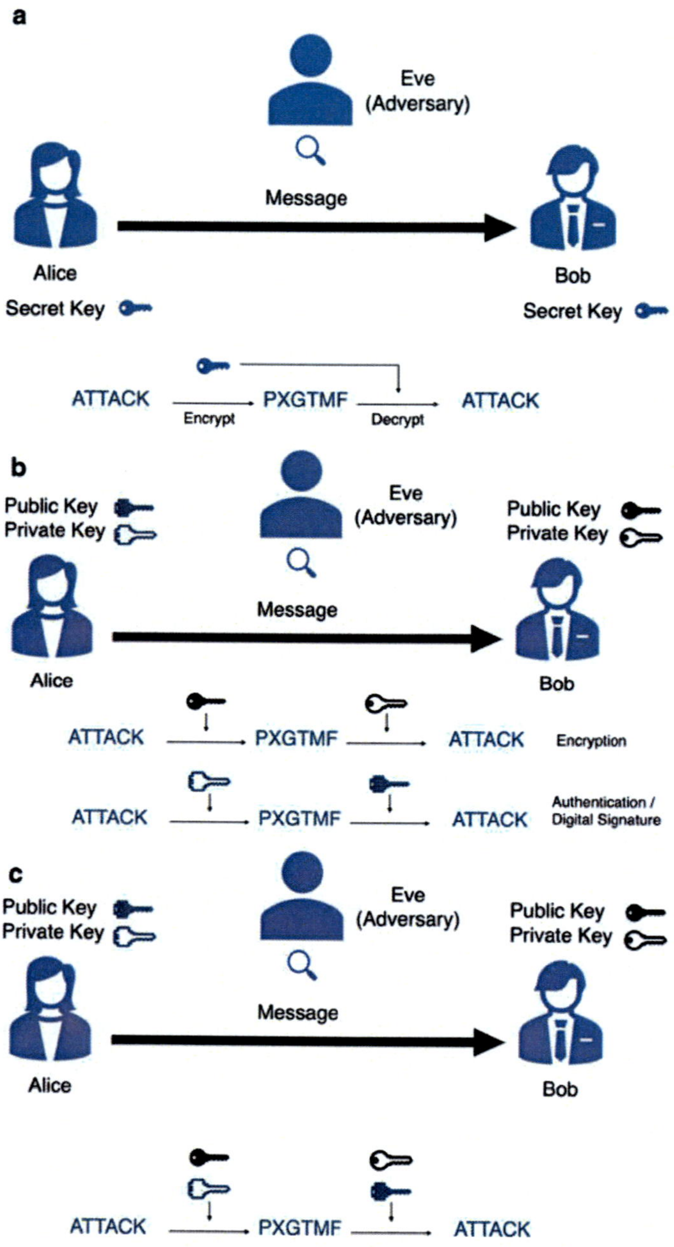

Fig. 3.2 Types of cryptographic algorithms: (**a**) symmetric or secret key cryptography, (**b**) and (**c**) asymmetric or public key cryptography. Asymmetric cryptography could be used for authentication or digital signature separately (**b**) or simultaneously (**c**). *Source* Author

mentioned before, as Alice and Bob do not need to meet each other to exchange the secret key. In fact, Alice and Bob do not even need to know each other before communicating their secret message.

Another important application of public key cryptography is authentication through digital signatures. Let us look at Fig. 3.2b again to explain. If Alice encrypts her message using her private key, then Bob can decrypt the message using Alice's public key, but so could Eve. In fact, anyone could decrypt Alice's message using her publicly available key. So, what is the point? This process allows anyone, including Bob, to ensure that the sender of the message is Alice. In other words, this is a digital signature of Alice that no one can fake (unless they have access to her private key). Things can get more interesting when we combine the two applications. If Alice encrypts her message twice, once with her private key and once with Bob's public key, then only Bon can decrypt the message, and he will immediately know that the message is authentic and from Alice (Fig. 3.2c). In a very simplistic form, this is how cryptography is used on the Blockchain to secure accounts and digital wallets.

3.5 Distributed and Decentralized Networks

A distributed system involves a set of distinct processes (e.g., computers) passing messages to one another and coordinating to accomplish a common objective (i.e., solving a computational problem) (Kasireddy 2019). In these systems, there is no central authority, and each node is connected to every other node and has the exact same authority (Fig. 3.3). In contrast, in a centralized system, there is one central authority or server, and all the other nodes act like clients or entities who accept messages from the server and enact them accordingly (Fig. 3.3).

A decentralized system is not centralized but may not be fully distributed (Fig. 3.3). In these systems, multiple servers receive messages from one central server. The individual nodes are connected to the secondary servers or each other. Decentralized

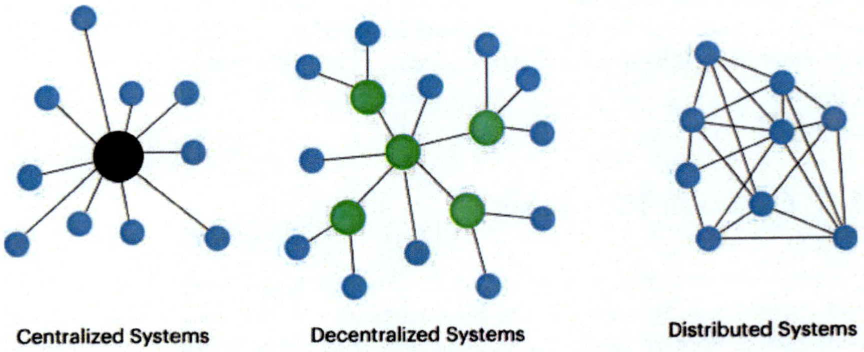

Fig. 3.3 Difference between centralized, decentralized, and distributed systems. *Source* Author

and distributed systems have specific properties that need to be considered during their design stage (Kasireddy 2019). In these systems, multiple events coincide (i.e., Concurrency). Also, time and order of events are hard to be synchronized as the nodes or computers are spatially separated (i.e., lack of a global clock). In addition, it is impossible to have a distributed/decentralized system without fault, be it non-malicious or malicious. Furthermore, message passing could happen synchronously or asynchronously. In a synchronous system, it is assumed that messages will be delivered within some fixed, known amount of time. In an asynchronous message-passing system, there is no fixed upper bound on how long a message will take to be received.

Blockchains are decentralized systems, and their design requires considering all these properties and finding technical solutions. One of the key issues that need to be addressed is the malicious attacks on the network. This brings us to a well-known problem in computer science called the Byzantine General Problem (Lamport et al. 1982). To define the problem, let's assume a group of Byzantine generals has surrounded and camped around a town, planning to attack and capture it (Fig. 3.4a). They need to communicate with each other to coordinate a simultaneous attack; otherwise, they may get defeated. What if some generals are traitors or spies? Public key cryptography, as described before, is insufficient because authentic and secure messages do not protect against traitors. The generals must have an algorithm to guarantee that (1) all royal generals decide upon the same plan of action, (2) a small number of traitors cannot cause the loyal generals to adopt a bad plan. In other words, in a distributed computing system, all participating nodes have to agree upon every message that is transmitted between the nodes. If a group of nodes is corrupt or the message they transmit is corrupt, then the whole network should not be affected by it and should resist this 'Attack' (Fig. 3.4b). In short, the network in its entirety has to agree upon every message transmitted in the network despite the foul play from malicious actors. This agreement is called a *consensus mechanism.*

In 1982, Lamport et al. proved that a solution to this problem was possible if two-thirds of the generals were royal (Lamport et al. 1982). Since then, different solutions have been developed by computer scientists. They are typically referred to as Byzantine Fault Tolerance systems, which tolerate the class of failures that belong to the Byzantine Generals Problem (Understanding Blockchain Fundamentals, Part 1: Byzantine Fault Tolerance | by Georgios Konstantopoulos | Loom Network | Medium, n.d.).

3.6 Consensus Mechanism

A consensus mechanism is an algorithm used to come to an agreement in distributed and decentralized systems, despite non-malicious and malicious faults. A typical consensus protocol is constituted of three steps (Kasireddy 2019):

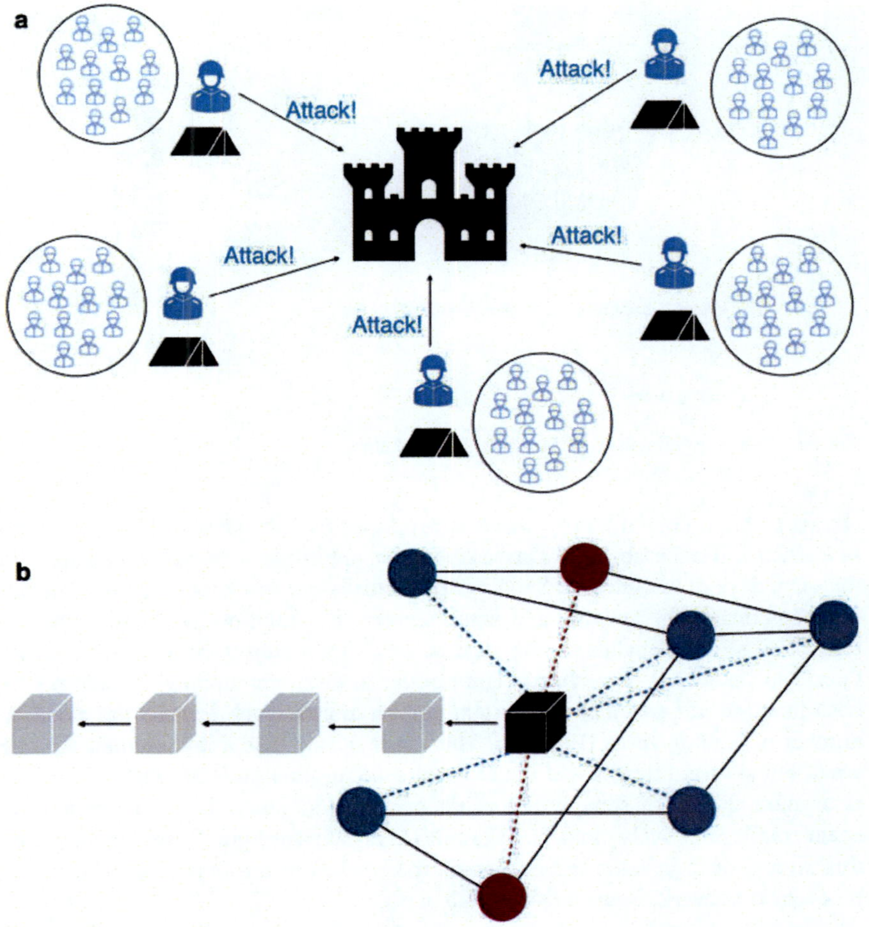

Fig. 3.4 **a** The classic Byzantine General Problem. **b** The same problem depicted in the blockchain context. Blockchain is a chain of data blocks. *Source* Author

1. Elect: the protocol should select a leader who proposes the next valid output (i.e., block)
2. Validate/Vote: the participants (nodes) will validate the proposed output (i.e., block)
3. Decide: the participants come to a consensus and finalize the output (i.e., block).

Different consensus protocols and mechanisms could be used in a Blockchain system. The first developed mechanism was Proof of Work. A Proof of Work (PoW) is computational proof that some work has been performed. This is usually implemented through a computational puzzle. For the puzzle to work, it needs to be (a) easy to state, (b) hard to solve, and (c) easy to verify (Understanding Blockchain Fundamentals, Part 2: Proof of Work & Proof of Stake | by Georgios Konstantopoulos | Loom

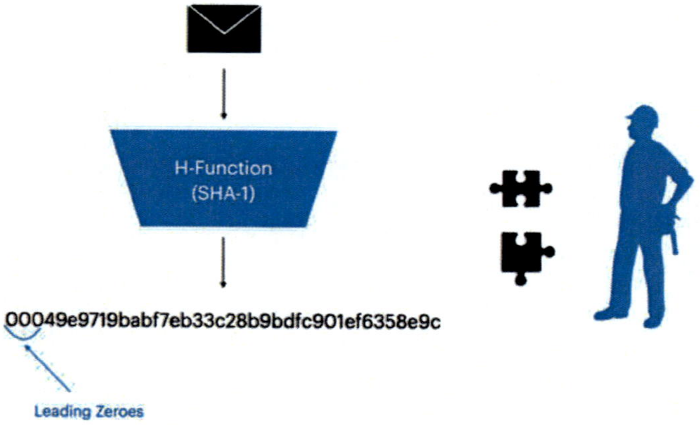

Fig. 3.5 PoW algorithm used in Hashcash. *Source* Author

Network | Medium, n.d.). PoW was first developed by Adam Back in 1997 and used in Hashcash (Hashcash, n.d.). Hashcash was an algorithm to prevent email spam in the early days of the internet. Sending spam emails is a low-cost malicious activity that can congest the network and email servers. The solution by Adam Back was to make it slightly challenging to send an email by requiring some computational puzzle to be solved beforehand. The puzzle, in fact, was finding an output of a hash function (digest) of the email text plus an arbitrary number that has a certain number of leading zeros (Fig. 3.5). This process may take a few seconds for one email but several minutes and hours when sending thousands of emails. The idea is to make the act of sending thousands of emails so costly to reduce or prevent spam emails effectively. Satoshi Nakamoto (a mysterious figure) subsequently used this algorithm to develop the consensus mechanism in the first successful public blockchain network, Bitcoin (Nakamoto, n.d.).

3.7 Blockchain Networks

Blockchain is a distributed ledger technology whose records are batched into timestamped blocks, and each block is identified by its cryptographic hash. Each block includes some data (e.g., transaction list) and references the hash of the block that came before it (Fig. 3.6a). This establishes a link between the blocks, thus creating a chain of blocks or blockchains (Fig. 3.6a). A blockchain network is a set of non-trusting writers sharing a digital database or ledger with no trusted middleman. Blockchain provides a robust, truly distributed peer-to-peer system that is tolerant of node failures.

The properties of Blockchain that makes it ideal for certain applications, including but not limited to digital currencies, are (Caldarelli 2020):

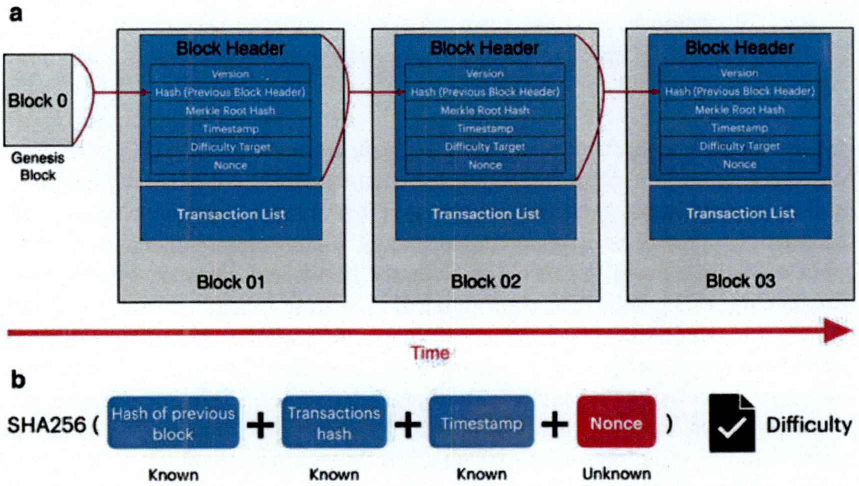

Fig. 3.6 **a** A schematic of the Bitcoin blockchain. Each block contains information about the block and a list of transactions. **b** PoW algorithm used in the Bitcoin network. *Source* Author

- Decentralization: There is the absence of an authority that constitutes a single point of trust/failure to approve transactions.
- Transparency: Records are auditable by all the participants in the network.
- Security and Immutability: Only private-key owners can start a transaction, and once added to the blockchain, forgery is very unlikely to happen.
- Censorship Resistance: The system is meant to prevent invalid transactions, not invalid users, so anyone—human, corporation, or even AI—may operate on the blockchain.
- Borderless: A blockchain network is not affected by distance or national borders. Even if the transaction happens in the same room or between two "poles," the rules remain the same.

As we mentioned earlier, the consensus mechanism is a critical blockchain element. In the first public blockchain network (Bitcoin), a SHA256 hash function is used for the mathematical puzzle of PoW, which takes the hash of the previous block, the transaction's hash of the current block, the timestamp of the current block, and a random number (nonce) as input (Fig. 3.6b). Miners in the bitcoin network are computers that participate in solving the mathematical puzzle by changing the nonce number (once at a time) to match the number of required leading zeroes in the final hash digest (Fig. 3.5). The first successful miner that solves the puzzle gets the reward (i.e., mined bitcoin) and broadcasts the block to other nodes to validate. The Bitcoin networks dynamically adjust the difficulty of the puzzle by changing the required leading zeroes to win the reward. The protocol's goal is to keep publishing a new block almost every 10 min regardless of the number of participants or the power of computers used.

The PoW consensus mechanism was an innovation that allowed for the emergence of Blockchain technology. However, this consensus mechanism suffers from significant shortcomings, including high energy consumption (discussed later in Chapter 5), relatively slow, and hard to scale. To solve some of these shortcomings, other modern consensus mechanisms have emerged. Another prominent mechanism is called Proof of Stake (PoS), where instead of requiring a mathematical puzzle (hash function) to be solved, the network requires the participant to stake something of value, typically the native token of the network (Understanding Blockchain Fundamentals, Part 2: Proof of Work & Proof of Stake | by Georgios Konstantopoulos | Loom Network | Medium, n.d.) (Fig. 3.7a). The algorithm will lock those valuable tokens and, in case of malicious or non-malicious (but harmful) activities, will be *slash* them. Slashing is a term used to represent taking skated tokens from participants by the algorithm when their negative activities are proven. To improve efficiency and reduce the chance of highest-stake validators taking disproportionate control of the network's consensus protocol, other variations, such as Delegated PoS and nominated PoS, have been proposed (Konstantopoulos 2020) (Fig. 3.7b). In these protocols, participants select a few nodes and delegate their staked tokens to them to participate in the consensus mechanism. The participant can change their nominees at any time if they are not happy with their performance.

Other consensus mechanisms have been developed in recent years, and each has its benefits and shortcomings, namely Proof of Authority, Proof of Capacity, Proof of Elapsed Time, Proof of Burn, etc. (A (Short) Guide to Blockchain Consensus Protocols – CoinDesk, n.d.). Knowing the advantages and disadvantages of each consensus mechanism is important when selecting a particular blockchain platform for an application. However, this is beyond the scope of this book, and there are numerous resources available on this topic for interested readers.

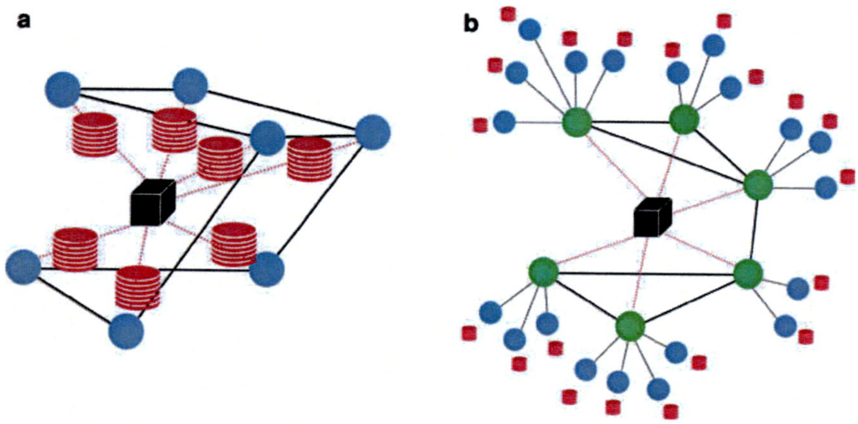

Fig. 3.7 **a** PoS consensus mechanism. **b** Delegated or nominated PoS consensus mechanism. *Source* Author

Regardless of the consensus mechanism used, a blockchain has the following typical workflow:

- New transactions are broadcast to all nodes
- Each node collects new transactions into a block
- In each "round," a random node gets to broadcast its block (leader)
- Other nodes accept the block if all transactions are valid (validate/vote)
- Nodes express acceptance by including its hash in the next block, lengthening the chain (reaching consensus).

3.8 Smart Contracts

Nick Szabo, a computer scientist, coined the term "Smart Contract" in 1995 before blockchains existed. In his words, a smart contract is "*a set of promises, specified in digital form, including protocols within which the parties perform on these promises*" (Smart Contracts, n.d.). The main aim of a smart contract is to automatically execute the terms of an agreement once the specified conditions are met. In the blockchain context, smart contracts are self-executing codes that reside on the blockchain and make changes in the ledger. They can be triggered automatically if a certain condition is met, such as if an agreement between the transacting parties is honored. This allows one to express business logic in code. Once deployed, the smart contract code is immutable, and the transaction history remains embedded in the blockchain on which it operates.

The main properties of smart contracts are (1) Trustlessnes: they are universally accessible, and no third parties or intermediaries are required, (2) Trackable: transactions can be traced and audited, (3) Irreversible (immutable): the security is paramount, and transactions are final, (4) Self-executing: this allows for reduce costs, increase speed, and use case variability, (5) Deterministic: the outcome is the same for everyone.

Smart contracts are another major innovation in the blockchain. They allow blockchain to become a platform for real-world applications. Ethereum was the first blockchain that was specifically developed to be able to run smart contracts. We have seen the application of these contracts recently result in blooming decentralized finance (DeFi). However, these innovations have more potential and possibilities in other areas, including sustainable energy. We will discuss those applications later in this book.

3.9 Blockchain Governance

Governance is defined as decision-making processes within an organization. The governance on blockchain could be conducted in a centralized or decentralized manner (Smits 2018). For example, the core developer team can decide to make changes in the blockchain architecture and execute it immediately (i.e., centralized), put their decision in a vote with the user community, or even allow users to propose new changes (i.e., decentralized). The examples are Bitcoin Improvement Proposal (BIP) and Ethereum Improvement Proposal (EIP). The voting can happen off-chain (on Discord, Twitter, Reddit, etc.) or on-chain by executing a voting smart contract.

Perhaps the most exciting application of blockchain-based governance so far has been Decentralized Autonomous Organizations (DAOs). DAOs are internet-native organizations collectively owned and managed by their members (Wright and Law 2021). They rely on blockchains, autonomous smart contracts, and digital assets to support organizations. Also, their decisions are governed by proposals and voting to ensure everyone in the organization has a voice.

The main characteristic of DAOs is that they are self-governing and are not influenced by outside forces. Its software operates on its own, with its by-laws were immutably written on the blockchain, not controlled by its creators. Typically, groups of like-minded individuals with specific projects and goals in mind form DAOs. Its identity is formed through consensus. Its authority is defined through voluntary endorsement and, ultimately, network effects.

3.10 Summary and Conclusion

Blockchain technology is a decentralized digital ledger that is secured by cryptography. It is also a multidisciplinary technology that benefits from game theoretic economic design. This chapter was designed to provide a brief introduction to this technology and how it works, intending to equip the readers with the information and technical understanding needed for the upcoming chapters. This chapter is not a comprehensive encyclopedia covering all the technical and historical development in this field. Therefore, many concepts and discussions are missing in this chapter. Interested readers are encouraged (but not required) to seek more technical information about this technology and its mechanics in computer science references.

References

11 Cryptographic Methods That Marked History: From the Caesar Cipher to Enigma Code and Beyond (n.d.) https://interestingengineering.com/innovation/11-cryptographic-methods-that-marked-history-from-the-caesar-cipher-to-enigma-code-and-beyond. Accessed 15 Jan 2023

A Brief History of Ledgers. Before Starting My Investigation Into… | by LLFOURN | Unraveling the Ouroboros | Medium (n.d.) https://medium.com/unraveling-the-ouroboros/a-brief-history-of-ledgers-b6ab84a7ff41. Accessed 15 Jan 2023

A Brief History of Money (2019) NAKAMOTO, December 30, 2019. https://nakamoto.com/a-brief-history-of-money/

A Cypherpunk's Manifesto (n.d.) https://www.activism.net/cypherpunk/manifesto.html. Accessed 1 Feb 2023

A (Short) Guide to Blockchain Consensus Protocols - CoinDesk (n.d.) https://www.coindesk.com/markets/2017/03/04/a-short-guide-to-blockchain-consensus-protocols/. Accessed 15 Jan 2023
Bit Gold | Satoshi Nakamoto Institute (n.d.) https://nakamotoinstitute.org/bit-gold/. Accessed 1 Feb 2023
B-Money | Satoshi Nakamoto Institute (n.d.) https://nakamotoinstitute.org/b-money/. Accessed 1 Feb 2023
Buterin V (n.d.) Ethereum: a next-generation smart contract and decentralized application platform, 36
Caldarelli G (2020) Real-world blockchain applications under the lens of the oracle problem. A systematic literature review. In: 2020 IEEE International Conference on Technology Management, Operations and Decisions (ICTMOD). IEEE, Marrakech, Morocco, pp 1–6. https://doi.org/10.1109/ICTMOD49425.2020.9380598
Genesis Block - Bitcoin Wiki (n.d.) https://en.bitcoin.it/wiki/Genesis_block. Accessed 1 Feb 2023
Hashcash (n.d.) https://nakamoto.com/hashcash/. Accessed 15 Jan 2023
How DigiCash Blew Everything (n.d.) https://cryptome.org/jya/digicrash.htm. Accessed 1 Feb 2023
Kasireddy P (2019) Let's take a crack at understanding distributed consensus. Medium (blog), October 29, 2019. https://medium.com/s/story/lets-take-a-crack-at-understanding-distributed-consensus-dad23d0dc95
Konstantopoulos G (2020) Understanding blockchain fundamentals, part 3: delegated proof of stake. Loom Network (blog), February 6, 2020. https://medium.com/loom-network/understanding-blockchain-fundamentals-part-3-delegated-proof-of-stake-b385a6b92ef
Lamport L, Shostak R, Pease M (1982) The byzantine generals problem. ACM Trans Program Lang Syst 4(3):382–401. https://doi.org/10.1145/357172.357176
Nakamoto S (2008) Bitcoin P2P E-cash paper, October 31, 2008. https://www.metzdowd.com/pipermail/cryptography/2008-October/014810.html
Nakamoto S (n.d.) Bitcoin: a peer-to-peer electronic cash system, 9
Rivest RL (1990) Cryptography. In: van Leeuwen J (ed) Handbook of theoretical computer science, volume A: algorithms and complexity. Elsevier and MIT Press, pp 717–755
Smart Contracts (n.d.) https://www.fon.hum.uva.nl/rob/Courses/InformationInSpeech/CDROM/Literature/LOTwinterschool2006/szabo.best.vwh.net/smart.contracts.html. Accessed 15 Jan 2023
Smits W-J (2018) Blockchain governance: what is it, what types are there and how does it work in practice? Watsonlaw (blog), October 24, 2018. https://watsonlaw.nl/en/blockchain-governance-what-is-it-what-types-are-there-and-how-does-it-work-in-practice/
Surowiecki J (2012) A brief history of money. IEEE Spectr 49(6):44–79. https://doi.org/10.1109/MSPEC.2012.6203967
The Cypherpunks (2019) NAKAMOTO, December 30, 2019. https://nakamoto.com/the-cypherpunks/
The Origins of Money Are Murky and Mysterious | Science News (2018, July 29) https://www.sciencenews.org/article/money-ancient-origins-debate-mystery
The Story of Cryptography: 20th Century Cryptography (n.d.) https://ghostvolt.com/articles/cryptography_20th_centuary.html. Accessed 15 Jan 2023
The Story of Cryptography: Historical Cryptography (n.d.) https://ghostvolt.com/articles/cryptography_history.html. Accessed 15 Jan 2023
The Story of Cryptography: Modern Cryptography (n.d.) https://ghostvolt.com/articles/cryptography_modern.html. Accessed 15 Jan 2023
Understanding Blockchain Fundamentals, Part 1: Byzantine Fault Tolerance | by Georgios Konstantopoulos | Loom Network | Medium (n.d.) https://medium.com/loom-network/understanding-blockchain-fundamentals-part-1-byzantine-fault-tolerance-245f46fe8419. Accessed 15 Jan 2023

Understanding Blockchain Fundamentals, Part 2: Proof of Work & Proof of Stake | by Georgios Konstantopoulos | Loom Network | Medium (n.d.) https://medium.com/loom-network/understanding-blockchain-fundamentals-part-2-proof-of-work-proof-of-stake-b6ae907c7edb. Accessed 15 Jan 2023

Whitfield D, Martin MH (1976) New directions in cryptography. IEEE Trans Infor Theor 22(6):644–654

Wood Dr G (n.d.) Polkadot: vision for a heterogeneous multi-chain framework

Wright A, Clinical Professor of Law at Benjamin N. Cardozo School of Law (2021) The rise of decentralized autonomous organizations: opportunities and challenges. Stanford Journal of Blockchain Law & Policy, June. https://stanford-jblp.pubpub.org/pub/rise-of-daos/release/1

Zetter K (n.d.) Bullion and bandits: the improbable rise and fall of E-Gold. Wired. https://www.wired.com/2009/06/e-gold/. Accessed 1 Feb 2023

Dr. Soheil Saraji is an Associate Professor of Energy and Petroleum Engineering, an Adjust Professor at the School of Energy Resources, and co-director of the Hydrocarbons Research Laboratory at the University of Wyoming. He has eighteen years of research experience and more than 35 peer-reviewed journal publications in subsurface energy extraction, storage, and carbon geo-sequestration. Furthermore, Dr. Saraji is a pioneer in applied blockchain research for the oil and gas industry. He has developed new courses and research initiatives on this topic at the University of Wyoming.

Chapter 4
Blockchain: Legal and Regulatory Issues

Karisma Karisma and Pardis Moslemzadeh Tehrani

4.1 Introduction

Blockchain technology (hereafter referred to as "blockchain" unless specifically mentioned) plays an increasingly significant role in multiple sectors, such as energy, banking, supply chain, and retail. The ubiquity of blockchain invokes disruptive and dynamic changes in industries as a decentralized, autonomous, and distributed ledger that operates rhythmically and efficiently without third parties. This chapter outlines the contours of legal and regulatory issues circumjacent to the application of blockchain technology by drawing on private law, criminal law, and public and administrative law when articulating the foundational and persistent challenges which have provoked commentary from many scholars. It pragmatically delineates the challenges tethered to a blockchain-enabled carbon economy by scrutinizing current debates and providing insights from a novel vantage point.

Section 4.2 of this chapter comprises three sub-parts, i.e., (i) contract law, (ii) liability and enforcement, and (iii) property law. First, we explore the plausible variance of blockchain with contract law attributes, particularly with the prevalence of blockchain-enabled smart contracts. Second, we consider the prevailing gaps in the attribution of liability and the proper enforcement of legislation in the ecosystem to provide safeguards for blockchain users. Third, we examine the property law facet by scrutinizing two concepts, chose in action and chose in possession underlying blockchain tokens. Section 4.3 of this chapter assesses potential cybersecurity attacks

K. Karisma (✉)
Faculty of Law, University of Malaya, Kuala Lumpur, Malaysia
e-mail: karisjay@outlook.com

P. Moslemzadeh Tehrani
School of Law, Faculty of Business, Law and Toursim, University of Sunderland, Sunderland, United Kingdom
e-mail: pardis.tehrani@sunderland.ac.uk

S. Saraji and S. Chen, *Sustainable Oil and Gas Using Blockchain*,
Lecture Notes in Energy 98, https://doi.org/10.1007/978-3-031-30697-6_4

and money-laundering activities that cripple the country's economy and infrastructure, further impacting investors' confidence in the blockchain landscape. Section 4.4 explores the facets of securities, commodities, taxation, and data protection laws. It first examines the general classification and legal status of blockchain tokens, which then advances to the complexity of imposing taxes on blockchain activities and assets. This part also delves into data protection laws that invoke the right to rectification and the right to be forgotten, which is at odds with the blockchain feature and functionality. Section 4.5 of this chapter explores blockchain-related legislation in the United States, Gibraltar, Malta, Estonia, Liechtenstein, and Switzerland. The legal and regulatory landscapes of many countries relating to blockchain remain rudimentary and fragmented. A meager and ill-designed framework might hinder the widespread application of blockchain, particularly when governance regimes shape the blockchain fora. Such governance regimes either provide a swifter implementation avenue for the technology or one that is unprogressive. A paradoxical event materializes that while countries reckon blockchain favorable, few have regulated blockchain by amending existing legislation or introducing industry-specific frameworks. This chapter navigates the governance framework of several countries, having advanced broad and compelling legislation to constitute global pacesetters in the blockchain race. Section 4.6 of this chapter is an extension of the contract law aspect in Section 4.2, which thoroughly evaluates how existing contract law principles align with the use of smart contracts. Section 4.6 also explores the eclectic challenges concerning the implementation of smart contracts, followed by a succinct precis of the legal and regulatory challenges presented earlier.

4.2 Private Law

4.2.1 Contract Law

Smart contracts underlying blockchain occupy a central position in business transactions. During smart contract creation, parties involved in the negotiation phase negotiate the terms, conditions, rights, and obligations underlying the contract between two or more parties. Contracting parties must reach a consensus on the content and description of the agreement. Subsequently, it requires translating the salient aspects of the contract into smart contractual codes. In codifying the legal prose, the aptness and functionality of the smart contract depends on the programming language to accurately reflect the contractual clauses. A new block comprising a smart contract is created and validated by a majority of nodes on the blockchain network. Here, the legality of smart contracts under existing contract law provisions becomes a discursive topic. The widespread applications of smart contracts are unlikely to oust the application of conventional principles, particularly concerning the performance of contractual clauses that cannot be coded onto the platform. However, it is pertinent to resolve the apparent friction between contract law and smart contracts by developing technology-driven principles to complement the foundational norms. We explore the area of smart contracts under Sect. 4.6 of this chapter.

4.2.2 *Allocation of Liability*

While multiple jurisdictions have devised appropriate legal and regulatory frameworks to circumvent the gaps in the blockchain mechanisms, other countries adopt a wait-and-see approach. The seminal question on the allocation of liability has received unsurprisingly little attention due to nascent blockchain systems. On the one hand, in a centralized landscape, third-party intermediaries are generally responsible for providing goods and services. On the other hand, in a decentralized and distributed environment, such as blockchain, participants engage in transactions autonomously, without the presence of third-party intermediaries. Blockchain landscapes trigger the question of the imposition of liability, mainly when an individual endures damage or harm in the case of failures, accidents, and errors from dysfunctional systems. In addition, blockchain supports international and inter-jurisdictional transactions and processes. Hence, as legal frameworks develop across multiple jurisdictions, the prerequisites and standards of civil and criminal liability may differ. The allocation of risk and liability is intricate in situations where systems operate simultaneously in online and offline landscapes, such as by combining conventional energy infrastructures with blockchain systems (Frommelt 2020).

Scholars present two distinct paradigms on the allocation of liability. First, developers, operators, and platform providers are accountable for the risk of harm and damage that materializes in the blockchain ecosystem due to the ambit of control and functions on operating systems, such as software and blockchain servers. Given the inherent difficulty in determining accountable actors in decentralized and distributed systems, developers and system operators may be jointly liable for breach of contract or tort claims (Kulms 2020). Second, liability can be distributed across the entire network of peers on a blockchain system to ensure equitable distribution of risks and benefits (de Almeida et al. 2021). Based on the precursory insights, "a system of pooled responsibilities" may be a viable solution in allocating liability between blockchain peers as it considers the speed and complexity of blockchain systems (de Almeida et al. 2021).

4.2.3 *Property Law*

This section explores whether digital assets constitute property law under the legal framework. In what follows, determining the legal status is particularly salient due to far-reaching implications in tort law and bankruptcy law, among others (Bacon et al. 2018). While digital assets may be defined as personal property, characterizing property interest to choses in possession and choses in action is cumbersome.

On the one hand, choses in possession (or tangible personalty) reflect the individuals' ability to claim or enforce their property rights through physical possession (Hosmer 1915). However, digital assets do not fit within the nuances of choses in possession. Additionally, courts in the United Kingdom do not recognize such characterization in light of long-standing principles. The case of Your Response Ltd. v Datateam Business Media Ltd. raised a pertinent question as to whether a party was entitled to exercise a lien over a digital database (Your Response Ltd v Datateam Business Media Ltd 2014). Lord Justice Moore-Bick held that considering digitalized materials as choses in possession is a significant departure from the deep-rooted law exemplified in OBG v Allan (OBG Limited and others v Allan 2007). His Lordship further asserted that this area was subject to parliamentary scrutiny and intervention by meaningfully engaging in legislative review (Your Response Ltd v Datateam Business Media Ltd 2014). Similarly, in Armstrong DLW GmbH v Winnington Networks Ltd., the EU emission allowances did not constitute choses in possession, given that the existing law did not allow the assumption or likening of the electronic form to a physical thing, of which possession is realizable (Armstrong DLW GmbH v Winnington Networks Ltd 2012). In essence, the legislative arm should take proactive measures to legislate this area by considering the strategic implications of recognizing digital assets as choses in possession.

On the other hand, choses in action are intangible property rights that are enforceable through legal action, as opposed to physical possession (Shaw 2009). Whether digital assets comprise a choses in action is contingent on the presence of clearly delineated rights embodying such digital assets against a specific issuing party or developer (Bacon et al. 2018). However, it is no easy feat to recognize digital assets as choses in action or possession. Countries have to navigate in uncharted territories due to the considerable confusion and uncertainties on the legal status of digital assets.

4.3 Criminal Law

4.3.1 Cybersecurity Attacks

Blockchain systems are subject to a variety of cybersecurity attacks that can hinder the smooth and secure operation of energy systems. This section identifies three key elements in assessing cybersecurity challenges, namely 'vulnerability,' 'threat,' and 'impact.' Blockchain systems are susceptible to technical vulnerabilities, organizational vulnerabilities, and systemic vulnerabilities owing to deficiencies in (a) blockchain designs, (b) weaknesses in operational policies and practices, and (c) interconnectivity with other systems, where exposing systems to cyber-risks can destabilize blockchain applications and processes in undesirable ways. In addition, adversaries can exploit blockchain systems by using the blockchain as a vantage point to augment and amplify existing attacks or launch new attacks (Saad et al.

2020). The proliferation of cyber threats imposes considerable societal and technological implications (specifically towards the viability, functionality, and effectiveness of blockchain) that are likely to compromise carbon economy applications. The following subsections explore governance initiatives and technology-driven regulatory instruments that have been developed in recent years in the form of standards and guidelines.

i. *German Federal Office for Information Security (BSI)*

In 2019, BSI, a national cybersecurity authority, published a comprehensive document titled Towards Secure Blockchains (TBS) to institute a dialogue regarding the nuances of blockchain technology. First, BSI acknowledges the widespread security challenges that generate unprecedented blockchain vulnerabilities, including (a) inherent hardware and software security issues and (b) new attack vectors that afflict security attacks on different interfaces and system components (Federal Office for Information Security [BSI] 2019). Second, as outlined by TBS, BSI plays an indispensable role in the investigation and assessment of blockchain technology. Further, TBS emphasizes the security certification of blockchain applications, which can benefit blockchain users. Attacks on the network may trigger severe implications for the security of the entire blockchain system. Therefore, TBS turns to a vast constellation of non-exhaustive measures, including security and contingency management, application and network security, and access control. TBS finds sufficient merit in adopting conventional security goals, such as integrity, authenticity, availability of services, confidentiality, anonymity, and pseudonymity for the blockchain landscape (Federal Office for Information Security [BSI] 2019).

ii. *National Institute of Standards and Technology (NIST)*

The NIST is a non-regulatory agency within the United States Department of Commerce that aims to "promote U.S. innovation and industrial competitiveness by advancing measurement science, standards, and technology" to augment economic security and the standard of living of its citizens. In 2018, the NIST published NISTIR 8202—Blockchain Technology Overview ("the Document"), which examines crucial information about blockchain technology, including its characterization, components, consensus models, forking, smart contracts, and common misconceptions about the blockchain landscape (Yaga et al. 2018). It highlights cybersecurity risks and the importance of entrenching risk assessment and management measures. In addition, the Document clarifies that the existing cybersecurity instruments in the form of standards and guidelines remain highly apropos for maintaining the security of systems that interface with and hinge on blockchain networks. While it may be necessary to fine-tune cybersecurity instruments in accordance with blockchain's specific attributes, existing cybersecurity frameworks create a solid foundation to safeguard blockchain networks from cyberattacks. It concludes that while the NIST Cybersecurity Framework is not specific to blockchain technology, it contains a broad set of standards and guidelines that are comprehensive enough to encompass blockchain applications and facilitate institutions to develop policies that ascertain and control cybersecurity risks.

iii. *European Union Agency for Cybersecurity (ENISA)*

ENISA published a paper titled Distributed Ledger Technology & Cybersecurity—Improving information security in the financial sector. ENISA explores different forms of Distributed Ledger Technology (DLT) and its integral components, including consensus protocols, cryptography, side chains, and smart contracts (European Union Agency for Cybersecurity 2017). In addition, this paper focuses specifically on traditional and DLT/blockchain-specific cybersecurity challenges. The publication delineates 'Good Practices' for addressing key cybersecurity challenges to assist commercial entities and organizations in securely implementing blockchain technology. While the paper focuses on circumventing security challenges in the financial sector, the best practice guidelines and standards are adaptable to non-financial sectors, owing to the generic nature of the challenges.

In a nutshell, blockchain technology offers multiple blockchain designs and use cases, and each blockchain model may exhibit unique characteristics. Regulators and policymakers ossify and fail to adapt security-related frameworks, guidelines, and standards when addressing distinct trajectories of blockchain. A brief assessment of cyber security authorities and organizations reveals the pertinence of governance regimes and authoritative bodies in circumventing security conflicts at the state, national, or regional levels.

4.3.2 Money Laundering Issues

Cryptocurrencies that operate on blockchain systems increasingly facilitate and orchestrate money laundering activities. The lack of regulatory oversight can cripple a country's economy and generate negative implications for users and other stakeholders. In this section, the pivot of our discussion is on cryptocurrencies. It is unfounded to associate the occurrence of such activities with blockchain technology, given its inherent design to operate applications lawfully. Due to the anonymity surrounding cryptocurrencies, regulatory bodies cannot effectively monitor all transactions.

As cryptocurrency transactions transcend jurisdictional boundaries, the existing fragmentation of international laws, standards, and guidelines may perpetuate criminal activities. The Financial Action Task Force (FATF) has prescribed the global policies and standards concerning Anti-money Laundering and Countering Terrorism Financing (AML/CFT) (Declaration of the Ministers of the Financial Action Task Force 2022). The FATF members are obliged to comply with such policies and transpose suitable national legal frameworks accordingly. The FATF recently updated its FATF Standards to streamline its adoption to "virtual assets" and "virtual assets service providers." This position led to the issuance of the FATF Guidance to shed light on the amendments made to the FATF Standards. The recommendations include augmenting the role of government bodies to impose sanctions and exercise enforcement actions in the event the service providers fail to observe the AML/CFT provisions. Government bodies should require service providers to adopt risk assessments and mitigation measures, such as the know-your-customer (KYC) verification process, screening of financial transactions, customer due diligence, and other prudent measures to hinder illegal use cases concerning cryptocurrencies. In addition, the 5th Anti-Money Laundering Directive (AMLD5) defines virtual currencies broadly to embrace all types of cryptocurrencies to combat illegal activities (Directive 2018/843 of 30 May 2018 on anti-money laundering and countering the financing of terrorism 2018). The AMLD5 represents a comprehensive and representative framework, under which platform operators are obliged to monitor the use of cryptocurrencies vis-à-vis technological measures and ensure transparency by maintaining records of customers.

4.4 Public and Administrative Law

4.4.1 Securities Law

With the proliferation of virtual currencies and digital tokens, many countries carefully probe such offerings. In this section, we explore security, utility, and currency tokens in greater depth. We explore the possible regulatory trajectory by discussing the approaches adopted by the United States (U.S.) and European Union (EU) surrounding the issuance and characterization of such tokens.

Security tokens commonly allude to investment, equity, or asset tokens, which underlie external tradeable and transferable physical assets. These tokens are analogous to conventional components, such as shares, debentures, or bonds (Guseva 2020). The United States employs the Howey Test to ascertain whether a given financial instrument constitutes an investment contract, a specific type of security (Securities and Exchange Commission v. Howey Co. 1946). Under the Howey test, a "transaction or scheme whereby a person invests his money in a common enterprise and is led to expect profits solely from the efforts of the promoter or a third party" would constitute a security, governed by securities legislation (Securities and Exchange Commission v. Howey Co. 1946). Focus is primarily on (a) placing dependence on others' efforts and (b) expectations of profits. The former concerns salient efforts that impact the success or failure of businesses, as opposed to merely administrative measures. The latter relates to the possibility of capital appreciation or earnings on transactions involving digital assets (Kaal 2022). William Hinman, the former Director of the U.S. Securities and Exchange Commission (SEC), highlights that not all digital assets constitute securities. It depends on the economic context and investors' expectations, prioritizing substance over form in conducting a wholesome legal analysis (Mendelson 2019). However, developers may make tenuous changes to blockchain applications to circumvent the second criterion of the Howey test and the application of securities regulations (Grennan 2022). In essence, the Howey test is not the be-all and end-all when characterizing tokens as regulators approach the situation on a case-to-case basis (Ferreira 2020). In 2019, the SEC issued a "Framework for 'Investment Contract' Analysis of Digital Assets" (the Framework), considering the arduousness of characterizing digital assets. The Framework reduces information asymmetries and ensures that market participants are familiar with the non-exhaustive factors in assessing digital assets as security (or otherwise) when partaking in the sale or distribution of such assets (U.S. Securities and Exchange Commission 2019). It is practically and theoretically possible that a digital asset that initially amounted to security may transform into something different that no longer constitutes security, and therefore securities legislation does not apply (Park 2018).

In the European Union, digital assets that comprise securities are bound by various regulations and directives, including but not limited to the Prospectus Regulation, Market Abuse Regulation, and Markets in Financial Instruments Directive (MiFID II). Similar to the U.S., in determining the applicability of securities directives and regulations, it is pertinent to examine whether the financial instruments constitute "transferable securities" defined under the MIFID II as "those classes of securities which are negotiable on the capital market," such as shares, bonds, and other comparable instruments (Council Directive 2014/65/EU on markets in financial instruments 2014). We consider the regulatory criteria that can determine the legal status of security tokens. First, the Prospectus Regulation provides for the transferability requirement, in that tokens must be transferable or assignable to other individuals (Council Regulation (EU) 2017/1129 on the prospectus to be published when securities are offered to the public or admitted to trading on a regulated market 2017). Most if not all digital tokens fulfill this requirement, primarily due to the ease in exercising such transfer, which leads us to the following prerequisite on negotiability. The MiFID requires securities to be "negotiable on the capital market," requiring the sale and purchase of securities to be facile and frictionless (Council Directive 2014/65/EU on markets in financial instruments 2014). The third requirement illustrated in the MiFID II necessitates the standardization of "transferable securities" that are homogeneous and mutually interchangeable. The existence of this requirement is implicit in the fulfillment of the negotiability criterion (Council Directive 2014/65/EU on markets in financial instruments 2014). In a nutshell, the EU Directives and Regulations prescribe three criteria assessed subjectively in determining whether tokens qualify as securities. Issuing operationalizable guidelines in this area may constitute a strategic approach for the EU to hinder ambiguity and regulatory fragmentation.

Utility tokens are means of exchange for products and services under the aegis of blockchain-based platforms that issue such tokens (Ferreira 2020). For instance, the ether token constituting the native token on the Ethereum platform possesses an intrinsic value that allows users to run smart contracts and execute transactions (Ferreira 2020). Accordingly, utility tokens do not constitute securities primarily because the price fluctuations in the goods and services market prejudice investors' profit instead of the organizational efforts of a company. Currency tokens are alternative means of exchange to facilitate the trade of goods and services which are generally exempted from securities regulation.

Based on the above discussion, the question of whether digital tokens constitute securities is remarkably difficult to answer. In light of the prevailing definitional ambiguity on the status of digital tokens, countries should develop guidelines to circumvent the nuances and challenges surrounding cryptocurrencies.

4.4.2 Commodities Law

In the U.S., the Commodity Futures Trading Commission (CFTC) is actively pursuing the treatment of virtual currencies as commodities under the Commodity Exchange Act 1936 (CEA). As a starting point, the CFTC modulates commodity futures contracts exerting control over "goods and articles [...] and all services, rights, and interests [...] in which contracts for future delivery are presently or in the future dealt in." Hence, natural persons and legal entities that trade digital assets through futures contracts fall under the purview of the CFTC. To further exemplify the area of digital assets and commodities, we draw on landmark decisions. In 2015, the CFTC brought an enforcement action against Coinflip for violating the CEA and CFTC's regulations by engaging in commodity options without prior registration. The CFTC held that virtual currencies constituted commodities under the Act and hence, are within the jurisdiction of the CFTC (In the Matter of: Coinflip, Inc., d/b/a Derivabit, and Francisco Riordan 2015).

In 2018, the CFTC brought a preliminary injunction against defendants, Patrick Mc Donnell and CabbageTech, for defrauding customers by engaging in deceptive and fraudulent schemes and conducting illegal activities concerning virtual currencies. In this case, the defendants urged customers to send money and virtual currencies for various services, but the provision of such services never materialized (Commodity Futures Trading Commission v. McDonnell 2018). Accordingly, the Court found that CTFC had jurisdiction to undertake enforcement measures vis-à-vis any fraudulent activities concerning virtual currencies transacted through channels of interstate commerce. Judge Jack B. Weinstein held that virtual currencies constitute "goods exchanged in a market for a uniform quality and value." Judge Weinstein further ruled that the "jurisdictional authority of CFTC to regulate virtual currencies as commodities does not preclude other agencies from exercising their regulatory power when virtual currencies function differently than derivative commodities." Hence, regulatory bodies exercising control over securities and commodities trading may possess "overlapping jurisdiction" in this area (Simmons 2021).

Based on the above, CFTC is demonstrating strong participation in the regulatory landscape by adopting robust enforcement measures to circumvent fraudulent activity and manipulation of the market. The CFTC is well-coordinated with other regulatory bodies, including the SEC and the Financial Crimes Enforcement Network (FinCEN), by taking collective action and providing oversight against illegal activities concerning digital assets (Commodity Futures Trading Commission 2019).

Unlike the nuanced debate by the U.S., other countries take a pessimistic stance in delineating the definitional scope of commodities and whether virtual currencies fall within the commodities legislation. Accordingly, legal guidance can unravel any regulatory uncertainties and ambiguities in existing laws, tighten enforcement measures, and circumvent any negative externalities from the proliferation and ubiquity of digital assets.

4.4.3 Privacy and Data Protection

The emergence of blockchain is a harbinger of technological development vis-à-vis decentralization, digitalization, automation of systems, and business processes, thus driving change in the energy sector. Scholars and academics widely use the term 'trustlessness' as a feature of blockchain to demonstrate the complete removal of reliance on third-party intermediaries to operate the ledger and verify transactions. Blockchain participants bestow trust in the integrity of the blockchain systems through consensus protocols and intelligent interplay of cryptography mechanisms.

Despite that, 'trustlessness,' as it stands, has emerged as a paradox or aporia for the rising threats and friction circumjacent to data protection underlying blockchain systems. While 'trustlessness' is at the core of blockchain technology, trust-building regulations from the outer periphery should advocate societal values, rights, and freedoms to hinder cascading and catastrophic effects of privacy and security risks.

Accordingly, we explore the core principles of significant pieces of legislation, namely the California Consumer Privacy Act of 2018 (CCPA) (California Consumer Privacy Act 2018) and the General Data Protection Regulations (GDPR) (General Data Protection Regulation 2016). Article 5 of the GDPR sets out seven fundamental principles for the processing of personal data, namely (a) lawfulness, fairness, and transparency, (b) purpose limitation, (c) data minimization, (d) accuracy, (e) storage limitation, (f) integrity and confidentiality, and (g) accountability. The CCPA advances similar principles to secure the California residents' data protection and privacy rights. To further refine, augment, and elevate data protection rights, California enacted the California Privacy Rights Act (CPRA) (California Privacy Rights Act 2020). The CCPA (including provisions under the CPRA) and GDPR are connoted as global standards in data protection landscapes and advance stringent safeguards to protect the personal data of data subjects (Lombino 2020).

It is salient to balance the competing interest of the data privacy rights with the freedom of information. In doing so, it is important to delineate the parameters of personal data. The GDPR defines personal data as information which involves an "identified or identifiable natural person" (General Data Protection Regulation 2016). Similarly, the CCPA refers to personal information as "information that identifies, relates to, describes, is capable of being associated with, or could reasonably be linked, directly or indirectly, with a particular consumer or household" (California Consumer Privacy Act 2018). Given the resulting privacy concerns and the real threats posed to personal data by the increasing use of blockchain, it is crucial to illustrate the significant concerns of technology users and regulators in the global dimension.

The subsequent sections explore the friction between blockchain and data protection regulations.

Storage limitation: Article 5(1)(e) of the GDPR introduces the principle of storage limitation which states that the personal data of the data subject shall be kept only for the duration necessary to attain the purposes for which they are processed (General Data Protection Regulation 2016). Likewise, Section 1798.100 (a)(3) of the CPRA imposes obligations of storage limitation, disallowing retention periods for longer than is reasonably necessary (California Privacy Rights Act 2020). Blockchain stores multiple copies of the full nodes on the ledger. While the storage of full copies of the data conduces data redundancy on the blockchain network, one can justify the storage of data for lengthy periods as necessary means to maintain the integrity and functionality of the blockchain. However, prolonged data retentions may compromise transactional data containing personally identifiable information.

Data minimization: This section is primarily concerned with exploring the discrepancy between the design of blockchain systems and the core principle of data minimization. Article 5(1)(c) of the GDPR states that personal data must be "adequate, relevant and limited" to what is necessary concerning the purposes for which they are processed (General Data Protection Regulation 2016). A similar regulatory texture is delineated under the CPRA, prohibiting organizations from collecting personal data that is disproportionate and not reasonably necessary. In a blockchain system, each full node on the ledger can replicate and store a complete and identical copy of the data to ensure transparency and data integrity, which are prominent features of blockchain systems. Storing full copies of personal data may precipitate conflictual interactions with the data minimization principle. The data of energy transactions attached to the blockchain structures are not limited to the perusal of the transacting parties but all full nodes on the blockchain systems. Hence, the immutability of the blockchain has some unintended consequences, as it is virtually impossible to delete or correct the data recorded on the blockchain ledger, resulting in such data being permanently present on the chain.

Purpose limitation: According to Article 5 (1)(b) of the GDPR, the "specified, explicit and legitimate purposes" for which data are collected must be determined, and personal data may not be further processed beyond the initially informed purpose without additional consent (General Data Protection Regulation 2016). Section 1798.100 (a)(1) and (2) of the CPRA presents a comparable purpose limitation obligation, disallowing the processing of personal data in a manner that is incompatible with the original and disclosed purpose. In the light of blockchain technology, this principle is indeed fundamental, as the purpose must be precisely defined to ensure compliance with the other principles (California Privacy Rights Act 2020). Unreasonably broad declarations for the processing of personal data are prohibited. The data controller determines the purpose(s) and the means for processing the personal data by specifying such purpose(s). Despite the prevailing GDPR requirements, the role of the data controller is distributed widely across the nodes in a decentralized network. Therefore, there is an absence of a unitary actor to determine the purposes and means of processing personal data. From a pragmatic perspective, it is essential to determine the existence of data controller(s) and whether blockchain nodes, miners, and users can be equally and jointly responsible as data controllers on a blockchain network.

Right to be forgotten: As outlined above, immutability is an appealing feature of blockchain technology. However, a significant constraint to introducing such a feature is the right to be forgotten (RTBF), an emerging legal concept under the GDPR and CCPA. Pursuant to Article 17, the data subject has the right to request from the data controller the erasure of personal data concerning him or her in a timely manner if one of the following grounds referred to in Article 17(1) applies (General Data Protection Regulation 2016). The RTBF propounded in CCPA is narrower, requiring businesses to delete personal data "which the business has collected from the consumer" (California Consumer Privacy Act 2018). Hence, businesses may be able to keep personal data received from third parties or data developed through prior transactions. Part of a technological breakthrough is recording the personal data of a data subject on a blockchain ledger in a tamper-proof manner (Tatar et al. 2020). However, there is a conflict between the inherent immutability of the blockchain ledger, the impossibility of deleting personal data from the blockchain network, and the RTBF.

Fulfilling the data subject's request to delete data in a blockchain ledger diminishes the utility of the blockchain because any modification or manipulation of data in a blockchain ledger distorts the entire chain of blocks, leading to undesirable consequences and inconsistencies (Tatar et al. 2020). In recent years, researchers have attempted to find breakthroughs or implement existing technological solutions in advanced ways.

Right to rectification: Under Article 5(1)(d) of the GDPR (further expanded in Article 16), if personal data are inaccurate, incomplete, or not up to date, reasonable steps shall be taken to ensure that such data are erased or rectified in a timely manner (General Data Protection Regulation 2016). The CPRA advances the right to rectification (RTR) akin to the GDPR. Several technical obstacles arise when implementing this right on a blockchain landscape. First, identifying and detecting errors across full nodes in a blockchain network is an arduous task. Even after the erroneous or incomplete data is identified, correction can be virtually impossible, especially on a permissionless blockchain. Data correction must be performed ubiquitously across multiple nodes, as modifying a local copy is insufficient, as it requires the consensus of a majority of participants on the blockchain network (Duarte 2019). Aside from the difficulty of coordinating and addressing the different nodes on the blockchain network, there is a likelihood that node disconnection may occur, making it challenging to create a fork on the existing blockchain to rectify the personal data (Duarte 2019). Compliance with the right of rectification is easier on a private or consortium blockchain because it includes a more centralized database than a permissionless blockchain (Duarte 2019). Underlying these challenges, failure to develop specific policies to implement the provision in the blockchain ecosystem effectively will hinder the exercise of the RTR by data subjects. Having described the contours of the challenges, which are hardly unprecedented, it is helpful to explore the practical possibilities for blockchain users and developers. The RTR can be exercised by making a supplementary declaration under Article 16 of the GDPR to reflect the current situation with energy participants, as it is impossible to delete previously inaccurate or incomplete personal data. Under the CPRA businesses relinquish their obligations by adopting "commercially responsible efforts" to confer the RTR to data subjects.

We consider a scenario in which individuals actively participate in blockchain-based carbon trading systems, engage in emissions reduction activities, and receive carbon credits. While blockchain applications can drive widespread societal and global change by (a) facilitating carbon trading systems and carbon credits transactions and (b) increasing reliability, transparency, and security of systems and processes, such applications can simultaneously trigger conflicting interactions and difficult trade-offs with data protection regimes. A well-designed blockchain system with new technical designs and architectural features can support multilateral goals and values embedded in data protection regulations.

Scholars Gupta & Mason elucidate that the "democratization driver of transparency" fortifies the freedom of information legislation and advances greater accountability (Mason and Gupta 2015). The Freedom of Information Act (FOIA) introduced in a record number of jurisdictions, provides public access to information maintained by the public authorities (PAs) to promote openness, accountability, and transparency. To achieve this, the legislative provisions highlight two aspects, namely (a) the obligation of the PAs to publish certain information in a manner that is open and accessible, and (b) to provide the public with the right to access documents and records held by them. However, the FOIA may contain numerous exemptions to withhold information from the public. For instance, where it involves "trade secrets and commercial or financial information obtained from a person [that is] privileged or confidential" (5 U.S.C. § 552 2018). The courts may continue to liberally interpret the Fourth Exemption of the FOIA, which favors the interests of corporations over the public interest, and thus preventing the disclosure of corporate information that may be otherwise annihilative to corporate entities (Lamdan 2014). However, to reduce carbon emissions in upstream, midstream, and downstream markets, countries should abstain from treating emissions data as highly confidential data or trade secrets. To properly align with climate change policies, regulators must be more proactive in advancing reflective provisions. For instance, the Clean Air Act provides that emission data is not confidential business information (42 U.S.C. § 7414(c) 2006). On the other hand, Aarhus Convention specifically establishes the right of the public to receive environmental information held by PAs. This reinforces that the principle of transparency, state of the environment, and public health and security outweighs commercial interest, by establishing the right to know and preventing the nondisclosure of emissions data. This has triggered other countries to enact legislative frameworks to provide open access to environmental data.

4.4.4 *Taxation Laws*

In this section, we answer prevailing questions regarding the imposition of capital gains tax and value-added tax (VAT) on digital assets. On the one hand, in most jurisdictions such as Norway, Finland, Germany, and Australia, taxation authorities impose capital gains tax on the income (gains) from the sale, trade, and exchange of digital assets or acquisition of goods or services using such assets (Solodan 2019). However, the payment of capital gains tax each time it is transacted may reduce the functionality of digital assets (Zetzsche et al. 2019). With the advent of cryptocurrency markets, governments have expressed concerns about the loss of revenue from the lack of a progressive framework for imposing taxes on cryptocurrencies. On the other hand, most countries do not levy VAT on the exchange of digital assets for legal tenders, fiat currencies, or other virtual assets. However, in relation to the payment of taxable goods and services via virtual currencies, the normal VAT provisions apply.

4.5 Regulatory Approaches to Blockchain Technology in Various Jurisdictions

In the following section, we explore countries that have adopted institutional or governance frameworks to regulate blockchain. These countries have an enabling policy environment for blockchain that can unlock and spur blockchain growth and remove any barriers of entry that generate disparities and hinder widespread blockchain adoption.

4.5.1 *United States*

4.5.1.1 Wyoming

Wyoming is at the forefront of the blockchain landscape, having enacted numerous legal and regulatory provisions, and industry players recognize the state as one of the most blockchain-friendly states. Such legislation has moved the state to establish a blockchain task force (Temte 2019), exempt digital tokens from state securities and money transmission laws (Wyo. Stat. Ann. § 17-4-206(a) 2018), establish a Select Committee on Blockchain, Financial Technology, and Digital Innovation Technology. These instruments have, in turn, developed and introduced necessary laws to advance blockchain and financial technology and define blockchain as a digital ledger that is chronological, consensus-based, and decentralized (Wyo. Stat. Ann. § 34-29-106 2017). In addition, the legislative framework authorizes corporations to use distributed electronic networks or databases to maintain corporations' electronic records (Wyo. Stat. Ann. § 17-16-1601 2018). Commercial entities may issue stock certificates via tokens, which are stored in an electronic format and entered onto a blockchain or other secure, auditable database (Wyo. Stat. Ann. § 17-16-625 2019). Briefly, state representatives have adopted a permissive legislative framework that underpins the growth of blockchain and accommodates emerging distributed ledger technologies.

4.5.1.2 Delaware

Delaware is making great strides in the blockchain fora, similar to Wyoming. The amendment to Delaware General Corporation Law (DGCL) allows corporations to utilize "electronic networks or databases" to maintain records in the regular course of the business (8 Del. Code Ann. § 224 2017). While state corporate laws impose various hurdles on blockchain stock ledgers, Delaware has taken a proactive approach by introducing amendments to Section 219(c) of the DGCL, advancing the definition of stock ledgers as "records administered by or on behalf of the corporation" (8 Del. Code Ann. § 219 2017). This amendment relinquishes the sole and direct maintenance of stockholder's records by corporations (or their agents). Instead, it recognizes efficient and cost-effective solutions and enables the widespread application of distributed ledgers for corporate recordkeeping. In addition, Section 232 amends electronic transmission to include "the use of, or participation in, 1 or more electronic networks or databases (including 1 or more distributed electronic networks or databases)" (8 Del. Code Ann. § 232 2017). In providing an enabling framework for corporations, legible paper form devised from electronic networks or databases "shall be valid and admissible in evidence" (8 Del. Code Ann. § 224 2017). In essence, the amendments are a starting point for the initial articulation and development of blockchain in corporate governance.

4.5.1.3 Illinois

Illinois enacted the Blockchain Technology Act (BTA) to embark on complex digital technology and drive change in the legal domain. BTA advances the definition of blockchain, cryptographic hash, electronic records, and smart contracts, equipped to (a) resolve legal uncertainty and ambiguity, (b) address governance gaps, and (c) modernize the legal and regulatory framework (Blockchain Technology Act 2019). By proactively adopting an ex-ante regulatory approach and integrating blockchain into legislative provisions, adjudication, and enforcement measures, Illinois delineates the scope and applicability of blockchain and clarifies unprecedented concepts. In addition, BTA augments the usage of blockchain and smart contracts by the local government to perform its functions and duties (Blockchain Technology Act 2019). In addition, the local government may not impose taxation or licensing requirements for the usage of blockchain and smart contracts (Blockchain Technology Act 2019). Interestingly, the Blockchain Business Development Act that came into force in 2020 requires the Department of Commerce and Economic Opportunity to consider the deployment of blockchain in banking facilities, appraise potential use cases, and provide feedback on salient regulatory amendments to facilitate the technological shift (Blockchain Technology Act 2019).

4.5.1.4 Nevada

Nevada's amendment to the Uniform Electronic Transactions Act (UETA) defines blockchain as an electronic record of transaction which is (a) "uniformly ordered," (b) "redundantly maintained or processed [...] to guarantee the consistency or nonrepudiation of the recorded transaction," and (c) "validated by the use of cryptography" (Nev. Rev. Stat. Ann. § 719.045 2019). Similar to Illinois, Nevada's UETA prohibits the local government from imposing taxes, certification, or licensing conditions concerning blockchain usage. In addition, Nevada's legislative framework implements regulatory sandbox regimes, such as the Regulatory Experimentation Program. It provides a safe and favorable environment for companies to test, validate, and improve blockchain-related products and services without legislative encumbrances. Legislation of Nevada facilitates frictionless business activities by authorizing corporations to maintain business records using blockchain technology (Nev. Rev. Stat. Ann. § 78.0297 2020). The above suggests that enabling legislative provisions connote beneficial outcomes in terms of profitability and efficacy and serve as an accelerator in blockchain uptake.

4.5.1.5 California

California's Assembly Bill 2658 (Calderon, Chapter 875) establishes a Blockchain Working Group (BWG). Following the legislative charge formulated by AB 2658, BWG advances policy recommendations and proposes amendments to the legal and regulatory framework in light of blockchain deployment. Additionally, the BWG adopts a succinct blockchain definition as a "domain of technology used to build decentralized systems that increase the verifiability of data shared among a group of participants that may not have a pre-existing trust relationship [...] Any such system must include one or more distributed ledgers, specialized datastores that provide a mathematically verifiable ordering of transactions recorded in the datastore" (Crittenden 2020). California adopts the term 'datastore' to demonstrate the verifiability of data shared between blockchain peers and further, many datastores could coexist simultaneously (Neitz 2021). In addition, BWG develops an ethical framework to balance technological innovation with any negative externalities arising from blockchain adoption. Therefore, regulators must pay special attention to fairness, equity, accessibility, trust, transparency, and sustainability and incentivize blockchain development congruous with the above principles.

In 2018, California enacted legislation amending Section 204 of the 2019 California Corporations Code, authorizing certain corporations to utilize blockchain technology to maintain specific corporate records, such as the information on the issuance and transfer of stocks (Corporations Code 1975). However, the legislation included a sunset clause effectively repealing the statutory provision on 1st January 2022.

4.5.1.6 Vermont

The legislative arm of Vermont amends the provisions on Court Procedure by appending blockchain enabling laws to existing rules of evidence. Accordingly, "[a] digital record electronically registered in a blockchain shall be self-authenticating according to Vermont Rule of Evidence 902 [...]" (Vermont Rules of Evidence 902 1983). However, unlike Nevada and Arizona, Vermont does not amend the Electronic Transactions Act but adds blockchain enabling provisions as evidentiary rules in court proceedings. In addition, Vermont passed the Senate Bill 269 of 2017, consisting of a statute on Blockchain Business Development to spur economic growth in Vermont and establish a blockchain hub by attracting blockchain business ventures and companies. The law advances two legal structures: blockchain-based limited liability companies (BBLLC) and personal information protection companies (PIPC) (11 V.S.A. § 4173 2018).

BBLLCs are businesses that not only adopt blockchain for a "material portion" of their commercial activities but opt for a BBLLC structure in their articles of organization and ensure compliance with legislative requirements (11 V.S.A. § 4172 2018). PIPCs are companies that provide personal data protection services and have fiduciary relationships with individual consumers to act in their best interests. The Act requires PIPCs to establish and maintain an overarching "information security program," which may include utilizing blockchain technology, sufficient to protect consumers' information.

4.5.2 Gibraltar

Gibraltar is a British Overseas Territory with autonomous legislative powers for "peace, order, and good government of Gibraltar" under the Gibraltar Constitution Order 2006 (Gibraltar Constitution Order 2006). Gibraltar is at the vanguard of DLT and crypto-related businesses. It enacted the "Financial Services (Distributed Ledger Technology Providers) Regulations 2020" (DLT Regulations), inaugurating the first DLT-specific legislation in the world (Financial Services (Distributed Ledger Technology Providers) Regulations 2020). This legislation adopts a principle-based approach by developing ten basic principles, such as (a) honesty and integrity, (b) customer welfare, (c) sufficient financial and non-financial resources, (d) risk management, (e) protection of customers' funds and assets, (f) corporate governance, (g) cyber security, (h) financial crime, (i) resilience, and (j) virtuous conduct, to advance regulatory goals and objectives. The existing taxation structure in Gibraltar omits to define the specific treatment of cryptocurrencies. Hence, general principles under existing legislation are applicable. Gibraltar does not impose capital gains, value-added, inheritance, or withholding tax. Hence, only corporation tax of 12.5% is payable provided the corporation's income is "accrued in or derived from" Gibraltar, which refers to the location of the activities that conduces profits (Income Tax Act 2010). This structure creates a favorable landscape for DLT-based activities. Cryptocurrencies in Gibraltar may constitute transferable securities or financial instruments, and natural or legal persons involved in their issuance, sale, or transfer are subject to legislation, such as the Financial Services Act 2019 (Garcia et al. 2022). In addition, a DLT firm is defined as a "relevant financial business" under the Proceeds of Crime Act 2015 (the Act) and is subject to anti-money laundering obligations. Therefore, such a firm should conform to due diligence requirements, adopt relevant procedures and systems to deter money laundering, and establish risk assessments and management measures (Financial Services (Distributed Ledger Technology Providers) Regulations 2020; Garcia et al. 2022).

4.5.3 Estonia

Like Gibraltar, Estonia is a pacesetter in crypto-related businesses and transactions by developing a comprehensive framework. Money Laundering and Terrorist Financing Prevention Act 2017 regulates virtual currency by defining it as "a value represented in the digital form, which is digitally transferable, preservable or tradable"(Money Laundering and Terrorist Financing Prevention Act 2017). The amendments to the MLTFPA, which came into effect in March 2020, stipulate that the legislative provisions imposed on financial institutions under this Act are similarly applicable to virtual currency service providers (VCSPs) (Money Laundering and Terrorist Financing Prevention Act 2017). VCSPs must revise internal procedures to address anti-money laundering provisions. As such, in establishing a commercial relationship with an e-resident or an individual from a country beyond the European Economic Area, VCSPs must identify the person and verify data with the assistance of information technology (Money Laundering and Terrorist Financing Prevention Act 2017). Besides that, such businesses must appoint a competent compliance officer with high educational standing, credentials, and experience (Money Laundering and Terrorist Financing Prevention Act 2017). The recent amendments which entered into force in March 2022 sets out further requirements, supplementing the current provisions and establishing rigorous safeguards. VCSPs are required to (a) formulate a business continuity plan of at least two years, (b) establish sound risk management strategies, and (c) demonstrate proper business reputation (Money Laundering and Terrorist Financing Prevention Act 2017). A VCSP must have sufficient share capital (i.e., at least 100,000 euros or 250,000 euros) contingent on the virtual currency services offered (Money Laundering and Terrorist Financing Prevention Act 2017). The amendments of the MLTFPA provide other significant changes and advance key details, besides the amendments highlighted above, that can govern the trajectory of business models and processes. Besides the financial sector, blockchain is widely used in healthcare, property, business, and succession registry, amongst others to uphold the integrity of the governance systems in Estonia (Estonian blockchain technology, n.d.).

4.5.4 Malta

In 2018, Malta enacted the Malta Digital Innovation Authority Act (MDIA Act), Innovative Technology Arrangements and Services Act (ITAS Act), and Virtual Financial Assets Act (VFAA) to ensure that Malta's legal and regulatory framework appeals to businesses and investors. The MDIA establishes the Malta Digital Innovation Authority (the Authority), responsible for the formal recognition of innovation technology arrangements or services (Malta Digital Innovation Authority Act 2018). Further, the Authority should promote and enforce "ethical and legitimate criteria" during arrangements or services' design and development phase (Malta Digital Innovation Authority Act 2018). The Authority and other regulatory bodies can engage in collaborative efforts and initiate initiative-taking action to prevent money laundering and other illegal activities (Malta Digital Innovation Authority Act 2018). Interestingly, the Authority is obliged to recognize and facilitate the right to opt-out, withdraw, or terminate participation in any arrangement (Malta Digital Innovation Authority Act 2018).

The ITAS Act regulates innovative technology arrangements (ITAs) and innovative technology services (ITSs). In addition, the ITAS Act sets out the various methods that the Authority may advance to recognize ITAs and ITSs. The Authority may certify the ITAs for at least one specified purpose by referring to quality, feature, attribute, behavior, or aspect (Innovative Technology Arrangements and Services Act 2018). The certificate issued to an ITA shall have an identifying public key or a brand name (Innovative Technology Arrangements and Services Act 2018). Further, the Authority may accede to the registration of service providers that provide ITSs (Innovative Technology Arrangements and Services Act 2018).

The VFAA regulates "Initial Virtual Financial Asset Offerings" and "Virtual Financial Assets" (VFAs). According to the Malta Financial Services Authority, the VFAA provides a framework to advance innovation and development of modern technologies while upholding the interests of consumers and investors. Individuals or business entities issuing VFAs or providing other related services must comply with the licensing requirements delineated under the Act (Virtual Financial Assets Act 2018). As such, in establishing whether the DLT asset falls within the purview of this Act, the MFSA has developed a "financial instrument test." Upon acquiring the license, service providers and operators can engage in the exchange of VFAs (Malta Financial Services Authority, n.d.). In essence, the VFAA ensures substantive legal certainty in the crypto landscape by delineating comprehensive regulatory provisions.

Along with the emerging technology and increasing digitalization, the Malta Business Registry, responsible for business information concerning new and existing companies, has recently shifted its operations toward blockchain-powered systems to increase transparency and efficiency and curb bureaucratic inertia (Malta Business Registry MBR 2022).

4.5.5 Switzerland

Switzerland's legislative arm enacted the "Adaptation of Federal Law to Developments in Distributed Ledger Technology" (DLT Act). In this section, we frame the events that triggered the development of the DLT Act (Adaptation of Federal Law to Developments in Distributed Ledger Technology 2020). In 2018, Switzerland issued a Federal Council report (FCR) entitled the "Legal framework for distributed ledger technology and blockchain in Switzerland." The FCR considered the legal position of crypto assets and tokens under civil legislation, financial market legislation, and anti-money laundering legislation and proposed legislative amendments to address the challenges arising from existing legislation (Federal Council of Switzerland 2018). In 2019, Federal Council issued a draft law regarding DLT and blockchain, which went through public scrutiny during the consultation process. The DLT Act, which came into effect in 2021, amends a multitude of legislation, such as the Code of Obligations, Federal Act of 1889 on Debt Enforcement and Bankruptcy, Federal Act of 1987 on International Private Law, Financial Services Act 2018, National Bank Act 2003, Banking Act 1934, Financial Institution Act 2018, Anti-Money Laundering Act 1997, and Federal Intermediated Securities Act, among others, to establish and leverage ledger-based securities. The amendments bring legal certainty, particularly with the augmentation of anti-money laundering provisions and the establishment of DLT trading facilities.

4.5.6 Liechtenstein

In 2019, Liechtenstein enacted the Tokens and Trusted Technology Service Providers Law (Token- und VT-Dienstleister-Gesetz), referred to as 'TVTG,' which came into force in 2020. Before its enactment, the Ministry for General Government Affairs and Finance of Liechtenstein published a consultation report that went through a public consultation process to enable a viable legal structure for tokenization processes. TVTG establishes the Token Container Model, where tokens "represent claims or rights of memberships against a person, rights to property or other absolute or relative rights and is assigned to one or more TT Identifiers" (Token- und VT-Dienstleister-Gesetz 2019). This definition recognizes any category or class of digitalized rights and assets placed within a highly flexible token model that can be owned or transferred. It further prompts the application of legal obligations and rules associated with such tokens, for instance, securities laws and financial markets laws imposed on security tokens. TVTG recognizes the role of a physical validator to enable synchronicity between the online and offline landscape (Token- und VT-Dienstleister-Gesetz 2019). In addition, the physical validator ensures the enforcement of blockchain users' rights, commensurate with the agreement. Liechtenstein advances technologically neutral legislation by an all-encompassing token economy framework and is a pacesetter toward effective tokenization efforts.

In essence, regulators have adopted novel strategies, approaches, and designs to regulate blockchain-based applications, ensuring resilience, efficacy, and coherency in managing complex interactions with technology. In other countries, the legal and regulatory framework remains fragmented and inadequate to fill governance gaps and address conflictual interactions. We provide an overview of the countries' regulatory approaches in Table 4.1.

4.5.7 Extraterritorial Jurisdiction and the Applicability of Local Laws

Blockchain and cryptocurrency applications operate across borders, effectuating a vast potential of transactional fluidity between multiple countries. In such circumstances, countries may assert regulatory authority and jurisdiction over the infringements of blockchain or cryptocurrency platforms. The absence of international comity on legislative, executive, or judicial acts may deepen the chasm and precipitate conflict between nations which could generate far-reaching effects. First, the legal uncertainty and ambiguity of what laws apply in transactional cases may present a dilemma for investors and other market actors (Spafford et al. 2019). Second, parallel actions in different courts may impose encumbrances on litigants from duplicative litigation (Spafford et al. 2019). Third, the lack of integration of blockchain and cryptocurrencies may contribute to widespread disruption and market inefficiencies, hampering the growth of innovative technology applications and business models (Spafford et al. 2019). The use of comity by courts and heightened deference to foreign states may be a global solution for transnational cases, having determined the states' interests. International organizations and countries have begun to adopt regulatory modalities on blockchain and cryptocurrency to achieve standardized benchmarks and parallel governance outcomes for regime harmonization and cooperation in addressing the latent conflicts of blockchain and cryptocurrency (Spafford et al. 2019).

For local laws to be applicable, there must be "sufficient nexus" with the transaction. In re Tezos Securities Litigation, courts considered the location of the blockchain nodes. In this case, the validation nodes for processing and confirming the transaction blocks were "clustered more densely in the United States that in any other country" (In Re Tezos Securities Litigation 2018). In conducting extraterritorial determination concerning the applicability of U.S. securities laws, the court also considered website hosting and marketing engagements that materialized in the U.S. (In Re Tezos Securities Litigation 2018). However, determining the applicable laws and

Table 4.1 Overview of the legal and regulatory framework of countries

Countries	Legal and regulatory facets						
	Distributed Ledger Technology or blockchain definition	Allow corporations /local government to use blockchain to maintain records	Admissibility of blockchain under the rules of evidence	Exemption(s) from taxation, fees, or licensing requirements	Regulations on cryptocurrency exchanges	Anti-money laundering provisions	Blockchain task force
Wyoming	●	●	○	●	●	●	●
Delaware	●	●	●	●	●	Federal and state laws	○
Illinois	●	●	●	●	●		●
Nevada	●	●	○	●	●		○
California	●	○	○	○	●		●
Vermont	●	●	●	○	●		○
Gibraltar	●	●	○	●	●	●	●
Estonia	●	●	○	●	●	●	●
Malta	●	●	○	○	●	●	●
Switzerland	●	●	○	○	●	●	●
Liechtenstein	●	●	○	○	●	●	○

● Presence of legislation, regulations, or policies
○ Absence of legislation, regulations, or policies based on publicly available information

forums to adjudicate disputes across geographical boundaries, in the blockchain and cryptocurrency landscapes is challenging. Therefore, courts would consider the governing laws and jurisdictions on a case-by-case basis depending on the contextual circumstances.

4.6 Smart Contracts

An entrepreneur or company can leverage smart contracts to circumvent any inefficiencies and uncertainties in commercial transactions. This sub-chapter attempts to provide a concise definition of smart contracts, acknowledging that the definition of smart contracts is vague and ambiguous. The following section examines the evolution and rise of smart contracts in the energy sector by delineating specific use cases and examples to demonstrate the promising domain of smart contracts. Subsequently, the authors examine the smart contracts' coherence (or lack thereof) and existing contract law elements (i.e., offer, acceptance, consideration, intention to create legal relations, and capacity). The authors then discuss the limits, barriers, and technological and legal challenges circumjacent to the usage of smart contracts. The shortcomings and challenges include the difficulty in translating plain contractual language into computer codes, security and privacy issues, and rigidity of smart contracts (in contrast with the fluidity of commercial transactions). Finally, the authors include a general assessment of the regulatory initiatives of countries in facilitating the widespread use and adoption of smart contracts.

4.6.1 Defining Smart Contracts

In 1994, an American computer scientist, Nick Szabo, formulated the term 'smart contract' as "a computerized transaction protocol that executes the terms of a contract" (Szabo 1996). At that time, the infrastructural gap and the lack of technological support stymied the widespread use of smart contracts (Perez and Zeadally 2022). The dawn of blockchain technology has piqued the interest of industry players by advancing immense opportunities to implement the technology in multiple domains and opened the doors to technological evolution with the dynamic integration of blockchain and smart contracts. Blockchain systems enhance the functionality and design of smart contracts through many blockchain attributes, including immutability, trustlessness, transparency, and reliability, thus furthering the ability to facilitate complex operations and contractual transactions.

There are definitional difficulties surrounding smart contracts, compounded by their inherent complexity in scope and breadth that obstructs regulatory actions in adopting an appropriate and comprehensive definition. Even though scholars and academics have acknowledged the existence of multiple definitions of smart contracts, there is no commonly accepted definition (Rühl 2021). The authors note that the definition should appreciate smart contracts' heterogeneity, diversity, and versatility, as a touchstone for disciplines integrating this mechanism. For the purposes of this chapter, it is sufficient to consider smart contracts as autonomous and self-enforceable instruments that execute agreed contractual terms when predefined conditions trigger output events aligned with contractual parameters. The use of smart contracts offers multiple benefits, including reducing transactional costs and solving complex coordination issues between parties that share common goals, ensuring robust and effective contractual performance.

4.6.2 Evolution and Rise of Smart Contracts in the Energy Sector

A smart contract operates on blockchain to facilitate energy transactions between parties, such as energy and flexibility trading. Primarily, it (a) automates and optimizes the matching of buyers and sellers by emulating the price of bids with the amount of energy generated, (b) verifies and validates the energy trade, and (c) processes the financial transactions between parties. In what follows, a smart contract serves as the basis for billing and charging solutions for energy transactions between participants and the energy grid. In addition, decentralized energy trading, can be effectively employed in retail applications, when consumers choose an energy supplier and enter a transaction. In such instances, a smart contract is a promising solution that monitors the energy generated and consumed and processes real-time energy settlements. Furthermore, a smart contract can determine energy costs and automatically implement payment policies.

In carbon trading operations, companies use blockchain-enabled smart contracts to record the carbon credits through offset projects and facilitate the distribution and trading of carbon allowances to substantially reduce transaction fees and overhead costs. In addition, appropriate pre-defined terms and conditions can be coded into smart contracts to ensure that the carbon offsetting and trading processes are carried out seamlessly without the intervention of central intermediaries. However, smart contracts present similar legal and regulatory challenges regardless of their application.

4.6.3 Formation of Smart Contracts

The formation of a legally enforceable contract depends on the determination of the following elements, namely (a) offer, (b) acceptance, (c) consideration, (d) intention to create legal relations, and (e) capacity. While these components are germane in most common law jurisdictions, other jurisdictions may prescribe different or additional requirements (Madir 2018). According to the UK Jurisdiction Task Force, a smart contract is equally capable of meeting the conditions for contract formation articulated in the laws of the land (Ferreira 2021). The elementary stages of contract law do not differ significantly for smart contracts, which are assessed objectively, not by mental assent, but by outward manifestation.

4.6.3.1 Offer

First and foremost, it must be a firm and definite offer addressed to a specific person(s), or the public at large, which, if accepted, can be converted into an agreement, pronouncing the offeror's willingness to undertake the contractual obligation on specified terms. Therefore, the formulation of an offer must be comprehensive because a mere declaration of intent is insufficient due to its exiguousness. Scholars and academics consider the deployment of a smart contract code on the blockchain network as the manifestation of an offer. According to J. Earls et al., the code uploaded to the network constitutes an offer if the participating nodes can implement the contractual code (Earls et al. 2018). Similarly, scholars Djurovic and Janssen consider the "binary computer code" that embodies a contract with definite terms uploaded to the blockchain ledger, as an 'offer' rather than a mere 'invitation to treat' (Djurovic and Janssen 2018). This smart contract code can be directed to a specific blockchain address or a wallet, or the public. Further, the use of cryptographic keys by the offeror and offeree demonstrates proof of the commitment that objectively represents an offer and communication thereof (Djurovic and Janssen 2018). Such an offer can be accepted once the contract code is entered into the general ledger and meets the requirements of a 'firm offer'.

4.6.3.2 Acceptance

Acceptance is premised upon the fact of acceptance and the communication of acceptance. Generally, acceptance of an offer by a party may be declared by formal acceptance or inferred from conduct. For instance, unequivocal conduct may clearly express acceptance, such as entering one's signature via a cryptographic private key (Earls et al. 2018). Additionally, the acceptance of a participating node can infer the transfer of cryptocurrencies or other forms of digital assets to the offeror. Although the design of smart contracts is to function as automated systems, the intention to enter a contract can only be expressed by the unqualified acceptance of all the terms and conditions contained in the offer, hence representing the intention of the parties to enter a contractual relationship. The smart contract rule (if X then Y) is consistent with the traditional elements of contract law, as acceptance infers the performance of the act contemplated in the offer.

4.6.3.3 Consideration

The notion of exchange of promises embeds in every definition of consideration (Waite 1918). While consideration is necessary for contract formation, recognizing the same is not an insurmountable obstacle in the context of smart contracts. It is firmly established that consideration must be sufficient but need not be adequate, which is indisputably the case with smart contracts, as consideration does not need to be economically equivalent to the promise (Djurovic and Janssen 2018). According to Durovic and Janssen, "10 Ether for a car" would suffice to provide some form of economic value (Djurovic and Janssen 2018). However, scholars Werbach & Cornell advance a thought-provoking proposition that smart contracts do not consist of "exchange of promises" (Werbach and Cornell 2017). Instead, "a contract to transfer one bitcoin upon such-and-such event occurring is not really a promise at all. It does not say 'I will pay you one bitcoin if such-and-such happens', but rather [...] 'You will be paid one bitcoin if such-and-such happens' [...]" which does not explicitly commit either party to anything (Werbach and Cornell 2017). One can circumvent this proposition in practical and commercial contexts, by assuming a mutual obligation through the expression of a common intention to enter into a contract (Andrews et al. 2017). Smart contracts can also constitute bilateral executory contracts analogous to instantaneous agreements (Andrews et al. 2017; Savelyev 2017). Since the courts interpret the doctrine of consideration based on morality, equity, or commercial convenience (Andrews et al. 2017), it is only appropriate to allow the same flexibility to smart contracts as that afforded to conventional contracts.

4.6.3.4 Intention to Create Legal Relations

In forming a legally binding contract, parties must intend to establish a legal relationship. This requirement is inextricably linked to the concept of offer, acceptance, and consideration. Established by Atkin LJ in 1919 in Balfour v Balfour, this requirement has undoubtedly made a remarkable impression. It stands as clear as day, resonating presumptively within the sphere of smart contracts, unless proven otherwise. Parties in a perilous state may be compelled to have in writing a contract, in its conventional form, to acknowledge that the smart contract is a valid agreement and to demonstrate an intention to create a legal relationship (Djurovic and Janssen 2018).

Savelyev proposes an alternative modulation, which states that by forming a smart contract, the parties intend to avail themselves of an alternative regulatory regime instead of conventional contract law principles (Savelyev 2017). Hence, this explains the lack of intention to create legal relations. In this context, however, Savelyev concurs that "if the result is in fact the same in substance as in the case of an ordinary contract [...] then it may be argued that the nature of the relations at its core are also the same" (Savelyev 2017). The fact that a contractual party is unlikely to resort to litigation based on the performance guarantee of smart contracts, is not indicative of a lack of intent to be legally bound (Djurovic and Janssen 2019). As aptly stated in Jones v Padavatton, "[t]he fact that a contracting party is in some circumstances unlikely to extract his pound of flesh does not mean that he has no right to it" (Jones v Padavatton 1969). In summary, if the agreement satisfies the elements of contract formation, it is almost evident that the parties intend it to be legally binding, particularly in a business setting.

4.6.3.5 Capacity

Parties must possess the total legal capacity to enter a contractual relationship. Although minors, the mentally disabled and drunkards lack sufficient capacity to understand the nature and extent of the transactions, these individuals would be able to enter contracts on available blockchain platforms, as these platforms do not inquire into, or filter legal incompetence. Moreover, it is challenging to identify blockchain users on such platforms because (a) cryptographic mechanisms are in place and (b) alphanumeric character strings represent blockchain addresses.

4.6.4 Modification and Performance of Smart Contracts

4.6.4.1 Interpretation of Smart Contracts

Difficulty in Translating Plain Language into Computer Code

Translating contractual language, including but not limited to rights, obligations and remedies into computer code is no easy feat. As scholars and academics astutely point out, the technical prose and nuances embedded in contractual languages are not easily replaceable by imperative programming languages (Carron and Botteron 2019). Additionally, the lack of directions and specifications embedded in smart contract codes can have far-reaching consequences for contracting parties. As more intricate computer scripting languages have emerged, it is pivotal to have a complete and thorough understanding of such languages for effortless drafting and implementation of codes (Lee and Khan 2020). The dearth of automated, user-oriented, and advanced translation tools that can express plain contractual languages into a programmed-coded format may be the basis for the passivity and disinclination of individuals and corporations to utilize smart contracts. Without such tools to facilitate drafting, interpretation, and enforcement, the adoption of smart contracts would therefore be less attractive, primarily due to the high costs associated with manual transcriptions. Moreover, manual transcriptions are prone to human errors that may materialize into interpretation disputes (Carron and Botteron 2019).

It is easy to program clear-cut smart contractual codes with clearly defined terms armed with sufficient precision and certainty. However, not all contract clauses can fulfill smart contracts' 'self-executory' feature (Henly et al. 2018). In light of heterogeneous contractual relationships, it can be an increasingly arduous and impractical undertaking to envisage a myriad of unforeseen circumstances and translate them into codes (Henly et al. 2018). One example is the difficulty in coding force majeure events, that may occur later beyond the reasonable control of the parties. Besides that, nebulous terms commonly adopted in legal clauses such as "good cause," "good faith" and "reasonable," are not easily transcribable into codes (Compagnucci et al. 2021).

Manifestation of a Shared Intention in Smart Contracts

The adoption of smart contracts as an embodiment of legal contracts may pose a challenge for individuals who are not technologically savvy, thus initiating various conflicts of interpretation. Given that there must be a mutual and unambiguous agreement between the parties for a legally binding contract to be formed, it necessitates an accurate expression of the contractual modalities to reflect the parties' agreement in codes (Lee and Khan 2020). Hence, parties must rely on a programmer to ensure the unambiguous manifestation of such an agreement, given that linguistic challenges may prevent parties from incorporating appropriate codes and verifying the precision of such codes (Compagnucci et al. 2021; Lee and Khan 2020). Scholars Compagnucci et al. recognize the need to draft a "term sheet" that instructs the programmer to specify the prescribed specifications and requirements. The parties may also (a) obtain a representation from the programmer that the code accurately reflects the parties' agreement or (b) enter an agreement with the programmer setting out specific requirements and conditions (Compagnucci et al. 2021).

4.6.4.2 Impossibility of Performance/ Non-performance

Failure to Perform Contract Due to Coding Error

Errors in smart contractual codes can have a detrimental effect on the performance of the contract and negate any benefits that might otherwise accrue. The characteristics of immutability and irreversibility may prompt unfavorable consequences by restricting the erasure, or amendments of any record embedded into a smart contract. In the event of a defect, error or malfunction in the codes, the distributed programs continue to run fully automated and self-executory manner without first seeking to rectify the code, contrary to the parties' contractual intention (Lee and Khan 2020). Errors in the drafting and execution of codes may result in delays, inaccurate billing or invoicing, and other deficiencies in transactions. Parties using smart contracts to automate transactions rely on the expertise of computer programmers to ensure the accuracy and precision of contractual codes. The common question is who might be responsible for coding errors that pose severe consequences for the contracting parties. In this instance, the contracting parties must understand the risks circumjacent to applying blockchain and smart contracts before entering a contract.

Smart contract applications are susceptible to bugs and malware and can be maliciously exploited and targeted by fraudulent users, leading to security concerns. These concerns may trigger imperfect contractual performance as taking remedial action is impossible, resulting in financial losses for both parties (Compagnucci et al. 2021). Smart contracts appear to have a paradoxical effect. While they purport to guarantee performance through self-enforceable and self-executable features, the system configurations impede modifications or adjustments, thus negating the flexibility of contractual performances.

Excuses for Non-performance

Modern societies have developed governance frameworks that define (a) what constitutes a contract, (b) who can enter a contract, and (c) whether there are legal sanctions that are clearly and unambiguously expressed if there is a breach of contract (or non-performance). In the first instance, the concept of force majeure relieves the contracting party of the responsibility for non-performance in the event of legal or physical constraints that are unforeseeable, irresistible, and impossible to fulfill. While programmers may attempt to translate force majeure clauses (in textual form) into contractual codes, it can be a laborious undertaking, particularly due to the inability to emulate such rules with sufficient precision. Specific causes of non-performance can be effortlessly programmed into smart contracts, while others are harder to envisage. Scholar Tai suggests two options: first, a simple default rule programmed into smart contracts that anticipate a range of possible circumstances that constitute force majeure events, and second, the reliance on expert oracles in more complex situations invoked by the contracting parties (Tai 2018). Therefore, having determined the attributability of the breach of contract and the presence of force majeure events, the affected party has the right to terminate the contract based on non-performance. Thus, inserting contractual codes that have such an effect absolves the affected party from future contractual performance (Tai 2018).

There is a recognizable fluidity in the law concerning the illegality of contracts and public policy for actual or intended law violations. It is increasingly onerous for programmers to code the alternative possibilities that may arise. Scholars suggest that parties invoke these doctrines by referring to expert oracles that act as arbitrators or judges (Tai 2018). The oracle has the unenviable task of accurately determining the illegality or public policy underlying the situation at hand and the consequences and remedies that flow from the nature and degree of illegality.

It is a challenge to translate situations unbeknown to the parties into codes. Similar to the traditional parameters of contractual performance, it would be absurd to require a party to perform its contractual obligations, especially if the performance of the contract is impossible in the commercial sense. In many scenarios, contracts that are tainted with illegality are void. Therefore, no obligation can be specifically enforced, contrary to the innate features of smart contracts as self-enforcing agreements.

4.6.5 Enforcement, Termination, Rescission and Dispute Resolution Mechanisms

Enforcement: In conventional parameters, contracting parties agree on dispute resolution and enforcement mechanisms. However, the emerging landscapes of smart contracts alter the reality of contractual performance by circumventing the need for such mechanisms, due to the self-executory and self-enforceability properties of smart contracts. Despite that, situations may arise where one party needs to enforce its contractual obligations on the other. In the first place, to enforce such obligations, it is essential to discern and establish the identity of the parties, without which enforcement remains a moot point. Unlike private or permissioned blockchains, public blockchains are open-sourced platforms that create uncertainty about the identity of the transacting parties beyond the blockchain addresses (McKinney et al. 2017).

Termination: Smart contracts are inherently immutable, and it is practically impossible to hinder the performance of such contracts. Therefore, contracting parties must rely on blockchain programmers to develop infrastructural designs and mechanisms that mandate the termination and modification of a contract embodied in a programmed-coded format.

Dispute resolution: The use of programmed-coded language in smart contracts may give rise to disputes. The inability to understand smart contractual terms with sufficient precision and certainty may lead to disagreements about the true intent of contractual provisions.

4.6.6 Selected Challenges in the Application of Smart Contracts

4.6.6.1 Static and Rigid Smart Contracts Versus Flexible Legal Contracts

A smart contract represents a sequence of 'if–then' instructions that are stringently defined and self-executable. However, such a contract advances insufficient weight to the commercial reality of the parties and negates any practical application of established contract law principles. Generally, smart contract codes deployed on a blockchain ledger cannot be modified. According to scholar Sklaroff, modification can only materialize if, in the first place, codes are built into smart contracts as "dormant alternatives," while incorporating the ability to 'turn on' the terms that are in an 'off' state (Sklaroff 2017). However, when drafting such a contract, parties must address their minds at the outset to factor in alternative functional terms, to amend the contract at a later stage (Sklaroff 2017). Traditional contract forms are flexible in that new transaction terms can be introduced by entering into a verbal agreement (at the minimum), to modify parties' performance requirements, rights or obligations set out in the original contract. However, because smart contracts do not have such flexibility, the parties may have to incorporate the modified terms into a new smart contract, which involves high transaction and drafting costs.

Moreover, a self-enforcing smart contract continues to operate regardless of the change in circumstances that appears to be commercially unreasonable or even absurd. Unlike conventional contracts where the terms of the contract are amendable to adapt to such changes that would otherwise render the performance inordinately onerous in the commercial landscape, smart contracts offer no such recourse.

In a commercially unattractive situation, a party of a traditional contract may deliberately breach the contract to espouse a more advantageous course of action (Mik 2017). Academics recognize this situation as efficient breaches of contract, advancing economic justifications such as wealth maximization and efficient resource allocation for deliberate non-performance (Collins 2003). Compensation is adequate for the aggrieved party resulting from such breach, where the party would be potentially better off (leaving no party worse-off). However, in a smart contractual setting, exercising a deliberate option *not to perform* is out of the question as rights and obligations under the agreement are exercised automatically (Collins 2003). In summary, self-enforceable and automated contracts are analogous to *performance obligations written in stone*, disallowing adjustments, revisions, and flexibility to make way for future contingencies and changing circumstances. Adhering to the maxim pacta sunt servanda (agreements must be kept) may prove extraneous and irrelevant considering the evolving and dynamic contractual relationship between parties.

4.6.7 Compatibility of Smart Contracts with Consumer Protection Laws—Applicability of Unfair Contract Terms Provisions

According to Savelyev, smart contracts have an egalitarian disposition. Therefore, consumer protection laws and provisions governing unfair contract terms ('UCT') are extraneous and irrelevant to smart contracts. This author respectfully takes a different view. Firstly, the term egalitarianism suggests that "equality that is present is not neutral [...] but is asserted" (Woodburn 1982). As advanced in egalitarian derivatives, the author posits that equality of interest and reinforces equal bargaining power envisaged in consumer protection legislation and UCT provisions. For instance, under Article 3.1 of the Directive on Unfair Terms in Consumer Contracts (DUT), the term 'unfair' is defined as "contrary to the requirement of good faith, it causes a significant imbalance in the parties' rights and obligations arising under the contract to the detriment of the consumer" (Unfair Terms in Consumer Contracts Directive 1993). In assessing the requirement of good faith, due regard must be given, inter alia, to the bargaining strength of parties and whether they are on an equal footing. Accordingly, consumer protection laws and UCT provisions, commonly applied in traditional contracts, may similarly apply to smart contracts. Therefore, what is required is a more practical and reasonable application of these provisions to smart contracts. Scholars have noted that "smart contracting is not and cannot be a blind spot for consumer protection law" (Djurovic and Janssen 2019).

Contrary to common claims, Durovic & Janssen state that the DUT, does not hold that contractual terms must be in textual form to be applicable (Djurovic and Janssen 2019). Otherwise, parties could translate terms from text into smart contract codes to circumvent UCT provisions. This position brings us to an additional point, in that it is required to draft terms of the DUT in "plain, intelligible language" (Article 5) and to construe the appropriate yardstick for determining whether such terms are plain and intelligible from the perspective of a typical or average consumer who is "reasonably well-informed and reasonably observant and circumspect" party (Office of Fair Trading v Abbey National 2008).

In business-to-consumer transactions, it is often the case that consumers rarely consider the need for legal advice. This consideration, coupled with the fact of limited autonomy, can exacerbate the issue of the legibility of computer language. In addition, companies may impose self-serving, exigent, and other unusual terms into smart contracts that consumers are unaware of, thus undermining free and informed consent and the equality of the bargaining position (Kasatkina 2021). This position brings us to another point, requiring businesses to provide translations for computer codes expressed in plain and intelligible language (Djurovic and Janssen 2019). Academics suggest the use of technology that allows codes to be presented in "textual form" to improve the understanding of the contracting parties (Forbes 2022).

There has been a gradual transition from a classical to a modern view of contract law in recent years. The classical view suggests that the imposition of unreasonable terms and unequal bargaining strength is not subject to judicial inquiry. Thus, the party signing the contract is bound irrespective of whether he or she has read the document. Such a position offers no protection to consumers in a smart contract landscape (L'Estrange v F Graucob Ltd 1934). The modern view considers 'bargaining strength' as a relevant consideration. In line with the enactment of legislation that safeguards the interests of consumers, courts should undertake a judicial inquiry on the transparency of such (unfair) terms, including whether they are unambiguous (Kasatkina 2021).

Furthermore, and as alluded to above, smart contracts can be considered legal contracts and the provisions of the UCT continue to apply to such contracts. This consideration provides much-needed protection for consumers by ensuring that the unfair term remains unapplied and that any remedy afforded is sufficiently compelling. However, a practical application to strike out unfair terms via UCT provisions remain a moot point, as the presence of unconscionable and unfair contractual terms within self-enforcing smart contracts continue to operate autonomously (Forbes 2022).

In essence, while contract law, by design, wittingly regulates market transactions, countries' legal and regulatory frameworks may prohibit the application of smart contracts, thereby precluding any form of relief or redress, in the event of a breach of contract or whiff of impropriety (Forbes 2022). Another potential concern is the enforcement of court judgments arising from smart contracts, which implies a lack of facilitation and protection of contractual relations and interests of the parties (Forbes 2022).

4.6.8 Security and Privacy Risks

Smart contracts are not free from a host of security and privacy issues. Given the recent advances in smart contracts, it is helpful to rehash the existing challenges that significantly plague the application of such contracts. One of the more deleterious attacks in smart contracts is the reentrancy attack, through which the attacker uses a recursive call function to perform repeated withdrawals to exhaust contract funds (Samreen and Alalfi 2020). Besides that, smart contract applications can trigger transaction-ordering dependence attacks, such that multiple transactions invoke functions in the same contract. Miners can impose an arbitrary order between transactions, which can prompt adversaries to launch attacks due to the inconsistent order and execution of transactions (Alkadi et al. 2020; Mavridou and Laszka 2018; Wang et al. 2018). Smart contracts are also vulnerable to timestamp dependency attacks. The timestamp reflects the time on the miners' local server used to perform critical operations. Malicious miners may manipulate the timestamps, to induce a favorable outcome for the miner and affect the performance of transactions on blockchain-enabled smart contracts (Praitheeshan et al. 2019). In circumventing the above vulnerabilities, technology developers can adopt security-by-design precepts to ensure the application of technical and organizational best practice guidelines for the entire lifecycle of the smart contract. Given the significant rise of smart contracts, policymakers should lead the way in designing policy and legal instrument designs guided by the ergonomics of security to ensure resilient and secure systems.

There is a discernible conflict between the application of smart contracts and data protection regulations that must be progressively curtailed. The core feature of blockchain-enabled smart contracts (i.e., immutability) prevents data subjects from exercising their right to rectification and erasure. Right to rectification refers to the ability to obtain from the data controller the correction of inaccurate personal data. In contrast, the right to erasure denotes the right to erase personal data without undue delay. The rectification, manipulation, and/or erasure of any data on the blockchain ledger may distort the entire chain of blocks resulting in undesirable consequences. In attempting to resolve the conflict between the data subject's right to the erasure of personal data and the immutability feature of smart contracts, scholars have proposed various technical measures. These measures include off-chain storage (Lima 2018), pruning transactions on blockchain ledgers (Florian et al. 2019), encryption (Politou et al. 2019), chameleon hash (Giessen 2019), consensus-based voting (Xu et al. 2021), self-destruct function (Herian 2021), block matrix (Kuhn et al. 2019), and private key (Al-Abdullah et al. 2020). As for the right to rectification, compliance is practically impossible in permissionless blockchain networks compared to private and consortium networks. Data correction must be performed across multiple nodes and requires the agreement of a majority of participants on the blockchain network. Regulators should codify data protection by design principles in legal and regulatory frameworks to ensure that technology developers, businesses and organizations implement a data privacy-centric approach during the engineering process.

4.6.9 Setting the Scene of Smart Contracts—Regulatory Initiatives in Various Countries on Smart Contracts

This section reviews multiple legal and regulatory initiatives spanning different jurisdictions, to provide an overview of the current legislative landscape.

4.6.9.1 United States (U.S.)

The states in the U.S. are at distinct stages of the legislative process. Wyoming, Arizona, Tennessee, Ohio, and New York, amongst other states, have introduced legislation to recognize smart contracts, navigate the nuances in enacting a comprehensive legislative framework and leverage the use of smart contracts across industrial domains. Certain states have enacted nearly identical provisions, advancing two common denominators in their legislative framework. These dominators include (a) authorizing smart contracts in the form of a code or programming language to automate transactions upon the occurrence of certain conditions by legally recording, storing, and transmitting information via distributed and decentralized ledgers, and (b) recognizing the existence of smart contracts in the commercial arena, including the legal effect, validity, and enforceability of smart contracts (Arcari 2018). While U.S. has been a leader in enacting technology-friendly regulations, the ever-turbulent evolution of technology has deterred other states from enacting specific legal and regulatory frameworks for smart contracts, prompting existing laws to bridge the gaps in enforceability and validity of smart contracts.

Wyoming

Wyoming has been spearheading the enactment of a smart contract framework by providing a favorable and promising landscape for the effective utilization of novel technologies. In the Sixty-Fifth General Session, Wyoming enacted Sections 34-29-101 through 105 on "digital assets" which came into force on 1st July 2019. According to Section 34-29-103(a), the "perfection of a security interest in digital securities" by a secured party may be attained through control, as defined under paragraph Section 34-29-103(e)(i) (Wyo. Stat. § 34-29-103 2019). Under subsection B, "control [...] means [...] a smart contract created by a secured party which has the exclusive legal authority to conduct a transaction relating to a digital asset [...]" (Wyo. Stat. § 34-29-103 2019). However, this provision is abstruse to provide any material effect to the entirety of the legislative provision particularly due to the all-encompassing effect of subsection (e)(i)(A). A smart contract is defined within the legislative provision as "an automated transaction [...] or any substantially similar analogue which is comprised of code, script or programming language that executes the terms of an agreement, and which may include taking custody of and transferring an asset, or issuing executable instructions for these actions, based on the occurrence or non-occurrence of specified conditions" (Wyo. Stat. § 34-29-103 2019). This provision seeks to give credence to the enforceability of smart contracts expressly.

Arizona

Arizona has surfaced as a pioneering state in embracing legislative provisions for smart contracts in commercial transactions, as early as 2017, encapsulating an amendment to the Arizona Electronic Transactions Act. Title 44, Chapter 26 of §44-7061(E) defines a smart contract as an "event-driven program, with state, which runs on a distributed, decentralized, shared and replicated ledger and that can take custody over and instruct transfer of assets on that ledger" (Ariz. Rev. Stat. § 44-7061 2017). Further, subsection (C) sheds light on the legal status of smart contracts by propounding that "[s]mart contracts may exist in commerce. "A contract relating to a transaction may not be denied legal effect, validity, or enforceability solely because that contract contains a smart contract term" (Ariz. Rev. Stat. § 44-7061 2017). The statute also provides that "a record or contract that is secured through blockchain technology is considered to be in an electronic form," and therefore legally recognized within the existing legal framework (Ariz. Rev. Stat. § 44-7061 2017). This definition advanced by the state legislature is technology-oriented and far-reaching, as it legitimizes the use and enforceability of smart contracts while encouraging the transfer of assets of various types and forms (Arcari 2018).

Tennessee

In 2019, Tennessee passed legislation that recognizes the legal effects, validity, and enforceability of executing a contract through a smart contract under Title 47, Chapter 10 § 47-10-201 (TN Code § 47-10-201 2019). While appearing to be like the legislative texts advanced by Arizona on this subject, there is a striking difference in the legal parameters of the two provisions.

Tennessee defines smart contracts as an "event-driven computer program, that executes on an electronic, distributed, decentralized, shared, and replicated ledger," to automate the following transactions, inter alia, (a) taking custody over and instructing transfer of assets; (b) engaging in the creation and distribution of electronic assets; (c) synchronizing information; and (d) managing identity and user access to applications (TN Code § 47-10-201 2019). In addition, Tennessee also provides that a "record or contract secured through blockchain" is an electronic record (TN Code § 47-10-201 2019). The authors note some similarities between Arizona and Tennessee in incorporating smart contracts within the current legislative frameworks. The absence of a harmonised and transparent regulatory framework propagates regulatory arbitrage, which may negatively affect the states in the U.S.

Ohio

Ohio amended the Uniform Electronic Transactions Act to include the following provisions. Firstly, "a contract may not be denied legal effect or enforceability solely because an electronic record was used in its formation" (Ohio Rev Code § 1306.06 2013). In addition, an electronic record or signature may not be denied legal effect or enforceability merely because it is in an electronic form (Ohio Rev Code § 1306.06 2013). Even though the term 'smart contract' is not explicit in legislative provisions, it provides sufficient legal recognition to smart contracts.

Arkansas

Arkansas passed HB 1944 into law in 2019. Arkansas Code Title 25, Chapter 32, Section 122(3) defines a smart contract as either "business logic that runs on a blockchain" or "a software program that stores rules on a shared and replicated ledger and uses the stored rules for: (i) Negotiating the terms of a contract; (ii) Automatically verifying the contract; and (iii) Executing the terms of a contract" (AR Code § 25-32-122 2019). As the statutory provision adopts the conjunction "and," all three elements must be satisfied to trigger the statutory application. While the latter two requirements are coherent with the feature of smart contracts, the first limb may be challenging to fulfill as the practical application of smart contracts in negotiating contractual terms is obscure and vague. Some scholars suggest that automated agents can be used to negotiate business terms and online transactions that smart contract applications can facilitate.

New York

Bills AB 3760 and SB 1801 were passed by Senate in early February, introducing new provisions into the state's Technology Law. § 302(7) defines a smart contract as an "event-driven program that runs on a distributed, decentralized, shared and replicated ledger and that can take custody over and instruct transfer of assets on that ledger" (NY State Tech L § 302 2014). This position mirrors the provisions advanced by the law of Arizona.

Illinois

Illinois passed the Public Act 101-514 Blockchain Technology Act defining a smart contract as "a contract stored as an electronic record which is verified by the use of a blockchain." Further, Section 10 stipulates that "[a] smart contract, record, or signature may not be denied legal effect or enforceability solely because a blockchain was used to create, store, or verify the smart contract, record, or signature" (Blockchain Technology Act 2019). Illinois recognizes the evidence of smart contracts, records, or signatures.

The Joint Economic Committee of the U.S. Congress notes that while the notion of smart contracts is new, the concept is rooted in the basic principles of contract law. Smart contracts facilitate the automatic enforcement of contractual provisions subsumed into codes, bypassing the judiciary's role in arbitrating contractual disputes (Joint Economic Committee 2018).

4.6.9.2 Gibraltar

While Gibraltar is at the forefront of introducing legislation on distributed ledger technology, there are no legal and regulatory frameworks and guidelines that guide the definition, enforceability, and validity of smart contracts. However, it is acceded by scholars and academics that smart contracts would be enforceable under Gibraltar law provided it demonstrates the cardinal components of offer, acceptance, consideration, and intention to create legal relations (Nagrani 2020).

4.6.9.3 Malta

Malta is a global player in the domain of blockchain and smart contracts. Under Maltese laws, smart contracts must comply with the provision set forth under the Malta Digital Innovation Authority Act (MDIA) and Innovation Technology Arrangements and Services Act (ITAS). A smart contract represents an innovative technology arrangement defined under the MDIA as a "computer protocol and/or an agreement concluded wholly or partly in an electronic form, which is automatable and enforceable by execution of computer code [...]" (Malta Digital Innovation Authority Act 2018). Further, MDIA recognizes that human input and control are not completely circumvented and can therefore be enforceable by ordinary legal methods. Malta adopts a progressive regulatory framework in implementing smart contracts to guarantee legal certainty in light of the emerging technology.

4.6.9.4 Liechtenstein

Strikingly similar to the position of Gibraltar, a smart contract is recognized as a legally binding contract in Liechtenstein provided that it meets prerequisites set forth under contract law are met. While there is a lack of a legislative and regulatory framework specifically designed for smart contracts, the Government Consultation Report on the Blockchain Act defines smart contracts briefly as "automated contracts that can also trigger transactions with tokens" (Ministry for General Government Affairs and Finance 2018).

4.6.9.5 Switzerland

Based on the Federal Council Report entitled the "Legal framework for Distributed Ledger Technology and Blockchain in Switzerland", "a smart contract is a computer protocol, usually based on a decentralized blockchain systems which allows automated contract execution between two or more parties with previously coded data" (Federal Council of Switzerland 2018). The Federal Council recognizes a smart contract as a technology for purposes of automated contractual execution and enforceability, in lieu of a 'legal contract' in the spirit of the Swiss Code of Obligations. Further, the Federal Council delineates three characteristics of smart contracts. First, there is no requirement for an intermediary to ensure contractual performance, as the agreeable terms are translated into a programmed-coded format to enable independent verification and enforcement. Second, a smart contract is immutable, in that it cannot be altered, modified, or deleted by contractual parties or participating nodes. Third, the Federal Council sets defined barriers for smart contracts applicable in the digital world vis-à-vis transactions involving electronic goods and services that can be the subject of smart contracts (Flühmann and Hsu 2021) The application of smart contracts in Switzerland is in the embryonic phase, and issues concerning liability, programming errors, and obstacles in implementing smart contracts warrant further discussion by the relevant parties.

Countries essentially set their policies and enact legal and regulatory provisions at their own pace, based on technology and market readiness levels. The countries with a regulatory framework on smart contracts have adopted technology-centric definitions using terms such as "event-driven programs" and "computer protocols." Despite that, the lack of enforcement and dispute resolution mechanisms are prodigious obstacles that may impede the application of smart contracts. While there are ongoing attempts to craft regulatory frameworks in line with the development of blockchain and smart contracts, countries' legislators are not deterred from upholding their validity, legality, and enforceability by applying existing policies and regulations.

The recent plunge in cryptocurrency, when algorithmic stablecoin TerraUSD and sister coin Luna collapsed, caused financial agony to many cryptocurrency enthusiasts and investors. While stablecoins, as their name suggest, seek to reduce volatility and fluctuations by maintaining value parity with the U.S. Dollars, they are highly susceptible to losing their peg, amid high inflation and economic downturn. In light of recent events, regulators have to consider whether stablecoins are a fit for and in existing legal and regulatory frameworks. Scholars have likened stablecoins to bank deposits as they are traded at fixed value relative to currency denominations

(Wilmarth 2021). The financial inclusion and regulation of stablecoins as deposit-like products and stablecoins issuers as banks, can provide regulatory oversight to address the inherent risks of stablecoins by effectively imposing equivalent standards as provided under the banking legislation such as applying liquidity requirements and deposit insurance. In essence, as financial stability has proved pivotal with the accelerated growth of stablecoins as means of payment, it is crucial to adopt regulatory landscapes that are fit for purpose.

4.7 Summary and Conclusion

This chapter explores the legal and regulatory challenges surrounding the implementation of blockchain technology. While the current regulatory trajectory of many countries is dubious and vague, other countries, such as the United States, Gibraltar, Malta, Estonia, Liechtenstein, and Switzerland, are frontrunners in terms of technological, operational, and legal maturity in the blockchain fora. Unlike the wait-and-see and ex-post (after the fact) approaches, which hinder the rapid rollout of blockchain applications in many industries, proper institutional and governance regimes provide far-reaching technological and societal advantages by advancing reasonable legal certainty. Specifically, this chapter examines the (a) liability gap in blockchain systems and plausible attribution of shared liability, (b) property rights and interests of digital assets, (c) applicability of securities, commodities, and taxation laws to digital assets, (d) susceptibility of the blockchain landscape to cybersecurity attacks and illegal activities, (e) potential breach of informational, decisional, and transactional privacy in the blockchain domain, and (f) legality of smart contracts under the existing legal landscape and associated challenges, to enrich our analysis on the importance of a resilient governance regime.

References

5 U.S.C. § 552 (2018)

8 Del. Code Ann. § 219 (2017)

8 Del. Code Ann. § 224 (2017)

8 Del. Code Ann. § 232 (2017)

11 V.S.A. § 4172 (2018)

11 V.S.A. § 4173 (2018)

42 U.S.C. § 7414(c) (2006)

Adaptation of Federal Law to Developments in Distributed Ledger Technology (2020)

Al-Abdullah M, Alsmadi I, AlAbdullah R, Farkas B (2020) Designing privacy-friendly data repositories: a framework for a blockchain that follows the GDPR. Digit Policy Regul Gov 22(5–6):389–411. https://doi.org/10.1108/DPRG-04-2020-0050

Alkadi O, Moustafa N, Turnbull B (2020) A review of intrusion detection and blockchain applications in the cloud: approaches, challenges and solutions. IEEE Access 8:104893–104917. https://doi.org/10.1109/ACCESS.2020.2999715

Andrews N, Brownsword R, Chuah J, Yihan G, Halson R, Khan A, Mik E, Stannard JE, Tolhurst GJ (2017) The law of contract (Common Law Series) (ed: Furmston M, Vol. 6). LexisNexis
AR Code § 25-32-122 (2019)
Arcari J (2018) Decoding smart contracts: technology, legitimacy, & legislative uniformity. Fordham J Corp Financ Law 24:363
Ariz. Rev. Stat. § 44-7061 (2017)
Armstrong DLW GmbH v Winnington Networks Ltd, EWHC 10 (2012)
Bacon J, Michels JD, Millard C, Singh J (2018) Blockchain demystified: a technical and legal introduction to distributed and centralized ledgers. Richmond J Law Technol 25:1
Blockchain Technology Act, 205 ILCS 730 (2019)
California Consumer Privacy Act, Cal. Legis. Serv. Ch. 55 (A.B. 375) (WEST) (2018)
California Privacy Rights Act (2020)
Carron B, Botteron V (2019) How smart can a contract be? In: Kraus D, Obrist T, Hari O (eds) Blockchains, smart contracts, decentralised autonomous organisations and the law. Edward Elgar Publishing, Gloucestershire, pp 101–143
Collins H (2003) The law of contract. Butterworths, London
Commodity Futures Trading Commission v. McDonnell, No. 18-CV-0361 (2018)
Commodity Futures Trading Commission (2019) CFTC, FinCEN, and SEC issue joint statement on activities involving digital assets. https://www.cftc.gov/PressRoom/SpeechesTestimony/cftcfincensecjointstatement101119. Accessed 14 June 2022
Compagnucci MC, Fenwick M, Haapio H (2021) Digital technology, future lawyers and the computable contract designer of tomorrow. In: Research handbook on contract design. Edward Elgar Publishing, Gloucestershire, pp 421–444
Corporations Code, Cal. Corp. Code § 204 (1975)
Council Directive 2014/65/EU on markets in financial instruments, OJ L 173 (2014)
Council Regulation (EU) 2017/1129 on the prospectus to be published when securities are offered to the public or admitted to trading on a regulated market (2017) Edited by European Parliament and the Council of the European Union
Crittenden C (2020) Blockchain in California: a roadmap. https://escholarship.org/uc/item/2j9596dp. Accessed 20 July 2022
de Almeida L, Cappelli V, Klausmann N, van Soest H (2021) Peer-to-peer trading and energy community in the electricity market: analysing the literature on law and regulation and looking ahead to future challenges. International Energy Agency. Accessed 18 July 2022
Declaration of the Ministers of the Financial Action Task Force (2022) https://www.fatf-gafi.org/media/fatf/documents/FATF-Ministerial-Declaration-April-2022.pdf. Accessed 8 Aug 2022
Directive 2018/843 of 30 May 2018 on anti-money laundering and countering the financing of terrorism (2018) Edited by European Parliament and the Council of the European Union
Djurovic M, Janssen A (2018) The formation of blockchain-based smart contracts in the light of contract law. Eur Rev Priv Law 26(6):753–771. https://doi.org/10.54648/erpl2018053
Djurovic M, Janssen A (2019) The formation of smart contracts and beyond: shaking the fundamentals of contract law. In: DiMatteo LA, Cannarsa M, Poncibò C (eds) Smart contracts and blockchain technology: role of contract law. Cambridge University Press, UK
Duarte D (2019) An introduction to blockchain technology from a legal perspective and its tensions with the GDPR. Cyberlaw Journal of the Cyberlaw Research Centre of the University of Lisbon School of Law. https://papers.ssrn.com/sol3/papers.cfm?abstract_id=3545331
Earls J, Smith M, Smith R (2018) Smart contracts: is the law ready. https://digitalchamber.s3.amazonaws.com/Smart-Contracts-Whitepaper-WEB.pdf. Accessed 8 Aug 2022
Estonian blockchain technology (n.d.) https://e-estonia.com/wp-content/uploads/2020mar-nochanges-faq-a4-v03-blockchain-1-1.pdf. Accessed 14 Aug 2022
European Union Agency for Cybersecurity (2017) Distributed ledger technology & cybersecurity - improving information security in the financial sector. https://www.enisa.europa.eu/publications/blockchain-security. Accessed 25 Aug 2022

Federal Council of Switzerland (2018) Legal framework for distributed ledger technology and blockchain in Switzerland - an overview with a focus on the financial sector. https://www.newsd.admin.ch/newsd/message/attachments/55153.pdf. Accessed 25 Aug 2022

Federal Office for Information Security (BSI) (2019) Towards Secure Blockchains. https://www.bsi.bund.de/SharedDocs/Downloads/EN/BSI/Crypto/Secure_Blockchain.pdf?__blob=publicationFile&v=3. Accessed 10 June 2022

Ferreira A (2020) Emerging regulatory approaches to blockchain based token economy. J Br Blockchain Assoc. https://doi.org/10.31585/jbba-3-1-(6)2020

Ferreira A (2021) Regulating smart contracts: legal revolution or simply evolution? Telecommunications Policy 45(2):102081. https://doi.org/10.1016/j.telpol.2020.102081. https://www.sciencedirect.com/science/article/pii/S0308596120301713

Financial Services (Distributed Ledger Technology Providers) Regulations, LN.2020/012 (2020)

Florian M, Henningsen S, Beaucamp S, Scheuermann B (2019) Erasing data from blockchain nodes. In: 2019 IEEE European Symposium on security and privacy workshops, pp 367–376. https://doi.org/10.1109/EuroSPW.2019.00047

Flühmann D, Hsu P (2021) Switzerland: blockchain. https://www.baerkarrer.ch/en/publications/switzerland-blockchain. Accessed 10 June 2022

Forbes L (2022) Consumer protection in the face of smart contracts. Loyola Consum Law Rev 34(1)

Frommelt E (2020) Liability challenges in the blockchain ecosystem. UC Davis Bus Law J 21:165

Garcia J, Garcia J, Collado J (2022) Blockchain & cryptocurrency laws and regulations global legal insights. https://www.globallegalinsights.com/practice-areas/blockchain-laws-and-regulations/gibraltar. Accessed 8 June 2022

General Data Protection Regulation, Regulation (EU) 2016/679 (2016)

Gibraltar Constitution Order (2006)

Giessen D (2019) Blockchain and the GDPR's right to erasure. University of Twente

Grennan J (2022) FinTech regulation in the United States: past, present, and future. https://doi.org/10.2139/ssrn.4045057

Guseva Y (2020) A conceptual framework for digital-asset securities: tokens and coins as debt and equity. Maryland Law Rev 80(1):166. https://digitalcommons.law.umaryland.edu/mlr/vol80/iss1/7

Henly C, Hartnett S, Endemann B, Tejblum B, Cohen DS (2018) Energizing the future with blockchain. Energy Law J 39:197

Herian R (2021) Smart contracts: a remedial analysis. Infor Commun Technol Law 30(1):17–34

Hosmer H (1915) The tax situs of personal property. Louis Law Rev 1:320

Income Tax Act, LN.2010/180 (2010)

Innovative Technology Arrangements and Services Act (2018)

In Re Tezos Securities Litigation, No. 17-cv-06779-RS (2018)

In the Matter of: Coinflip, Inc., d/b/a Derivabit, and Francisco Riordan, CFTC Docket No. 15-29 (2015)

Joint Economic Committee (2018) Report of the Joint Economic Committee Congress of the United States on the 2018 Economic Report of the President. US Government Printing Office

Jones v Padavatton, 2 All ER 616 (1969)

Kaal, Wulf A. 2022. "Title." Available at SSRN

Kasatkina M (2021) Consumer protection in the light of smart contracts. ELTE Law J 1:95

Kuhn R, Yaga D, Voas J (2019) Rethinking distributed ledger technology. Comput Commun 52(2):68–72

Kulms R (2020) Blockchains: private law matters special feature. Singap J Leg Stud 2020:63

Lamdan SS (2014) Sunshine for sale: environmental contractors and the freedom of information act. Vermont Law Rev 15(2):227–261

Lee J, Khan VM (2020) Blockchain and smart contract for peer-to-peer energy trading platform: legal obstacles and regulatory solutions. UIC Rev Intellect Property Law 19(4):1

L'Estrange v F Graucob Ltd 2 KB 394 (1934)

Lima C (2018) Blockchain GDPR privacy by design. IEEE Blockchain Group. https://blockchain.ieee.org/images/files/pdf/blockchain-gdpr-privacy-by-design.pdf

Lombino C (2020) New privacy laws require changed operations on commercial websites. In: The LegalTech book: the legal technology handbook for investors, entrepreneurs and FinTech visionaries. Wiley, pp 78–82

Madir J (2018) Smart contracts: (how) do they fit under existing legal frameworks? https://doi.org/10.2139/ssrn.3301463

Malta Business Registry MBR (2022) https://www.legal-malta.com/articles/malta-business-registry-mbr. Accessed 14 Aug 2022

Malta Digital Innovation Authority Act, Act No. XXXI (2018)

Malta Financial Services Authority (n.d.) Guidance note to the financial instrument test. https://www.mfsa.mt/wp-content/uploads/2019/04/20190405_GuidanceFITest.pdf. Accessed 8 Aug 2022

Mason M, Gupta A (2015) Transparency. In: Research handbook on climate governance. Edward Elgar, Gloucestershire, pp 446–457

Mavridou A, Laszka A (2018) Designing secure ethereum smart contracts: a finite state machine based approach. In: International conference on financial cryptography and data security, pp 523–540

McKinney SA, Landy R, Wilka R (2017) Smart contracts, blockchain, and the next frontier of transactional law. Wash JL Tech Arts 13:313

Mendelson M (2019) From initial coin offerings to security tokens: a US Federal Securities law analysis. Stanford Technol Law Rev 22:52

Mik E (2017) Smart contracts: terminology, technical limitations and real world complexity. Law Innov Technol 9(2):269–300

Ministry for General Government Affairs and Finance (2018) Government consultation report on the creation of a law on transaction systems based on Trustworthy Technologies (TT). https://www.naegele.law/files/Downloads/2018-10-05-Unofficial-Translation-of-the-Draft-Blockchain-Act.pdf. Accessed 8 Aug 2022

Money Laundering and Terrorist Financing Prevention Act, Decree of the President of the Republic No. 174 (2017)

Nagrani V (2020) Gibraltar: blockchain comparative guide. https://www.mondaq.com/gibraltar/technology/935292/blockchain-comparative-guide. Accessed 2 May 2022

Neitz MB (2021) How to regulate blockchain's real-life applications: lessons from the California blockchain working group. Jurimetrics J 61:185–217

Nev. Rev. Stat. Ann. § 719.045 (2019)

Nev. Rev. Stat. Ann. § 78.0297 (2020)

NY State Tech L § 302 (2014)

OBG Limited and others v Allan, UKHL 21 (2007)

Office of Fair Trading v Abbey National, EWHC 875 (2008)

Ohio Rev Code § 1306.06 (2013)

Park JJ (2018) When are tokens securities? Some questions from the perplexed. UCLA School of Law, Law-Econ Research Paper No. 18-13. Lowell Milken Institute Policy Report. https://corpgov.law.harvard.edu/2018/12/20/when-are-tokens-securities-some-questions-from-the-perplexed/. Accessed 8 Aug 2022

Perez AJ, Zeadally S (2022) Secure and privacy-preserving crowdsensing using smart contracts: issues and solutions. Comp Sci Rev 43:100450

Politou E, Casino F, Alepis E, Patsakis C (2019) Blockchain mutability: challenges and proposed solutions. IEEE Trans Emerg Top Comput. https://doi.org/10.1109/TETC.2019.2949510

Praitheeshan P, Pan L, Yu J, Liu J, Doss R (2019) Security analysis methods on ethereum smart contract vulnerabilities: a survey. https://doi.org/10.48550/arxiv.1908.08605

Rühl G (2021) Smart (legal) contracts, or: which (contract) law for smart contracts? In: Cappiello B, Carullo G (eds) Blockchain, law and governance. Springer, Switzerland, pp 159–180

Saad M, Spaulding J, Njilla L, Kamhoua C, Shetty S, Nyang DH, Mohaisen D (2020) Exploring the attack surface of blockchain: a comprehensive survey. IEEE Commun Surv Tutorials 22(3):1977–2008. https://doi.org/10.1109/COMST.2020.2975999

Samreen NF, Alalfi MH (2020) Reentrancy vulnerability identification in Ethereum smart contracts. 2020 IEEE International Workshop on Blockchain Oriented Software Engineering (IWBOSE)

Savelyev A (2017) Contract law 2.0: 'smart' contracts as the beginning of the end of classic contract law. Inf Commu Technol law 26(2):116–134

Securities and Exchange Commission v. Howey Co., 328 U.S. 293 (1946)

Shaw SLK (2009) Conversion of intangible property: a modest, but principled extension? A historical perspective. Victoria Univ Wellington Law Rev 40(2):419–440

Simmons A (2021) Regulating Libra: will legal and regulatory uncertainty prevent the launch of Facebook's cryptocurrency project? J Bus Technol Law 16:83. https://digitalcommons.law.umaryland.edu/jbtl/vol16/iss1/4

Sklaroff JM (2017) Smart contracts and the cost of inflexibility. Univ Pennsylvania Law Rev 166:263. https://scholarship.law.upenn.edu/cgi/viewcontent.cgi?article=1009&context=prize_papers

Solodan K (2019) Legal regulation of cryptocurrency taxation in European countries. Eur J Law Public Adm 6(1):64–74. https://doi.org/10.18662/eljpa/64

Spafford ML, Stanaway DF, Chung S (2019) Blockchain and cryptocurrencies: a cross-border conundrum. J Invest Compliance 20(3):10–19. https://doi.org/10.1108/JOIC-05-2019-0027

Szabo N (1996) Nick Szabo–Smart contracts: building blocks for digital markets. https://www.fon.hum.uva.nl/rob/Courses/InformationInSpeech/CDROM/Literature/LOTwinterschool2006/szabo.best.vwh.net/smart_contracts_2.html

Tai ETT (2018) Force majeure and excuses in smart contracts. Eur Rev Priv Law 26(6)

Tatar U, Gokce Y, Nussbaum B (2020) Law versus technology: blockchain, GDPR, and tough tradeoffs. Comput Law Secur Rev 38. https://doi.org/10.1016/j.clsr.2020.105454

Temte MN (2019) Blockchain challenges traditional contract law: just how smart are smart contracts. Wyoming Law Rev 19:87

TN Code § 47-10-201 (2019)

Token- und VT-Dienstleister-Gesetz, No. 301 (2019)

Unfair Terms in Consumer Contracts Directive (1993) Edited by European Parliament and the Council of the European Union

U.S. Securities and Exchange Commission (2019) Framework for "investment contract" analysis of digital assets. https://www.sec.gov/corpfin/framework-investment-contract-analysis-digital-assets#_edn1. Accessed 25 Aug 2022

Vermont Rules of Evidence 902 (1983)

Virtual Financial Assets Act (2018)

Waite JB (1918) Performance of an existing obligation as consideration for a promise. University of Michigan Law School. https://repository.law.umich.edu/articles/903

Wang S, Yuan Y, Wang X, Li J, Qin R, Wang F-W (2018) An overview of smart contract: architecture, applications, and future trends. 2018 IEEE Intelligent Vehicles Symposium (IV)

Werbach K, Cornell N (2017) Contracts ex machina. Duke Law J 67:313

Wilmarth AE (2021) It's time to regulate stablecoins as deposits and require their issuers to be FDIC-insured banks. GWU Law School Public Law Research Paper No. 2022-01. https://papers.ssrn.com/sol3/papers.cfm?abstract_id=4000795. Accessed 25 Aug 2022

Woodburn J (1982) Egalitarian societies. Roy Anthropol Inst Great Br Irel 17(3):431–451. https://doi.org/10.2307/2801707

Wyo. Stat. § 34-29-103 (2019)

Wyo. Stat. Ann. § 17-4-206(a) (2018)

Wyo. Stat. Ann. § 17-16-625 (2019)

Wyo. Stat. Ann. § 17-16-1601 (2018)

Wyo. Stat. Ann. § 34-29-106 (2017)

Xu S, Ning J, Ma J, Huang X, Deng RH (2021) K-time modifiable and epoch-based redactable blockchain. IEEE Trans Inf Forensics Secur Commun Netw 16:4507–4520. https://doi.org/10.1109/TIFS.2021.3107146

Yaga D, Mell P, Roby N, Scarfone K (2018) Blockchain Technology Overview. https://nvlpubs.nist.gov/nistpubs/ir/2018/nist.ir.8202.pdf. Accessed 10 June 2022

Your Response Ltd v Datateam Business Media Ltd, EWCA Civ 281 (2014)

Zetzsche DA, Buckley RP, Arner DW (2019) Regulating LIBRA: the transformative potential of Facebook's cryptocurrency and possible regulatory responses. University of Hong Kong Faculty of Law Research Paper No. 2019/042

Karisma Karisma is a PhD scholar at the Faculty of Law at University Malaya. Her research interests center around law and technology from an interdisciplinary and comparative perspective and the viability of blockchain technology in the energy sector, particularly in the legal and regulatory aspects. Before pursuing her PhD, she was a law lecturer at a private higher educational institution in Malaysia.

Pardis Moslemzadeh Tehrani is the Associate Head of the Law School, at the Faculty of Business, Law and Tourism, University of Sunderland. Before joining the Sunderland University, she was a Senior Lecturer and Visiting Associate Professor at the Faculty of Law, University of Malaya. Her research spans the area of IT law, international law and legal research methodology. She is the author and co-author of more than 70 research papers, books, book chapters and conference papers. She has served on many conferences, seminars, forums and workshop programme committees and delivered talks at various panels, NGOs, research centres and international bodies such as ICRC and other Human Rights organisations.

Part II
Applications

Chapter 5
Blockchain and Sustainable Energy

Soheil Saraji

5.1 Introduction

Blockchain is an emerging and disruptive technology in the energy sector with potential applications in recording and tracking data exchanges, utilizing a distributed system to verify transactions, improving energy efficiency, allowing shared governance, facilitating the startup process for financial companies, reducing overhead costs, increasing energy security, boosting climate actionability, and enhancing international energy competitiveness. The energy sector can benefit greatly from the emerging blockchain applications that are currently under development. For instance, blockchain could make it possible to track low-carbon energy and associated certificates from their origin through every stage of transaction. It could also transform the way parties across a value chain interact, make new collaborations possible, and create transactive energy within both renewable and nonrenewable energy businesses.

In the oil and gas industry, blockchain has emerged as a promising innovation that could play an essential role in delivering the technological and commercial capabilities that the sector will need to achieve its goals of modernization and digitization. It could offer operational cost reductions, increased efficiency, fast and automated processes, and transparency. Other applications could include blockchain-enabled transactional digital platforms for trading (Andoni et al. 2019), the design and construction of wells and facilities, the tracking of drilling equipment history and maintenance, the automation of drilling as well as the optimization of drilling operations, and supply-chain management (Elijah et al. 2021). Current examples of employed applications include guaranteeing the authenticity of wellbore rock and fluid samples and creating a shared consensus about the progress of drilling campaigns (Perrons and Cosby 2020).

S. Saraji (✉)
Energy and Petroleum Engineering, University of Wyoming, Laramie, WY, USA
e-mail: ssaraji@uwyo.edu

S. Saraji and S. Chen, *Sustainable Oil and Gas Using Blockchain*,
Lecture Notes in Energy 98, https://doi.org/10.1007/978-3-031-30697-6_5

We surveyed a few experts familiar with this field about the potential applications of blockchain for the sustainable oil and gas industry. John Westerheide Senior Director of Customer Solutions, Project Canary) told us: "*I think there are a whole lot of opportunities for blockchain in oil and gas traditional energy supply chains. I think as it relates to the energy transition, and driving capital efficient markets, especially around the rise of ESG, investing in that type of deal, the ability to bring low cost, high fidelity understanding of oil and gas supply chains, and being able to differentiate producers, midstream operators, Petrochem plant, utilities,* etc., *based off of that information from both a consumer perspective, but also from capital markets and investment perspective, is definitely one of the main use cases as it relates to ESG in the energy transition.*" He then continued: "*One that I've always been fascinated with is fractional ownership of minerals and production and being able to open direct investment from non-accredited investors into oil and gas production. Now, if you think about it, your ability to get in and invest in the development of an oil and gas well, or mineral rights usually takes large capital and many intermediaries. And I think Blockchain's ability to enable fractional ownership, reduce the administrative cost associated with that type of play is very interesting. And it creates a diversity of exposure to assets, when you don't have to go in on long large chunks of minerals or large chunks of wells, you can diversify at a finer point and make up a portfolio that may be more risk tolerant, resilient,* etc."

Dan Cearnau, Co-Founder and CTO of EarnDLT, thinks transparency and traceability are the key properties of blockchain when it comes to applying it to the energy sector. "*I think it's mostly about transparency. And this is what blockchain is bringing to the table. The blockchain has two big advantages. One is transparency and the fact that you can't hide things on a chain. Once something is issued, it is distributed over the network. You can't cook the books, you can't erase documents, so it increases the transparency. And I think that will matter a lot in this area and this market. The second is, of course, the traceability; the fact that when you're buying a token that represents certified gas, you know whom you're buying it from. And that it has that environmental impact, is validated, and is accredited to do that. Because otherwise, they think there's going to be a lot of fake carbon credits, people faking in order to reduce their environmental impact.*" When asked about other potential applications, he replied: "*in the financing sector of the oil and gas industry, there's definitely a lot of potential. The second [application] is in supply chain management, because they have a lot of heavy machines, that are manufactured and distributed and sold from one manufacturer to distributor, and so on. You can also use it for inventory management.*"

Sriram Srinivasan, Chair of Mining and Minerals ESG WG at OriginBX, explained: "*The killer application of the internet was email. When we consider all of the different forms of communication, such as WeChat, WhatsApp, and others, we tend to think of email as if it hasn't changed over the years. In fact, that is absolutely true. I mean, different variations of email have some sort of communication between parties. The killer application of blockchain is finance. You'll see that often applications that become successful will typically be variations of the killer application. So you have to think about it in the petroleum business, there is data, there is process, and then there is payment. Typically in large companies, treasury and procurement are in two silos. What a blockchain does is finally figures out there is a real linkage between those two and in those 30 days and 60 days on those payment cycles, the linkage happens. And many of the deals that happen today happen because of the nature of our intermediaries need to charge because they got to hold that inventory for some time. And then there is a risk associated with this transaction and the cost of the risk is added to the cost of the goods. Blockchain does not remove it. What it does is it makes the risk transparently known in a suitable way so that the right parties can take the course of actions needed. So eventually, the killer app for the blockchain, I think, is the finance.*"

As discussed in Chap. 1, the oil and gas industry is undergoing a transformation to adapt itself to the global energy transition and secure a place in the future of the energy sector. Reducing carbon emissions and improving environmental, social, and governance (ESG) activities will be the focus of the industry. In addition, the markets seem to be interested in investing and rewarding ESG-compliant activities within the industry. Therefore, the author also believes the intersection of blockchain, and sustainability is a ripe area for rapid development and growth in the industry. Blockchain has been proposed as a key technology enabling reliable and trustless tracking of greenhouse gas emissions and other environmental data (Cannon et al. 2022) and blockchain-based life cycle assessment (Zhang et al. 2020). This will allow the stakeholders to make purchasing decisions based on the environmental performance of any product. Furthermore, this allows for ESG attributes to be tokenized and traded. In the proceeding sections, we provide a few examples to highlight the potential applications of blockchain for sustainable energy development. However, first, let us discuss if blockchain, itself, is a sustainable technology.

5.2 Blockchain Sustainability

Blockchain technology and, more specifically, the Bitcoin network have been criticized because of its large energy consumption and associated greenhouse gas emissions. It is estimated that the annual electricity consumption and carbon emissions of Bitcoin had reached around 45.8 TWh and 22 MtCO_2 by 2018 (Stoll et al. 2019). These emission levels are equivalent to the levels produced by Jordan, Sri Lanka, and Kansas City. The other top 20 blockchains (from a total value perspective) combined emit about half of this amount (Gallersdörfer et al. 2020). However, it should be noted that Bitcoin and other 13 (out of the top 20) Blockchains, studied at the time, use proof-of-work consensus. This consensus mechanism, by nature, is very energy-intensive as participants with higher computing power and higher energy consumption have a better chance of mining the next block and receiving the rewards. This naturally incentivizes investing in more advanced hardware and consuming more electricity.

To address this inefficiency, the blockchain industry has been shifting its sources of energy from fossil fuels to renewable energy to operate more sustainably. Another proposal towards more sustainable mining operations is the development of mobile mining units to use methane waste in the oil and gas fields, which is otherwise either released into the atmosphere or burnt on the spot due to a lack of distribution pipeline and uneconomical volumes. Recently, several startups have developed mobile mining units for this purpose. This new development is beneficial as the new mining units move to off-grid sources and reduce the load from the electric grid. Although burning unwanted natural gas to generate electricity helps with avoiding energy waste and the release of methane into the atmosphere, it still releases carbon dioxide all the same. These operations seem to be mostly motivated by the economics of cheap or free gas streams (i.e., electricity) from the oil and gas fields, rather than any real efforts toward mitigating the environmental impact of the mining operation.

The most important activity in the blockchain industry toward sustainable operation is the effort to change the engine of the blockchain (i.e., consensus mechanism), which is the main source of energy consumption. It is helpful, here, to borrow an analogy from Justin Drake (researcher at the Ethereum Foundation), who used a comparison between a car and a blockchain network. In this analogy, the engine of the car is the consensus mechanism, which is Proof of Work (PoW) in the Bitcoin network. We can consider PoW as a diesel engine that is energy inefficient and emits a considerable amount of greenhouse gas. By redesigning the car (i.e., blockchain), we can replace the diesel engine (PoW) with better, more scalable, and more energy-efficient engines. In this other and more modern engine (i.e., consensus mechanisms), the security of the network is guaranteed by different mechanisms such as staking valuable currency (i.e., Poof of Staking, PoS), trusted validators (PoA), etc., as discussed in Chap. 3. The PoW is perceived as the gold standard of network security in the blockchain industry, and replacing it with other mechanisms may decrease the security or decentralization of the network. However, years of academic research on variations of PoS shows that it could provide relatively comparable security and decentralization while improving on sustainability and scalability.

In October 2022, the second largest blockchain (in market cap), Ethereum, successfully switched its engine from traditional PoW to PoS, which is known in the community as the *Merge*. The operation was described by Justine Drake as replacing the engine of a moving car. This was an impressive and seamless operation following a long-term plan within the Ethereum community, whose founder (Vitalik Buterin) and developers had recognized the sustainability and scalability vulnerabilities of PoW early on and conceived a multi-year plan to replace it with PoS. The initial assessments after the merge show 99.9% reduction in energy consumption and, by proxy, carbon emissions of the Ethereum network as a result.

This was an important development, as Ethereum is the second historic blockchain with an active developer community and the first smart contract platform. More importantly, this is not an isolated attempt, as many other prominent smart contract platforms that followed the success of Ethereum in bringing real-world applications to the Blockchain designed and implemented their protocols using energy-efficient consensus mechanisms, such as Cardano, Polkadot, Solana, Cosmos, and etc. Today, the top smart contract blockchain networks are using energy-efficient and scalable security models for their networks, such as variations of PoS. Although this is mostly driven by the blockchain industry's own need for greater efficiency, lower costs, and faster throughput, it also shows that the blockchain industry is aware of the criticism of its carbon footprint, and there is a systematic movement toward a sustainable solution.

5.3 Sustainable Energy Applications

In this section, we explore applications of blockchain in sustainable energy. This is a fast-growing area, and new applications emerge every year. The list of applications provided here is a sample of existing or developing applications and is not intended to be a comprehensive list. This section starts by focusing on applications for sustainable oil and gas, consistent with the title of this book. Other applications in the broader energy sector are also provided for interested readers.

5.3.1 Monitoring, Verifying, and Reporting (MRV)

Carbon is an interesting element. It is a part of our everyday life on the earth, and it exists everywhere, in the ocean, in the soil, in the air, and in animal and plant bodies. More importantly, it has been the main source of our energy (coal, oil, and gas), allowing for exponential improvements in the quality of human life in the past two centuries. On the other hand, the excess use of fossil-based fuel and carbon emissions to the atmosphere (in the form of carbon dioxide and methane) has shifted the balance between various natural sinks and sources of carbon on the earth. There is a growing concern that this shift, if continued, may push this equilibrium/balance past the point of return. This can cause a tectonic change in the climate and negatively impact plant, animal, and human life on the earth. To manage these changes in the short and medium term, there seems to be an urgent need to manage the carbon in our world. This is where the concept of carbon management comes into play.

In the past decade, companies across different industries, including the oil and gas industry, have been adopting mandatory or voluntary Greenhouse Gas (GHG) emission reporting. The force behind this trend is either mandates by governments or market push by the customers. In order to provide accurate GHG emission reporting, the process of measuring (monitoring), verifying, and reporting (MVR) has emerged. The goal of MRV, when applied to emission data, is to develop a process and standard for gathering and reporting the GHG data. The integrity of this data is important, especially when applied to the entire supply chain of a product in which multiple entities are involved with competing interests. For example, to produce and deliver a barrel of oil, different companies and processes are involved, including drilling, production, transportation, etc. Blockchain technology could help with providing security, transparency, and integrity to the data for the entire process. Since the data recorded on the blockchain is immutable and transparent, the track record will also be accessible to authorities at any time to review. Furthermore, the data digest on blockchain could be tokenized for sale to interested parties. This is discussed in the next section.

5.3.2 Certified and Tokenized Differentiated Fuels

Another advantage of developing MRV processes using blockchain technology is that it allows for the monetization of this data. Why would a customer pay extra for ESG data attached to fuel? John Westerheide (Senior Director of Customer Solutions, Project Canary) says: "*Regardless of the industry that you look at right now, but specifically in B2C industries, this idea of what we consume, being aligned to our personal values, and differentiated along the lines of provenance, sustainability,* etc., *is important. We saw this in the diamond industry when they started talking about ethical sourcing. We see it in the building materials industry, the food industry,* etc. *One of my favorite examples is I picked up a water bottle the other day and it told me where it came from, and that the plastic was sustainably sourced or recycled. What is more fungible in this world than two hydrogen molecules attached to an oxygen molecule? But we're differentiating those hydrogen and oxygen molecules based on their provenance, how they're delivered, how they're packaged,* etc. *Where we haven't seen that yet is in the energy industry. And what I mean by that is, we've seen it between the world of traditional fuels and renewables, or Green Energy Fuels. ... So, the things that we're most dependent on in this world like the sources of energy: electricity, gasoline,* etc., *are the things we have the lowest visibility and understanding of their provenance, where it comes from, how it was produced, and the supply chain by which it was delivered to us.*"

There have been some initiatives in the industry to develop fuels with an attached EGS data layer as premium fuels. The initial market response has been positive to these developments. An example is the concept of digital fuels: "*Digital fuels are basically digitalizing the actual commodity in itself. The first product that we kicked off in digital fuels was DNG (Digital Natural Gas). Think of it like a birth certificate that comes with each MMBtu of gas. It's still the gas; it is still being produced the same exact way. But now we're just putting data and attaching that data that has those environmental attributes into an immutable digital file. So, each molecule comes with a birth certificate that's digitally captured, that gives you the environmental attributes like carbon, methane, VOCs, time it was produced, which facility it was produced [from], and things like that, just basic, preliminary primary data.*" Explained Jasmine Zhu, VP of Market Development at Xpansiv.

However, lack of trust between involved parties, misalignment between economic incentives versus environmental impact, transparency, and auditability requirements are the main challenges in this space. Blockchain technology is a great platform for addressing these challenges and becoming the backbone for certifying differentiated fuels. This information could then be used to tokenize the emission and other ESG-related attributes of the fuel for sale to the markets. The entities that are interested in achieving their ESG compliance goals are then given the option to buy these tokens at a price and include them in their ESG reports. The owners of these tokens, on the other hand, will receive extra proceeds by selling the tokens. This provides monetary incentives for companies to invest in reducing their emissions and becoming ESG-compliant. This topic is further discussed through a real-world demonstration in the case study discussed later in Sect. 5.4.

5.3.3 Carbon Credits and Offsets

About two decades ago (December 1997), countries around the world got together in Kyoto (Japan) and signed an agreement with the purpose of managing greenhouse gas (GHG) emissions. The Kyoto protocol is the first international application of the cap-and-trade emission rights system. One indirect consequence of this protocol was a shift in viewing carbon instead of a problem as a commodity to be traded. If there is a limit on emitting carbon into the atmosphere by each country, then carbon emissions (or more precisely, avoiding carbon emissions) become a commodity with significant value. Since then, carbon credits and offsets have been gaining traction as a carbon management tool with economic incentives. This is an area that seems to be primed for innovation and development in the coming years, as countries around the world adopt voluntary and mandatory emission targets.

Blockchain technology provides a safe, reliable, efficient, convenient, open, and inclusive platform that is uniquely suited for implementing Carbon Credit Markets. The immutable cryptographically-secured distributed ledger on the Blockchain allows for reliable issuance and tracking of carbon credits. Public blockchains are easily accessible to small and medium-sized enterprises, reducing the entry threshold for the carbon trading market. Furthermore, the information provided by companies is transparent and accessible to everyone. Recently, free automated market makers (AMMs) have been developed on blockchains, allowing for the trading of digitized assets directly on the Blockchain with minimal algorithmic fees and without intermediaries. This provides the infrastructure required to create a digital carbon credit ecosystem to engage all interested stakeholders.

One proposed scheme is shown in Fig. 5.1 (Saraji and Borowczak 2021). In this scheme, different stakeholders involved are *Generators* of carbon credit (i.e., wind farms, tree-planting operations, CO_2 sequestration projects, etc.) and *Consumers* of carbon credit (i.e., carbon emitters or polluters of any kind such as the energy industry) as well as other stakeholders such as regulators, concerned citizens, and validators. *Validators* are an essential part of this ecosystem. They are accredited, globally distributed, technically competent consultants incentivized to parameterize appropriately and onboard projects to an open architecture marketplace that matches interested parties generating and retiring carbon credits.

The carbon credits will be minted natively on the blockchain or transferred to the Blockchain by converting them into digital tokens distributed to carbon credit generators after properly validating their projects. Buyers and sellers of carbon credit

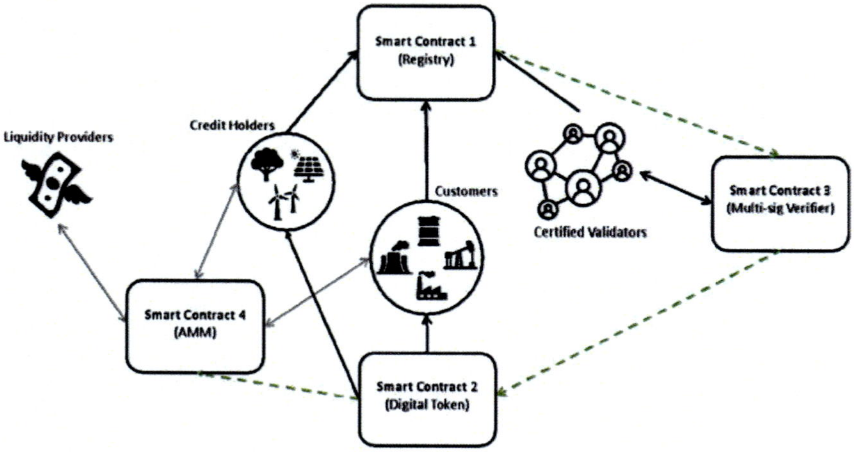

Fig. 5.1 Flow diagram showing the proposed Carbon credit ecosystem on a blockchain (Saraji and Borowczak 2021)

will use a decentralized exchange platform on Blockchain to trade Carbon credits. The price will be determined by market dynamics driven by supply and demand. The Carbon Tokens would be retired via a "buy and burn" model by sending the given Carbon Tokens to a smart contract or defined blockchain address whose private key is not known by any party and can be visible to the collective of validators as well as regulators or other stakeholders. The companies and individuals who successfully burn their Carbon Tokens will be issued non-fungible tokens as a carbon removal certificate.

5.3.4 *Renewable Energy Certificates*

A Renewable Energy Certificate (REC) is created when 1 MWh of renewable energy (such as hydro solar, wind, biomass, or geothermal) is produced. The resulting RECs may be sold separately and apart from the electricity in what is known as "unbundling." When the buyer of the REC uses the credit to demonstrate an "offset" of 1 MW of electrical usage, then the REC is retired and can no longer be used by any other power user (Mutually Evolving Technologies: Blockchain, Renewable Energy, and Energy Storage, n.d.). With the purchase of a REC, the buyer is adopting the renewable attributes of the specified amount and type of renewable energy generation. RECs are tradeable, non-tangible energy commodities. The suppliers of RECs are mandated by law to disclose the quantity, type (hydro, solar, wind, etc.), and geographical source of the certificates. The renewable energy certificate (REC) market, in particular, can be revolutionized using blockchain technology. The recording of transactions between electric generation suppliers and electric distribution companies can be automated and easily accessed with this new technology. Blockchain produces an easy-to-understand ledger, graphical user interface, and generates data that is easily extractable to be used by other, preexisting enterprise-level software. Lastly, and perhaps most importantly, the blockchain can include all the critical data points required by law in the United States for REC generation, registration, and recording (Revolutionizing Renewable Energy Certificate Markets with Tokenization IBM Supply Chain and Blockchain Blog, n.d.).

5.3.5 Energy Internet of Things (e-IoT)

The Internet of things (IoT) comprises various mechanical devices and sensors, which are connected to each other through a gateway (Muhanji et al. 2019). With the rapid development of wireless sensor networks, smart devices, and traditional information and communication technologies, there is tremendous growth in the use of IoT applications and services in our everyday life (Pal et al. 2021). Research has shown that decentralization in IoT access control may overcome the limitations of traditional centralized access control models, such as single-point of failure, reliance on trusted third parties, internal attacks, and central leaks (Song et al. 2021). Blockchain can provide a secure, distributed, and anonymous framework for IoT devices, enabling secure access and data management. Recently, the energy Internet of Things (e-IoT) has been proposed as an energy management solution in modern energy grids (Muhanji et al. 2019). In the modern digital energy grid of the future, all devices that consume electricity are internet-enabled and, consequently, can coordinate their energy consumption with the rest of the grid in or near real-time (Muhanji et al. 2019). In addition, the integration of blockchain-enabled IoT technology into the smart electricity grid will promote the decentralized generation, transmission, distribution, and management of energy (Muhanji et al. 2019).

For instance, *Ondiflo* is a Texas-based startup that leverages sensor data and blockchain technology to fully automate conversations between humans, devices (IoTs), and business systems in the oil field. For example, by using blockchain, they provide visibility to tank levels and locations. Operators can provide truckers with the ability to manage dispatch and scheduling better. Also, by providing transparency to track movements, the trucker provides the operator with the benefit of planning and risk avoidance (Ondiflo Is Using Blockchain to Revolutionize the Oil and Gas Industry I AWS Startups Blog 2019).

5.3.6 Integrated Distributed Electricity Network

Integrated distributed electricity networks consist of small-scaled operations that provide trackable, transparent transactions at lower costs (Blockchain in Energy Markets, n.d.). The network makes use of information technology and management technology to relate larger, more complex devices and systems (Cioara et al. 2020). Digital technologies like blockchain are creating new options for streamlining and simplifying paper processes, and for disrupting long-established business models (Blockchain for Commodities: Trading Opportunities in a Digital Age | S&P Global, n.d.). Decentralized management and coordination of energy systems are emerging trends facilitated by the uptake of the Internet of Things and Blockchain, offering new opportunities for more secure, resilient, and efficient energy distribution. Even though the use of distributed ledger technology in the energy domain is promising, decentralized smart grid management solutions are in the early stages. Smart grids coordinate the requirements and capabilities of all grid operators, generators, end-users, and electricity market stakeholders. A smart grid network encompasses numerous connected IoT devices, which can react to electricity grid signals by adjusting their power consumption accordingly, as and when required. Battery storage is seen as being a key ingredient in helping to maintain this grid stability and pricing flexibility, soaking up excesses during periods of high energy production and returning it to the grid when demand outstrips supply (Cioara et al. 2020). Blockchains could assist in management of decentralized networks, flexibility services or asset management. Blockchains could achieve integrated flexibility trading platforms and optimize flexible resources, which might otherwise lead to expensive network upgrades. As a result, blockchains might also affect revenues and tariffs for network use (Andoni et al. 2019).

A good example is *Energy Web* Foundation. Energy Web (EW) is a global nonprofit organization accelerating a low-carbon, customer-centric electricity system by unleashing the potential of open-source, decentralized technologies. In 2019, EW launched the Energy Web Chain, an open-source enterprise blockchain platform tailored to the retail energy sector. This platform aims at providing services and integrating a wide range of distributed energy resources. There are a variety of projects and initiatives that have launched since the inception of this platform by EW, including a decentralized finance crowdfunding platform to accelerate energy access in sub-Saharan Africa (Bebat Launches EasyBat, an Open-Source, Decentralized Solution for Battery Lifecycle Management | by Energy Web | Energy Web | Medium, n.d.), energy market exchanges to accelerate the development of renewable energy projects in fragile, energy-poor countries (Energy Web and Energy Peace Partners Announce Peace Renewable Energy Credit (P-REC) Digital Marketplace Platform | by Energy Web | Energy Web | Medium, n.d.), a platform for decentralized lifecycle battery asset management (ENGIE Energy Access and Energy Web Announce DeFi Crowdfunding Platform to Help Scale Solar, Mini Grids in Sub-Saharan Africa | by Energy Web | Energy Web | Medium, n.d.), etc.

5.3.7 *Energy Commodity Trading*

Commodity trading typically relies on traditional documentation and written records to process and accomplish a transaction (Blockchain for Commodities: Trading Opportunities in a Digital Age | S&P Global, n.d.). Commodity trading systems built on decentralized ledgers provide the required security, immutability, and real-time view of pricing and transaction status (Blockchain in Energy Markets, n.d.). Replacing expensive proprietary systems can open the door to a wider variety of market participants. Applying blockchain technology to commodity trading would be cheaper, more efficient, provide more security, and rapidity than existing proprietary systems. The benefit of blockchain in commodity trading is that it can eliminate the slow adaptability of large-scale proprietary systems, which is one of the major limiting factors. The unique advantage for this use case in the commodity and energy trading market is creating an ecosystem that encompasses the start-to-finish transaction life cycle, in essence, a private blockchain network. Finally, this results in potential cost savings coupled with improving processes more efficiently. Commodity trading within specifically the energy sector includes, but is not limited to, crude oil, natural gas, gasoline, and petrochemical products.

5.3.8 *Peer-to-Peer Energy Trading*

Peer-to-peer trading is one of the early, universal blockchain applications regarding decentralized energy transmission, widely considered to be a viable solution for the future. Peer-to-peer, or P2P, is a direct energy trade structure that includes a group of participants, i.e., generators, consumers, and prosumers (Soto et al. 2021). The direct energy trade structure encourages trade between consumers and prosumers without intermediating conventional energy suppliers, typically within a local grid network (Zhang et al. 2018). In addition to trading, peer-to-peer allows energy generated to be stored, exported, and sold. Peer-to-peer trading contains sub-applications, including wholesale trading and utility trading, each following the same basic principles.

A successful example of peer-to-peer energy trading is the *Brooklyn Microgrid* project (Can the Brooklyn Microgrid Project Revolutionise the Energy Market? | by The Beam | TheBeamMagazine | Medium, n.d.). Brooklyn Microgrid, implemented by *LO3 Energy*, is a demonstration project where citizens can buy and sell locally produced solar electricity from one another. The project started in early 2015, and in April 2016, the first community activity took place when three residents of President Street in Park Slope participated in the first-ever peer-to-peer energy transactions (Can the Brooklyn Microgrid Project Revolutionise the Energy Market? | by The Beam | TheBeamMagazine | Medium, n.d.). The project started first on a private blockchain to remove intermediaries and facilitate peer-to-peer transactions. In 2020, following the success of the Brooklyn Microgrid project, LO3 Energy released its open-source blockchain solution for utilities called *Pando*.

5.3.9 Internet of Vehicles

Electric vehicles (EVs) are significant drivers of the global energy transition, and their use is expected to grow steadily in the next 15 years (Mackenzie 2018). EVs emerge as new unconventional and highly disruptive participants in the grid that can add significant benefits and flexibility (Jurdak et al. 2021). These vehicles are conventionally viewed as energy consumers that are charged at charging stations to support their movement. EVs' charging and discharging behavior have a significant impact on the power system stability and power market operation (Huang et al. 2021). For example, the electricity exchange between EVs and the power grid, wireless charging among EVs, and EV route optimization based on the availability of vehicles and static charging stations, are new challenges that need to be addressed (Jurdak et al. 2021). Blockchain has attracted tremendous attention as a potential solution to many of these challenges due to its salient features, including decentralization, security, anonymity, auditability, and transparency (Jurdak et al. 2021; Sharma 2019).

5.3.10 Energy Data Management

Concerning data management, the blockchain could become a support technology that can perfectly manage all data collected through electricity meters and facilitate real-time consumption monitoring. Furthermore, blockchain offers consumers the possibility to control their energy sources and energy data. This offers real-time and secure updates of their used energy and the ability to securely share subsets of data with the market. The global energy markets have experimented with these concepts, including blockchain infrastructure connected to water meters that can trace flows and identify issues in need of repair (Blockchain in Energy Markets, n.d.).

5.3.11 Supply-Chain Management

Blockchain is a potentially disruptive technology for the design, organization, operations, and general management of supply chains. Blockchain's ability to guarantee the reliability, traceability, and authenticity of the information, along with smart contractual relationships for a trustless environment, all portend a significant rethinking of supply chains and supply chain management (Esmaeilian et al. 2020). Blockchain has the potential to advance supply chain processes and supply chain management towards sustainability. Blockchain plays a particularly important sustainability role. Four main Blockchain capabilities can support sustainable supply chains: (1) they help reduce the product recall and rework due to its tracking capabilities; (2) they make it easy to trace the actual footprint of products and determine the accurate amount of carbon tax that each company should be charged; (3) they facilitate recycling behavior by incentivizing individuals to participate in deposit-based recycling programs; (4) they improve the efficiency of emission trading schemes by reducing fraud and improving the fidelity of the system (Esmaeilian et al. 2020).

5.4 Case Study—Tokenizing Emission Data and Other Verifiable Environmental Attributes

"*PureWest Grows Market for Certified Gas Through Partnership with Earndlt and Project Canary to Tokenize Verifiable Environmental Attributes*" (PR Newswire 2022). This is the headline that summarizes the case study presented in this section. It involves a gas producer, a standard-setting and evaluating party, a Blockchain solution company, and a digital exchange platform. We have interviewed individuals within each of these entities to better understand the role that blockchain plays in the process. We included excerpts from those interviews in this section. The complete interviews are available in the last part of the book.

5.4.1 Monitoring and Certification

It all starts with a small gas producer in Wyoming. "*We are a fossil fuel company. And so, we have to find a way to communicate that natural gas is part of the solution toward the energy transition and find a way to present our product as being part of that solution. We've set a high bar for ourselves; part of it is regulatory-driven. We're in the Upper Green River Basin, which is an ozone non-attainment area. And that's driven a lot of our improvements, for sure. But we've tried to push that even further.*" Said Kelly Bott (SVP of ESG, land, and regulatory, PureWest Energy).

Being aware of the low level of their methane emissions compared to national levels reported by the EPA, PureWest Energy decided to explore monetizing their low emission levels. "*When we started thinking about certifying our gas, there were a couple of things that we were trying to do. It's very helpful to have a third party come in and audit and give a blessing to what we're doing, but also to give some recommendations for improvements. So, our methane intensity rate is one of the lowest in the country for 2020, according to the most recent EPA verification, and it is 0.05 percent.*" Explained Kelly Bott.

There is only a handful of standard-setting and certifying bodies for methane emissions active in the oil and gas industry, such as MiQ, Equitable Origins, and Project canary. PureWest landed on one of the options, as Kelly Bott mentioned: "*when we discovered Project Canary, one of the first things we noticed was that they were comprehensive in their certification program. They're looking at everything from our emissions profile and performance to how we protect the land, our reclamation activities, our safety metrics, and how we interact in the community. As a result, it was a complete package. And that's what we were after, because in the Upper Green River Basin, we've got protected viewsheds and wildlife, air quality issues, and world-class waters and recreation. And people are very, very interested in making sure that we're doing things right.*"

Project Canary provides TrustWell™ certification, which covers a variety of criteria, including emissions, environmental, spill prevention, waste management, etc. Their process is explained by Kelly Botts: "*To get those certifications, we had to provide everything, from every piece of documentation and every planning document that we put together, to Project Canary, who was very interested in knowing if we did things correctly from the start when it came to planning the wellbore, planning the surface, planning for risk, identifying risk, and mitigating risk. They went out and did a field tour to make sure that what they saw in person matched what we said we saw in person. And then they go through and evaluate the risk specific to each site and how we're mitigating that risk. And then they provide this certification to us that gives us our rating and tells us how we're doing right now, along with some suggestions for things that we can do better.*"

In this process, the producer is responsible for monitoring and collecting the environmental attributes, including methane emissions. This data is typically collected via sensors installed in the field (mostly accumulated around well pads) in real-time. This is essential because "*if my application is to provide immutable data associated with a natural gas marketing contract for a certified gas deal where emissions intensity is written into the structure of that contract, you're going to have continuous monitoring on the ground. You need that high fidelity and low or high intermittency data.*" Said John Westerheide (Senior Director of Customer Solutions, Project Canary). However, there are newer technologies that allow for larger-scale monitoring of emissions. John continued: "*Now, if I'm just trying to get an understanding of an aerial basis, where leaks might be occurring, let's say 100 m by 100 m or a kilometer by kilometer, satellites are great for that. But in a place like the Permian, where I may have checkerboarded, lease holdings, and operations, being able to tie that back to a specific action on a specific piece of equipment, for a specific operator is just not realistic. Drone flyovers or just flyovers in general are a great way of improving the fidelity and, therefore actionability of operations. But still, they're discrete, They're intermittent,* etc. *So, it's this idea that it's going to take a combination of all three, sort of what I'll call bottoms up continuous monitoring to tops down the satellite, and then getting the data linked on both a spatial as well as its own portal temporarily, to correlate what the bottoms up to tops down actually is, and to build a holistic picture.*"

This multi-level monitoring scheme will generate a massive amount of data from multiple sources that eventually need to be reconciled. Kelly Bott agrees: "*That's part of the pilot we're working on; we also have another pilot with a company called Validere. Validere takes multiple data streams. So, they're pulling in everything that we use to build our calculated emission inventory. Our equipment, our throughputs, and all of our emission factors that's one data stream; they're looking at data from monitors and are also technology agnostic. So, they're looking at multiple different sensor technologies that are coming in. And then they're also doing mass balance calculations.*"

5.4.2 *Monetization and Markets*

As we mentioned earlier, the process of emission monitoring requires setting up an array of sensors in the field, and developing platforms and processes to collect, store, and analyze this data. Furthermore, a third-party verifier needs to be hired to review and certify the data. Why would a small producer with a small profit margin invest in such a process? Bryan Hassler (Chairman, PureWest Energy) explained: "*When you look at PureWest, or Jonah Energy on the other side of the fault divide [Jonah Field/Pinedale Anticline, Wyoming], we operate mostly on federal lands, close to national parks, ozone nonattainment areas. So, our license to operate was built around the lowest cost emissions so that we could continue to operate. It's tough in a $2 [per MMBTU] environment; it's much easier to take a bit of your profitability and invest it in continuous monitoring, independent third-party evaluation, and find a way to tokenize, say, the certified gas that you ultimately produce. I would hope that there's a breakeven point associated with what we're doing. We're also seeing markets request more and more; what's your methane intensity?*".

Can these companies tap into the environmental commodities (i.e., ESG-compliant commodities) markets to sell their products at a premium price? This is a fast-growing market, and there are a few marketplaces enabling these trades. An example is Xpansiv, a registry for environmental commodities, which has partnered with Project Canary to bring transparency and differentiation for digital fuels. Jasmine Zhu (VP of Market Development, Xpansiv) explained: "*Xpansiv is the global platform for environmental commodities. And what that is, is basically a marketplace where we help facilitate buyers and sellers around environmental attributes. We cover several markets, one being carbon, and another being water. Then there's also the RECs market, which is the renewable energy credits, i.e., solar, wind, and differentiated fuels, which is the certified gas. And what it is really just helping to bring transparency to the market. There's price discovery that gets shown on a daily basis.*"

How does certified gas accrue premium value in the market? "*It's paying a premium for the data or certificate that shows that the gas that you're buying is cleaner than the gas that has no data. The EPA puts out a methane emission intensity number every year, and the most recent one from 2020, I believe, was 0.437 percent. So if you historically have never measured what it is that came with your natural gas, you could presume that your footprint is the national average (0.347) for methane emissions. But if you start buying, now, gas that has a certificate that tells you that you buy those environmental attributes that basically pairs up with your physical molecule that says, you're buying gas with the 0.08 or 0.05 type of methane intensity, then you could voluntarily make those reduction claims because now you've got data that backs up to the actual commodity that you're purchasing versus just buying the commodities with no transparency of what it is. The benefit of it is really, again, providing that fugitive methane number to help with the scope 3 [emissions] as a buyer. And let's solve the ability to offset the fossil fuel that you're still buying at the end of the day.*"—Jasmine Zhu.

There is a real movement in the markets and industry to accommodate this new type of premium fuels. "*Kinder Morgan Tennessee gas pipeline, just three days ago, the July 4 [2002] weekend, got FERC approved to offer two different pools, one they call producer certify gas, which is certified gas, and the other is regular gas. ...you got these producers who's going to be 100% certified, and there are these big producers like Chesapeake and EQT, who, at some point, are not going to be able to just offer you regular gas and they want to be either distinguished from a monetary standpoint or incentivize or, align themselves differently than everyone else who isn't certified so then you got midstream guys like the pipelines who said okay, you know what, we'll put you guys in a different pool.*"—Jasmine Zhu.

5.4.3 Blockchain and Tokenization

So far, we have learned that the process of developing differentiated oil or gas (i.e., Digital Fuels, Certified Gas, Low-carbon oil, etc.) and monetizing it happens in three steps:

1. Monitoring emissions, which generates a huge amount of data from different sources that needs to be stored securely.
2. Certification by a third party, who reviews and analyzes the data before issuing the certificates.
3. Monetization by listing those certificates on registries for price discovery and purchase by the customers.

The security and transparency of the generated data, as well as the reputation of the certifying party, in other words, trust and integrity, is the key element of this entire process. This is where Blockchain technology could help with providing security, transparency, and integrity to the entire process. Furthermore, digitizing these certificates in the form of tokens will allow for more flexibility in trading digital assets (i.e., Digital Fuels) and more robust tracking and monitoring by the relevant authorities.

"*I think what blockchain promises to provide is transparency to the world and humankind's greatest and most complex infrastructure, which is our energy infrastructure. It also provides transparency to the consumer about such information as; where did my energy come from? How is it produced? Do I like the value statements of who produced it and how they produced it? and am I willing to make trade-offs on cost, price, reliability, and resiliency associated with those verified attributes of the energy that I'm consuming? I think the biggest thing that blockchain does is lower the administrative cost of creating data-backed transparency in those complex supply chains. To follow the molecule from the reservoir to the burner tip, to the electron, to the wire, to somebody turning on a light bulb using it, to the number of intermediaries or the number of people involved in that transaction for that molecule, how many different assets it hit, who owns those assets,* etc., *is amazingly complex. And before advancements in technology, like blockchain, the administrative cost of that tracking and transparency and collating and curating the data that goes along with it was almost impossible. Although it would be possible, it'd be costly.*" *Said* John Westerheide.

Bryan Hassler argued: "*I think there is a role for blockchain in terms of attaching attributes to physical production that's continuously monitored, to give folks assurance that what we say we're doing and what the independent third-party validator says they're doing is completely documented.*" Kelly Bott agrees: "*we talk about differentiated gas, whether it's certified gas or some other attributes that we're attaching to it, we need to be able to have a mechanism that assures a buyer that they're buying it that this attribute,* per se, *hasn't been sold multiple times, that it's real, that it's valid, and that there was a unit of gas produced. … The cool thing about blockchain is that we can have a unit of gas that we can sell on the market. Then there are the environmental characteristics, which we may be able to attribute to a completely different person or entity. And it's cool because we can have a molecule or a unit of gas coming from a platinum pad, and maybe we can attach methane performance to it, or we're one of the first companies in the Rockies to get our freshwater-friendly, verified attribute from Project Canary; that attribute can be attached to that blockchain token, and so, it's really exciting that we'll be able to do that.*"

This is the point at which companies working on blockchain solutions for the oil and gas industry come into play. EarnDLT is a small startup company who, in collaboration with PureWest and Project Canary, minted (perhaps the first ever) digital tokens representing low methane emission and other environmental attributes associated with produced gas by PureWest in Wyoming. "*What EarnDLT provides is the ability for a user and energy producer in this case, to document all of their green energy production in a data room that has been authenticated by the cryptography and the blockchain. Further to that we allow for that data room, or that cluster, or that group of data, to be unitized, and tokenized. So that the owner of that data can transfer the data and even sell that data to another user. And the way that energy companies will use it is basically by monitoring their energy production on wells that have been certified by a third-party science-based organization as low methane intensity or as certified gas on that well, and what our platform will allow for is the producer to report directly to the blockchain on their production numbers, but then allow for that third-party scientific organization to report directly on that unit of energy regarding the methane intensity and the certification of that as green energy. And so, in combination with the blockchain, what you have is that immutable, permanent record around that unit of energy, that MMBtu of gas, certified gas, and the methane that was produced or leaked in the production of that unit of energy.*"

This is a very exciting development in the area of Blockchain and sustainable energy. This experiment may open the path for other companies to pursue differentiated digital fuel markets. However, there are serious challenges to scaling these developments at a national and international level, which will be discussed next.

5.4.4 Regulations and Standardization

Regulations and standardization are two major areas of uncertainty in the commercialization of differentiated premium fuels. "*Standardization of rules, regulations, and protocol, puts everybody out on a common playing field. And that typically needs to be the kind of government mandates. As much as we don't want government in our business, standardization across states, as opposed to state by state, makes things more efficient and more audible. We need to be able to audit what's going on, And I think it brings costs down across the board.*" Said Bryan Hassler, he continued: "*The Freeport is down, so we've lost two BCF a day of export capacity through LNG. But a lot of the International LNG players being European and Asian of origin, are starting to press their LNG totaling facilitators to find the lowest methane intensity gas out there, because they're pressed by European standards and developing Asian standards.*"

There is a push for further development in this area by the markets, explains Jasmine Zhu: "*The market has been stressing the whole concept of an independent third-party certifier. I do think that's critical. The market could handle a handful of standards, just like we have three rating agencies. It is like Googles and Apples of the world, they do similar things, they have different methodologies, and the market can handle that. It is just we need to eliminate a lot of that noise that's coming out from the sidelines. And any type of direction that the governing agencies could help would be also very impactful because now we're not just thinking domestic implications, we want to make sure that what we measure here and how we measure those emissions, also pass the smell test when we send those LNG cargoes to Europe. And they recognize how we measure and they recognize our practices, so that there is that uniformity from an international standpoint as well.*"

John Westerheide agrees: "*I think standards are going to be jurisdiction-specific. So, I don't see us getting to be a global standard. I don't think that's realistic. Nor do I think it's practical, because different paces at different places, you may need to understand the need locally to determine what the standard should be. That way, what we're doing here in the US, doesn't dictate what they must do somewhere else that isn't the US, and may not be facing the same challenges. Now what I think the commonality is, regardless of jurisdiction, and location, is the ability to use data to inform the standards, right? I think one of the challenges that we always face is the setting of standards without a true understanding; operational, real-world understanding versus an academic or theoretical understanding. And so, what I hope to see is that when standards are set, those standards are set based on good quality, real-world data, and information that allow us to understand both the pros and cons of the standards that are to be adopted as it relates to where it is in the certified gas market right now or looking at ESG profiles associated with hydrocarbons.*"

5.5 Challenges and Risks of Blockchain in the Energy

The advantages that blockchains can provide the energy sector are numerous and can potentially outweigh the challenges and risks. Blockchain data are essentially unchangeable, and highly reliable, providing fault tolerance and strong integrity (Teufel et al. 2019). Since anyone can generate blocks, create, and keep copies, an attack on one or more nodes does not impact the entire network. Notably, blockchains do not require an intermediary. The decentralized architecture of blockchains also can provide many benefits, such as fraud prevention, faster transaction times, and no single point of failure (Teufel et al. 2019). However, there is a wide range of challenges that need to be addressed for blockchain to become widely adopted in the industry, ranging from technical challenges to unclear regulations and legal frameworks. In the following sections, we will highlight some of these challenges.

5.5.1 Workforce Digital Skills

A recent online survey of oil and gas executives representing a cross-section of oil producers and service companies showed that there are many challenges to the adoption of digital technologies within the industry, including organizational and cultural barriers, the ability to change in a timely manner, a lack of workers with the right skills and the challenge of training workers to use new technologies (How Do You Reshape When Today's Future May Not Be Tomorrow's Reality? - Oil and Gas Digital Transformation and the Workforce Survey 2020, n.d.). Blockchain is not a typical topic taught at universities, especially outside of the computer science field as an applied tool. Although academia should take the blame for falling behind in teaching applied digital technology, the industry itself may need to take the lead to collaborate with academia in developing courses and teaching materials on topics such as Blockchain. Being cognizant of this challenge, the author has developed a new course titled "Blockchain in Energy" at the University of Wyoming with the aim to train engineering and other students on a basic understanding of blockchain and its applications in the energy sector. Furthermore, we plan to develop more applied courses and workshops on this topic in collaboration with the industry.

5.5.2 Standardization and Interoperability

The lack of standardization and interoperability among technology platforms within the blockchain technology space makes it difficult for different entities developing blockchain-based solutions to interoperate and communicate (Perrons and Cosby 2020). In an interview with the author, Sriram Srinivasan (Chair of Mining and Minerals ESG WG at OriginBX) said: "*To share data you need a system that protects your privacy. This system needs to interoperate with other blockchains. …Everyone will likely have a blockchain integration of their own. It's not clear that everyone uses the same blockchain. From what we know about history, that may not be the case. So, you need a way for blockchains to talk to each other. This could mean that large refiners each have their own blockchain, buyers have their own,* etc. *So, there needs to be a way for blockchains to talk to each other. That's where the IBC (Inter Blockchain Protocol) comes in. Different blockchains have thought about it differently, the inter-ledger foundation has spent a lot of time thinking about it, and how all of us will build technology that essentially is interoperable natively. So, if we interoperate, the next thing we need is standards.*" He then continued: "*Meaning if I'm going to just work with you, then we can all use a database. But if I am getting data, not just from you to recapture but from somebody else and from a third person, then I need to build systems that get from all three people. If there was a standard and I just adapted to that standard then any new party can just send me the information and I can send information out. So what I'm trying to make is a case for why blockchain, and if you think blockchain, you automatically need standards, otherwise the blockchain becomes not terribly useful.*"

5.5.3 The Oracle Problem

The immutable cryptographically-secured distributed ledger on the Blockchain allows for reliable tracking and monitoring of physical assets. However, real-world applications of Blockchain suffer from a fundamental shortcoming known as the *Oracle Problem* (Caldarelli 2020). That means blockchains are blind to the real world, so they depend on *Oracles*. Typically, Oracles are centralized and trusted third parties that constitute the interface between blockchains and the real world. As oracles reintroduce the concepts of trusted third parties and centralization, their implementation is often seen as a problem. They introduce a single point of failure breaking the trustlessness of the system. A recent systematic literature review on the subject showed that from a sample of 142 journal papers discussing blockchain real-world applications, only 15% considered the role of oracles, and less than 10% underlined the limitations of the oracle problem (Caldarelli 2020). To bring applications of Blockchain and smart contracts to the energy sector, we need to address this problem.

Solving the oracle problem is a multidisciplinary and active area of research. For example, a considerable attempt to limit the oracle problem was made by *Chainlink*. They proposed a system of decentralized oracles based on reputation to reproduce a blockchain's consensus mechanism. However, in many applications, decentralization is not sufficient or plausible to address the oracle problem, and data authenticity cannot be objectively verified. In these cases, a trust model is needed for the smart contract environment to keep a certain degree of reliability. A trust model is an intuitive scheme that outlines why the smart contract application should be trusted. There are a multitude of potential solutions for this problem being developed that are specific to particular applications ranging from the incorporation of temper-proof IoT devices to developing a network of experts through a DAO, etc.

5.5.4 *Legal and Regulatory Issues*

The most significant obstacles against blockchain energy commercialization are legal and regulatory issues, as these issues include contract law, energy law, data protection, liability, consumer rights, and sovereignty (Teufel et al. 2019). This topic is further discussed in a dedicated chapter in this book (Chapter 4).

5.6 Summary and Conclusion

Blockchain technology is in the early stages of research, implementation, development, and commercialization. Much research and discussion have been conducted on the emerging possibilities of this technology in the energy industry. In combination with the current global energy transition, this trend provides a unique opportunity to disrupt the traditional energy sector. Blockchain technology can provide a secure digital platform facilitating digitization, decarbonization, and decentralization of energy systems. It is especially suited for monitoring, verifying, and reporting GHG emissions and other ESG attributes of energy products. In this chapter, we provide a list of potential blockchain applications for sustainable energy development. More detailed discussions of some applications, including Methane emission abatement, carbon capture and storage, sustainable aviation fuels, and sustainable plastics, are provided in the upcoming chapters.

References

Andoni M, Robu V, Flynn D, Abram S, Geach D, Jenkins D, McCallum P, Peacock A (2019) Blockchain technology in the energy sector: a systematic review of challenges and opportunities. Renew Sustain Energy Rev 100:143–174. https://doi.org/10.1016/j.rser.2018.10.014

Bebat Launches EasyBat, an Open-Source, Decentralized Solution for Battery Lifecycle Management | by Energy Web | Energy Web | Medium (n.d.) https://medium.com/energy-web-insights/bebat-launches-easybat-an-open-source-decentralized-solution-for-battery-lifecycle-management-281f2ace61e9. Accessed 15 Jan 2023

Blockchain for Commodities: Trading Opportunities in a Digital Age | S&P Global (n.d.) https://www.spglobal.com/en/research-insights/featured/special-editorial/blockchain-for-commodities-trading-opportunities-in-a-digital-age. Accessed 15 Jan 2023

Blockchain in Energy Markets (n.d.) https://lisk.com/blog/posts/blockchain-energy-markets. Accessed 15 Jan 2023

Caldarelli G (2020) Understanding the blockchain oracle problem: a call for action. Information 11(11):509. https://doi.org/10.3390/info11110509

Can the Brooklyn Microgrid Project Revolutionise the Energy Market? | by The Beam | TheBeam-Magazine | Medium (n.d.) https://medium.com/thebeammagazine/can-the-brooklyn-microgrid-project-revolutionise-the-energy-market-ae2c13ec0341. Accessed 15 Jan 2023

Cannon C, Hughes S, Johnson M, Long E, Natali P (2022) Principles for blockchain-based emissions reporting. RMI

Cioara T, Pop C, Zanc R, Anghel I, Antal M, Salomie I (2020) Smart grid management using blockchain: future scenarios and challenges. In: 2020 19th RoEduNet conference: networking in education and research (RoEduNet), pp 1–5. https://doi.org/10.1109/RoEduNet51892.2020.9324874

Elijah O, Ling PA, Rahim SKA, Geok TK, Arsad A, Kadir EA, Abdurrahman M, Junin R, Agi A, Abdulfatah MY (2021) A survey on industry 4.0 for the oil and gas industry: upstream sector. IEEE Access 9:144438–144468. https://doi.org/10.1109/ACCESS.2021.3121302

Energy Web and Energy Peace Partners Announce Peace Renewable Energy Credit (P-REC) Digital Marketplace Platform | by Energy Web | Energy Web | Medium (n.d.) https://medium.com/energy-web-insights/energy-web-and-energy-peace-partners-announce-peace-renewable-energy-credit-p-rec-digital-6295ea233218. Accessed 15 Jan 2023

ENGIE Energy Access and Energy Web Announce DeFi Crowdfunding Platform to Help Scale Solar, Mini Grids in Sub-Saharan Africa | by Energy Web | Energy Web | Medium (n.d.) https://medium.com/energy-web-insights/engie-energy-access-and-energy-web-announce-defi-crowdfunding-platform-to-help-scale-solar-mini-2142029ad84f. Accessed 15 Jan 2023

Esmaeilian B, Sarkis J, Lewis K, Behdad S (2020) Blockchain for the future of sustainable supply chain management in industry 4.0. Resour Conserv Recycl 163:105064. https://doi.org/10.1016/j.resconrec.2020.105064

Gallersdörfer U, Klaaßen L, Stoll C (2020) Energy consumption of cryptocurrencies beyond bitcoin. Joule 4(9):1843–1846. https://doi.org/10.1016/j.joule.2020.07.013

How Do You Reshape When Today's Future May Not Be Tomorrow's Reality? - Oil and Gas Digital Transformation and the Workforce Survey 2020 (n.d.) TRUE Global Intelligence. ey.com/oilandgas/digitalskills

Huang Z, Li Z, Lai CS, Zhao Z, Wu X, Li X, Tong N, Lai LL (2021) A novel power market mechanism based on blockchain for electric vehicle charging stations. Electronics 10(3):307. https://doi.org/10.3390/electronics10030307

Jurdak R, Dorri A, Vilathgamuwa M (2021) A trusted and privacy-preserving internet of mobile energy. IEEE Commun Mag 59(6):89–95. https://doi.org/10.1109/MCOM.001.2000754

Mackenzie W (2018) When, why and how is the global energy transition going to happen? October 16, 2018. https://www.woodmac.com/news/feature/global-energy-transition/

Muhanji SO, Flint AE, Farid AM (2019) The development of IoT within energy infrastructure. In: Muhanji SO, Flint AE, Farid AM (eds) EIoT: the development of the energy Internet of Things in energy infrastructure. Springer International Publishing, Cham, pp 27–90. https://doi.org/10.1007/978-3-030-10427-6_3

Mutually Evolving Technologies: Blockchain, Renewable Energy, and Energy Storage (n.d.) https://www.americanbar.org/groups/business_law/publications/blt/2019/12/evolving-tech/. Accessed 15 Jan 2023

Ondiflo Is Using Blockchain to Revolutionize the Oil and Gas Industry | AWS Startups Blog (2019, August 28) https://aws.amazon.com/blogs/startups/ondiflo-is-using-blockchain-to-revolutionize-the-oil-and-gas-industry/

Pal S, Dorri A, Jurdak R (2021) Blockchain for IoT access control: recent trends and future research directions. arXiv. https://doi.org/10.48550/arXiv.2106.04808

Perrons RK, Cosby T (2020) Applying blockchain in the geoenergy domain: the road to interoperability and standards. Appl Energy 262:114545. https://doi.org/10.1016/j.apenergy.2020.114545

PR Newswire (2022 November 15) "Purewest Grows Market For Certified Gas Through Partnership With Earndlt And Project Canary To Tokenize Verifiable Environmental Attributes." https://www.yahoo.com/now/purewest-growsmarket-certified-gas-140000925.html
Revolutionizing Renewable Energy Certificate Markets with Tokenization IBM Supply Chain and Blockchain Blog (n.d.) https://www.ibm.com/blogs/blockchain/2021/08/revolutionizing-renewable-energy-certificate-markets-with-tokenization/. Accessed 15 Jan 2023
Saraji S, Borowczak M (2021) A blockchain-based carbon credit ecosystem [White Paper]. http://arxiv.org/abs/2107.00185
Sharma V (2019) An energy-efficient transaction model for the blockchain-enabled Internet of Vehicles (IoV). IEEE Commun Lett 23(2):246–249. https://doi.org/10.1109/LCOMM.2018.2883629
Song L, Ju X, Zhu Z, Li M (2021) An access control model for the Internet of Things based on zero-knowledge token and blockchain. EURASIP J Wirel Commun Netw 2021(1):105. https://doi.org/10.1186/s13638-021-01986-4
Soto EA, Bosman LB, Wollega E, Leon-Salas WD (2021) Peer-to-Peer energy trading: a review of the literature. Appl Energy 283:116268. https://doi.org/10.1016/j.apenergy.2020.116268
Stoll C, Klaaßen L, Gallersdörfer U (2019) The carbon footprint of bitcoin. Joule 3(7):1647–1661. https://doi.org/10.1016/j.joule.2019.05.012
Teufel B, Sentic A, Barmet M (2019) Blockchain energy: blockchain in future energy systems. J Electron Sci Technol 17(4):100011. https://doi.org/10.1016/j.jnlest.2020.100011
Zhang C, Jianzhong Wu, Zhou Y, Cheng M, Long C (2018) Peer-to-peer energy trading in a microgrid. Appl Energy 220:1–12. https://doi.org/10.1016/j.apenergy.2018.03.010
Zhang A, Zhong RY, Farooque M, Kang K, Venkatesh VG (2020) Blockchain-based life cycle assessment: an implementation framework and system architecture. Resour Conserv Recycl 152:104512. https://doi.org/10.1016/j.resconrec.2019.104512

Dr. Soheil Saraji is an associate professor of Energy and Petroleum Engineering, an adjust professor at the School of Energy Resources, and co-director of the Hydrocarbons Research Laboratory at the University of Wyoming. He has eighteen years of research experience and more than 35 peer-reviewed journal publications in subsurface energy extraction, storage, and carbon geosequestration. Furthermore, Dr. Saraji is a pioneer in applied blockchain research for the oil and gas industry. He has developed new courses and research initiatives on this topic at the University of Wyoming.

Chapter 6
Reducing Methane Emissions

Bertrand Williams Rioux

6.1 Introduction

The Global Methane Tracker published by the International Energy Agency (IEA) reports that around 30% of global warming is a result of methane emissions, with the energy sector representing a significant share (IEA 2022a).[1] Methane is a potent GHG with a much stronger warming potential than Carbon Dioxide (CO_2). Wasted methane also represents a wasted energy commodity as the primary component of commercial natural gas. It is estimated that the methane wasted by the oil and gas industry annually has a market value of tens of billions of U.S. dollars. However, this requires investing in existing and new technologies that come with real financial risks.

A range of incentives is needed to address this waste, requiring cooperation across several stakeholders in the oil and gas industry. Data providers specializing in measuring and monitoring emissions provide essential metrics to coordinate action, from devices installed within oil and gas facilities to independent satellite monitoring systems. Standard-setting agencies and professional auditors provide independent verification services. They impart trust onto metrics to impact the production, trade, and investment in oil and gas commodities. The actors and market forces that can leverage these metrics are:

- Providers of emission performance certificates used in voluntary markets for sustainable commodities. S&P Global Methane Performance Certificates & the MiQ standard.

[1] Other important sources of methane emissions include agriculture and human waste.

B. Williams Rioux (✉)
Two Ravens Energy & Climate Consulting Ltd, Calgary, Canada
e-mail: bertrand@tworavens.consulting

S. Saraji and S. Chen, *Sustainable Oil and Gas Using Blockchain*,
Lecture Notes in Energy 98, https://doi.org/10.1007/978-3-031-30697-6_6

- Investors in the oil and gas industry that prioritize Environmental, Social, and Governance (ESG) factors linked to emission performance metrics, such as Sustainability Linked Bonds issued by energy producers.
- Governments, legal bodies, and international organizations that regulate suppliers and coordinate voluntary reduction efforts.

This chapter speaks directly to the interests of these stakeholders and the oil and gas producers. It will cover three topics related to the creation of a measurement economy built on environmental data to create actionable metrics for methane waste reduction. First, we define the scale of the problem and the different organizations working on it. Second, what is being done to solve the problem will be explored, including technical innovation, voluntary initiatives, government regulation and legal action. Finally, the where and how elements and features of Blockchain can be applied to the governance of emissions data, and the development of the measurement economy to help reduce methane waste, will be discussed.

Blockchain offers a framework that can support shared accounting and transaction networks built on independently audited emissions data. We discuss applications of blockchain and the broader Distributed Ledger Technology (DLT) space to identity management, verifiable credentials, and smart programmable contracts used by stakeholders in the oil and gas supply chain. As discussed in earlier chapters, a distributed network is used to execute "smart contracts" between stakeholders without a central counterparty. This can help coordinate decisions across groups with different, and potentially, competing interests, while supporting transparency and traceability.

We discuss how blockchain can be applied to the aggregation of emission data from a growing number of sources. This includes improving access to and interoperability of data related to the operation of oil and gas facilities. Furthermore, we explore how Decentralized Autonomous Organization (DAO), a tool for peer-to-peer governance and decision-making, can be applied to the verification of emission data for the measurement economy. Finally, we discuss applying blockchain to the tokenization of emission data that can be integrated into sustainable commodity markets and financing.

6.1.1 Oil and Gas in a Net Zero Future

The path to a net zero future is complicated by the dependency between the global economy and oil and gas supplies that make up a large share of GHG emissions. They are vital sources of energy and value creation for the global economy needed for the energy transition. In addition, they provide essential feedstocks required by industry to build the physical infrastructure for a net zero future, in particular, energy-intensive industries like cement, steel, aluminum, and fertilizer production that cannot be easily decarbonized, but also to produce solar panels, wind turbines, and other sources of clean energy. Figure 6.1 shows the projected demand for oil and gas until 2050 in the U.S. from the reference scenario of the Annual Energy Outlook published by the

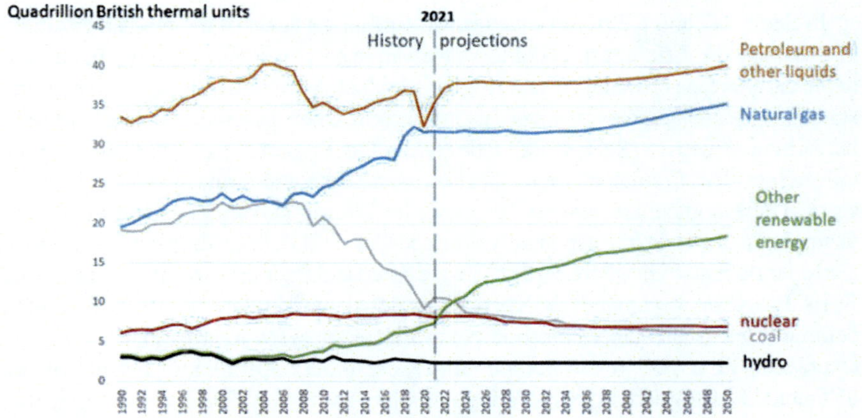

Fig. 6.1 U.S. energy consumption from the Annual Energy Outlook 2022 reference scenario. *Source* Annual Energy Outlook 2022 EIA (2022)

Energy Information Administration (EIA). Given the strategic importance and role of oil and gas in the coming decades, the first step in moving towards Net Zero is reducing as much waste from their production as possible.

As discussed by Daniel Yergin, the Pulitzer Prize-winning author of books about the global energy landscape, decoupling from oil and gas will not be straightforward (Yergin 2021). The net-zero targets set by many countries require optimizing the use of existing oil and gas assets. As discussed in this chapter, blockchain can help coordinate collective action by stakeholders to achieve this effort.

6.1.2 Oil and Gas in the Measurement Economy

The measurement economy is playing an increasing role in shaping the function of the oil and gas industry. Measuring and tracking the primary physical attributes of fuels was vital to the formation of oil and gas commodity markets in the twentieth century. This included grading fuels based on impurities and density, leading to terms like light sweet crude oil and sour gas. These characteristics are important in describing how fuels can be used and are valued. Moving into the twenty-first century, secondary environmental attributes are playing an increasing role in the trade of commodities. These attributes include the GHG emission intensity of fuels, such as the percentage of methane emitted per volume of natural gas produced or flaring intensity per barrel of oil. Since these qualities are not directly transferred with the physical commodity, a new set of standards has emerged to verify claims within the measurement economy.

In Sects. 6.2 and 6.3 of this chapter, we outline the scale of the methane problem and identify the different stakeholders involved in solving it, respectively. In addition to the various data providers measuring the problem (Sect. 6.3.1), we discuss different standard-setting agencies involved in certificate markets based on methane emission performance metrics (Sect. 6.3.2). Companies like Project Canary and MiQ provide tailored services to support the verification of methane emissions by oil and gas operators. These companies provide standards for issuing performance certificates, for example, in the trade of Responsibly Sourced Gas (RSG).[2] Standard-setting agencies also operate registries where certificates are listed and traded; for example, the Digital Fuels Programs organized by Xpansiv, a global marketplace for ESG-compliant commodities. Sustainable commodity certificates provide metrics that can also be leveraged by the ESG investor community (Sect. 6.6.3). Specifically, for the creation of financial instruments incorporating emission-based Key Performance Indicators (KPIs) that impact the access to and cost of capital for projects. The ESG community needs reliable and consistent metrics to make informed decisions about funding the oil and gas industry. Raising capital for sustainable fuel projects extends across the value chain into the production of low carbon or carbon-neutral Liquified Natural Gas (LNG) and hydrogen fuel. While these projects promise cleaner fuel sources, they face the challenge of verifying emission reduction across the supply chain. For example, the liquefaction and regasification of LNG or hydrogen production by steam methane reforming. New certification practices and standards are crucial to advancing these efforts.

The enforcement of existing and new policies by regulators are crucial to achieving global methane reduction targets, the topic of Sect. 6.3.5. By leveraging existing legal agreements, landowners can also help pressure the oil and gas industry to reduce methane waste (Sect. 6.3.6). Other voluntary initiatives with interest in oil and gas operations also support this effort (Sect. 6.3.7), such as the World Bank's Global Gas Reduction Flaring Partnership, or the Global Methane Pledge. The United Nations Environment Programme has also introduced The Oil & Gas Methane Partnership 2.0, a measurement based reporting framework used by producers to compare progress across the oil and gas industry.

The solutions available to the oil and gas industry to prevent methane waste are covered in Sect. 6.4. Many of these can be deployed at no net cost by selling the captured methane as natural gas. However, this often comes with market barriers and economic risks that both raise the marginal production cost of producers and complicate the financing of reduction efforts.

Section 6.5 discusses how advances in blockchain-based governance systems and transaction networks can help advance the measurement economy for methane abatement by converting secondary environmental attributes into digital assets verified by trusted professionals. Section 6.5.1 Looks specifically at the use of blockchain for

[2] RSG is a term originally trademarked by Project Canary describing efforts to validate steps by producers to align with or outperform industry benchmarks by reducing methane emissions, flaring operations, as well as other impacts on land and water across the fuel supply chain. Project Canary also announced that it was giving the trademark to the large energy community (Project Canary 2022).

methane data governance in three areas: (a) the creation of verifiable credentials assigned to trusted data providers and verification professionals, (b) creating a more efficient marketplace for secondary environmental attributes of oil & gas products, (c) the automation of verification and monitoring systems. The second area addresses challenges in improving accessibility and transparency while protecting the commercial interests of energy companies through the aggregation of emission and energy production data.

Section 6.5.2 discusses how the participation of the government, including laws and regulations adapted to and developed for blockchain applications, impacts its use in the measurement economy. For example, tokenizing verified emission data for use in sustainable commodity markets. In Section 6.6, we discuss how digital emission assets can be constructed to facilitate the trade and tracking of RSG certificates. Institutional oversight and enforcement are crucial to addressing greenwashing claims, which attempts to hide or underestimate the environmental impacts of a project. Institutional investors that apply strict ESG reporting requirements could leverage the transparency and immutability features supported by blockchain. Section 6.6.3 covers how digital emission assets could be used by ESG investors to monitor oil and gas projects. This includes organizing financial instruments that target emission reductions, such as sustainability-linked loans and bonds, to influence a company's access to and cost of capital.

6.2 The Scale of the Methane Emissions Problem

There are two major sources of waste GHG emissions by the oil and gas industry associated with unused methane during the production, processing, and distribution of oil and natural gas. These are emissions not associated with the extraction of useful work, such as carbon dioxide (CO_2) produced during the combustion of fuel. Our definition of waste methane emissions by the oil and gas industry excludes the sector's own use of fuel, direct or indirect. The first and largest source of waste emissions comes from the leakage, and intentional venting, of methane gas. Methane is a potent GHG with a global warming potential (GWP) that is multiple times that of CO_2.[3]

The second source comes from routine flaring, most often during the extraction of associated natural gas from crude oil wells. When there is insufficient infrastructure to process and market the natural gas to consumers, producers will often reinject it into the oil field to improve oil recovery. When the natural gas cannot be reinjected or processed, it is combusted in a flare stack as a waste product. This emits large amounts of carbon dioxide (CO_2), but significantly reduces the global warming potential of

[3] Depending on the time scale considered, the emission impact of 1 kg of methane (CH_4) relative to 1 kg of CO_2 can range from 80 times over a 20 year period, to 30 times over 100 years. While methane is a more potent GHG, it has a much shorter residency time. Each molecule will naturally convert into CO_2 in the atmosphere in 8 years on average.

venting the methane. Depending on the combustion efficiency, flaring also leads to direct methane emissions. According to the IEA, global combustion efficiency of flare stacks is 92% (IEA 2022b). However, poor performance standards by some producers and weak enforcement of regulation by the government can result in flares going unlit for extended periods leaking methane gas into the atmosphere.

The World Bank and the IEA estimate in 2022 show that more than 242 billion cubic meters (bcm) of methane extracted by the oil and gas industry were released into the atmosphere as waste. This includes gas flaring operations estimated at 144 bcm by the World Bank in 2021 (World Bank 2022a). In addition, 98 bcm of direct methane emissions (venting and leakage) are reported by the IEA (IEA 2021). As the primary component of natural gas, a valuable energy commodity, this represents a wasted input to the global economy and a lost commercial opportunity. As reported by the Environmental Defense Fund (EDF), the market value of all flared gas in 2020 was roughly $15 billion, based on a natural gas price of $3 per million British thermal units (MMBtu) (~$0.10 $/m^3) reported in U.S. Henry Hub spot market (EDF 2021). This price can vary significantly, exceeding $8/mmbtu in the second quarter of 2022 (YCharts 2022). Spot prices in the international trade of Liquified Natural Gas (LNG) exceed $10/MMBtu and have reached peaks of more than $50/MMBtu in the Asian markets (International Monetary Fund 2022). The takeaway here is that methane emissions from both leakage and flaring represent billions of dollars of wasted economic resources.

While reducing methane emissions has a cost, in many cases, these can be covered by the revenues generated from selling natural gas. Based on the IEA's Global Methane Tracker, the abatement potential of methane emissions by the oil and gas industry is 71%, or 57 million tons (mt) (IEA 2022c). It is estimated that 58% of the technical abatement, or 33 mt can be achieved at no net cost. In other words, the value associated with substituting gas purchases, can more than compensate for the cost of capturing the methane.[4]

Gas flaring also has real abatement opportunities. While 25% of flaring occurs as routine safety measures, the remaining 75% are non-routine events that can be avoided with the right investment. The World Bank estimates that the cost of ending routine flaring could exceed $100 billion (World Bank 2022b). The number varies depending on assumptions on location and accessibility to natural gas markets; however, it is estimated that around 40% of total gas flares could be avoided at no net cost. For example, the EDF found that 30% of oil leases in the Permian Basin in the state of Texas could deploy methane recovery solutions at minimal to no net cost, and 32% of leases are responsible for more than 95% of vented and flared gas (EDF 2017). Figure 6.2 illustrates the total methane emissions (bars) and intensity (diamond) among some of the largest oil and gas producers. Figure 6.3 offers similar data but for flaring volumes from oil and gas operations. As reported by the World Bank, the top seven countries (Russia, Iraq, Iran, U.S., Algeria, Venezuela,

[4] The IEA adjusted the abatement cost in different countries by the 5-year moving average of natural gas import prices to determine the net cost.

and Nigeria) account for about 65% of global gas flaring and 40% of oil production (World Bank 2021).

The challenge is that methane venting and flaring are distributed across many geographic locations and governance structures. Some of the highest percentages of flared and vented methane occur in Eurasia and Africa, where regulation is either absent or not enforced. Venezuela, with one of the largest proven reserves of oil, has some of the worst flaring performance, with aging infrastructure and poorly enforced regulation (Berg 2021). In 2020, flaring intensity in Venezuela was more than double the pre-2018 peak. Meanwhile, Norway, a highly developed oil and gas exporting country, has some of the most advanced methane regulations and the lowest reported emission intensities. Achieving the goals of the Global Methane Pledge and the zero routine flaring initiative will require rapidly closing these gaps.

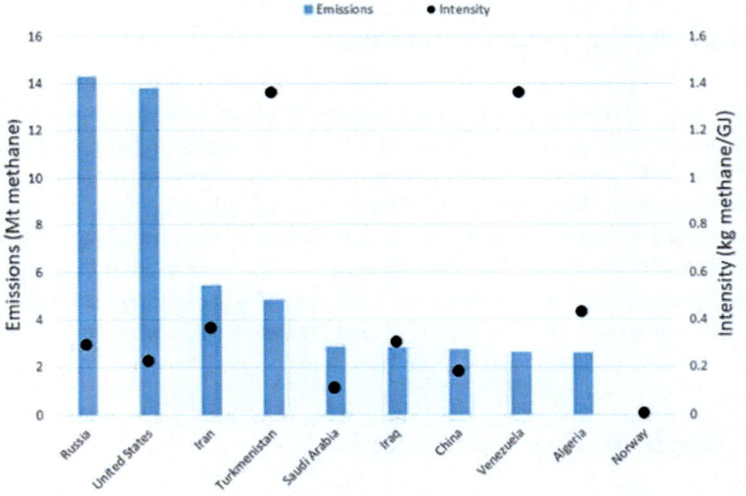

Fig. 6.2 Total methane emissions in million tones (mt) and methane intensity of oil and gas and intensity in kg in methane per gigajoule (GJ) of production. *Source* IEA (2022d)

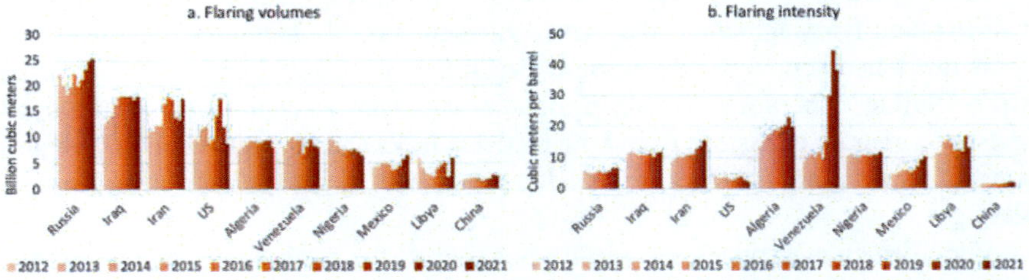

Fig. 6.3 Flaring volumes (**a**) in billion cubic meter and intensity (**b**) in cubic meter per barrel of oil in 10 countries with the largest annual flaring volumes. *Source* Global Flaring Data World Bank (2022a)

Analysis by Flare-Intel, a provider of flare detection services using satellite data, highlights the role that flared methane could play in addressing concerns over shortages in global energy supplies. A report focuses on how natural gas captured from flare stacks in North Africa could supply import demand in Europe using existing infrastructure (Davis et al. 2022). This can contribute to diversifying sources of gas to improve Europe's energy security. The EU relies on Russia for roughly 40% of its natural gas imports (Fisher 2022). Concerns of elevated dependency on Russian natural gas were elevated following the war launched in Ukraine in the winter of 2022. Realizing such opportunities will require a balanced mix of incentives and deterrents: voluntary action by the private sector, government policies, a shift in capital allocation by lenders and international investors, and pressure from public advocacy groups.

6.3 Solving the Methane Problem

A major step forward in solving the methane problem is how we quantify it. A variety of new data service providers have been introduced in just the past decade to improve both internal measurement and independent monitoring of oil and gas industry emissions. These efforts, detailed below, feed into a variety of initiatives and organizations focused on reducing methane emissions by the oil and gas industry. We outline several of these, ranging from the data providers and standard-setting bodies tasked with verifying emission data, and legal action by government and landowners to voluntary programs and markets to advance reduction efforts.

6.3.1 Methane Measurement Tools

Advances in measurement tools for both direct methane emissions and flaring have played an essential role in highlighting the severity of the problem. This includes improvements in onsite detection technologies deployed by oil and gas companies to optimize operations. Independent observation and monitoring tools also play a central role in quantifying aggregate and point source emissions. One of the most powerful tools that has emerged in measuring emissions by the oil and gas sector has come from satellite technology. Visible, ultraviolet, and shortwave infrared spectrograph is used to identify emissions from both methane leaks and flaring. Table 6.1 provides a list of various satellites that are tuned to detect both direct methane emissions and flaring.

Flare detection services use visible imaging of the earth at night to assess the volume of gas being flared globally on a near real-time basis. Night-time detection of flaring data is made possible by the Visible Infrared Imaging Radiometer Suite (VIIRS) instrument aboard joint NASA/NOAA polar-orbiting satellites. The visible

Table 6.1 Satellites for the detection of methane emissions and flaring

Name	Emissions	Resolution
TROPOMI	Venting & Leakage	~5 km (global mapping)
MethaneSAT		~500 m (area mapping)
GHGSat		~50 m (location mapping)
VIIRS	Flaring	~500 m

intensity of the flare can be used to determine the volume of gas at a very high resolution. This data is used by Flare-Intel, a provider of global flare-detection services, to help oil and gas companies to detect and reduce emissions. Satellite-based instrumentation is also used for observing direct methane emissions from intentional venting and unintentional leaks. Methane cannot be seen using the visible spectrum; however, infrared and ultraviolet spectral analysis can detect emissions from space. Coverage and resolution are not quite as advanced as visible flare detection, but new satellites are quickly filling the gaps in global methane surveillance from space.

The European Space Agency (ESA) launched the TROPOMI instrument that uses a combination of different detection bands to detect methane emissions globally with a resolution of around 50 km^2. While not sufficient to support facility-level emission reporting, it is good for regional detection and global mapping. GHGSat, a Canadian cleantech company, has deployed a fleet of satellites capable of tracking methane emissions with a resolution of up to 25 m. With an increasing number of satellites in orbit, GHGSat can provide vital data for real-time measurement of emissions. It is more suitable for facility-scale detection, rather than global mapping. It provides lower detection precision than GHG and focuses on quantifying higher-emission concentrations. Both satellite solutions have been successful in detecting large unreported leaks. MethaneSAT is another program set up as a subsidiary of EDF. With satellites first launched in 2022, it is designed to provide high global precision, proposed as an Area Mapping tool that can bridge gaps between TROPOMI and GHGSat. It has been developed as a public good that can be used for free.

Satellite surveillance tools provide a useful high-level perspective on global emissions that are helping to identify major problem areas and advance global action. Climate TRACE (Tracking Real-time Atmospheric Carbon Emissions) is leveraging such instrumentation, combined with data analytics, to improve the speed and accuracy of detecting man-made GHG emissions (Climate Trace 2021). Their objective is to accelerate the speed and accuracy of climate action.

Ground and air-based emissions detection systems are also used both for regional and facility-level measurement. Permian Methane Analysis Project (PermianMAP), another project set up by the EDF, provides regional estimates of the methane emission in the Permian Basin, one of the largest oil and gas production basins located primarily in the State of Texas. It uses a combination of satellites, helicopters, aircrafts, autonomous vehicles, and towers to assess methane emissions across the region, providing a public source of data (EDF 2022).

Oil and gas producers also deploy their own methane detection and monitoring systems at the well site. Typically, this data is used for operational and safety management, emission inventory book-keeping, and to comply with the regulation. Instrumentation can be operated directly by the producer or by service providers specializing in emission detection. Producers are increasingly integrating production data into cloud data analytical services. This helps producers both track the attributes of marketed products, but also to monitor emission performance, such as combustion efficiency of flare stacks, and leak detection. There are two companies leading the North American market in managing production data. Validere provides cloud-based data management services related to the quality and characteristics of oil and gas production volumes. They also provide solutions to help companies manage emissions performance data linked to their production operations. Project Canary provides tailored environmental assessment solutions using instrumentation they install and operate on-site. This includes monitoring methane emission, as well as land and water impacts used in RSG certification.

6.3.2 Standard-Setting Agencies and Certificate Registries

Given the growing amount of measurement data, a major challenge is addressing differences and potential discrepancies between self and independent monitoring. In many cases, self-reported values are simply not available or have not been subject to third-party review. Standard-setting agencies help fill this gap by playing the role of intermediary at the interface of voluntary action and investor relations. This includes verifying emission data that can then be linked to markets for RSG certificates and financial instruments linked to sustainable energy production (see Sect. 6.3.3). These verification networks can be quite complex and involve coordinating data across multiple data providers to quantify different sustainability metrics. For reference, the providers of various RSG metrics discussed below are listed in Table 6.2.

Table 6.2 RSG certificate programs organized by different providers

Provider	Standard name	Specifications
Project Canary	TrustWell	A variety of criteria covering (e.g., rating, environmental, spill prevention, waste management, etc.) are used to assign a Silver, Gold or Platinum Grade
MiQ	MiQ Standard	A to F grading systems for methane intensity within a specific threshold: (0.05, 0.1, 0.2, 0.5, 1.2%)
S&P Global/Xpansiv	Methane Performance Certificate (MPC)	Methane intensity below a threshold of 0.1% relative to an industry average of 0.437%

While still in the early development stage, these markets have significant growth expectations. According to S&P Global Commodity Insights, an energy commodity information provider, the market for certified RSG in the U.S. was expected to nearly double from an estimated 7 bcf per day (bcf/d) in 2021 to 12.3 bcf/d in 2022, roughly 14% of total U.S. production, with others predicting volumes up to 20% (Hallahan and Corral 2021).

Some of the larger standard-setting agencies have already been introduced in earlier chapters, including Verra and Gold Standard serving a range of emission reduction projects. This section highlights the major providers that focus on emissions by the oil and gas sector. In addition to operating instruments that measure emissions at upstream oil and gas facilities, Project Canary participates in the RSG by offering the TrustWell™ service for independent third-party verification and monitoring (Project Canary 2020). While Project Canary enables precise onsite verification of emission data, it does not operate a registry where RSG certificates are listed and traded. However, it has partnered with Xpansiv, a global market provider for data-driven commodity trading, which operates a Digital Fuels Registry where RSG certificates can be listed and traded.

S&P Global Commodity Insights has also partnered with Xpansiv to provide its own Methane Performance Certificate (MPC) (S&P Global 2021). An MPC is issued to natural gas producers with a methane intensity below the 0.1% threshold. The methane intensity metrics are independently verified by Xpansiv with the MPC issued on their Digital Fuels registry. Methane intensity premiums are assigned proportionately to the percentage reduction in methane intensity relative to an industry average, defined as 0.437% (S&P Global 2022a).

MiQ is another leading provider of methane emission standards that also provides a registry for the certificates independently verified by its own verification network. However, unlike Project Canary, MiQ does not provide onsite methane measurement services, focusing instead on coordinating independent data providers and auditors. It issues a range of methane intensity performance grades (A–F) based on available data verified by its network, with the highest grade awarded to producers within the 0.05% threshold.

As the demand for these certificate markets grows, oil and gas producers have an opportunity to monetize their commitment to sustainable production. The trade of certified natural gas has seen premiums ranging from 1 to 2% of market prices for natural gas (Freitas 2022). S&P performs daily price assessments for the MPCs; prices are converted to a \$/mtCO2e basis by assessing the volume of emissions compared to the industry average. Using the daily MPC prices, S&P reports methane intensity premiums for the cost of reducing methane intensity in different production basins of the U.S. (S&P Global 2022b).

6.3.3 ESG Investors and Sustainable Finance

In addition to using methane performance for RSG certification, the metrics can also be used by ESG investors for sustainable financing. This includes adjusting the cost of capital through KPI-based financial instruments known as Sustainability Linked Loans (SLL) and bonds (SLB). Interest rates are increased or decreased based on emission permanence metrics linked to ESG goals, with interest margins varying from 5 to 10%. These instruments differ from the use of proceeds-based instruments that require capital to be spent on specific investments. SLBs differ from SLLs in that they also require funds to be used for specific investments, with the latter offering greater flexibility. SLBs are typically applied on the downside with a failure to achieve a KPI target resulting in rate increases, while SLLs will be used to adjust rates up and down. Both are labor intensive and can have a noticeable increase in operating costs to comply with rigorous measurement and reporting requirements. However, with proper management of the underlying metrics, improvements in a producer's access to and cost of capital can outweigh the additional costs. The first SLB in the oil and gas industry was issued by Eni, an Italian multinational oil and gas company. It was linked to a target pledging to reduce upstream emissions of the oil and gas producer by 7.4 million tons (Credit Agricole 2021).

6.3.4 Low-Carbon Fuels

In addition to reducing emissions from oil and gas extraction, sustainable finance is playing a role in developing new supplies of low-carbon fuels. Low-carbon fuel standard is a policy instrument that requires visibility across the lifecycle emissions of a fuel supply chain.[5] While all hydrocarbon fuels emit GHGs, some have lower impacts than others. The standards target reductions in the carbon intensity of conventional and new transportation fuels.

Compressed and liquified natural gas has lower emission factors than traditional diesel and gasoline. Biofuels sourced from renewable biomass can significantly reduce lifecycle emissions. In addition to achieving lower direct emissions, the amount of CO_2 produced during combustion, embodied emissions can be reduced during upstream production, including methane venting and flaring, or by recycling CO_2 from the atmosphere downstream (biomass production). Reducing upstream emissions is also essential to produce low-carbon hydrogen. There is growing interest in using hydrogen both as a fuel for transport and as a low-carbon feedstock for hard-to-decarbonize industries.

The challenge is providing verified metrics to reliably document the lifecycle emissions of low-carbon fuel. Standard-setting bodies are tasked with filling the information gaps. Organizations like Verra and the Gold Standard employ career

[5] Low carbon fuel standards were first introduced in California in 2007.

practitioners who are loosely governed by standards established by the International Organization for Standardization (ISO) to verify data and apply Life Cycle Assessment (LCA) methodologies to assess environmental impacts. Each component accounted for in the LCA should have credible information. However, this is often not the case, with many studies relying on historical data that assess conditions reported several years in the past. Plevin et al. (2017) review the LCA methods used to assign emission intensity standards and regulations on fuels in the US, Canada, and the EU. They find that the different models, estimates, and projections often lead to subjective and unverifiable results that compromise the effectiveness of low-carbon fuel standards.

Oil and gas companies looking to increase the production of low-carbon fuels will need to look for solutions to address these gaps. Oil and gas companies are already supplying LNG and compressed natural gas to the global markets. Marketing them as low-carbon or even carbon neutral by packaging carbon offsets with their distribution will require value chain certification of emissions to fully confirm their role as low-carbon fuels. LNG production, processing, and distribution can account for up to 33% of total lifecycle emissions. In the case of hydrogen production, 100% of GHG emission occurs upstream during the steam methane reforming process.

6.3.5 Methane Regulation

Government efforts to regulate methane emissions have faced mixed success. Among oil and gas producing countries with more advanced economic development (e.g., wealth per capita), strict enforcement of methane regulation is not the norm. The oil and gas sector in Norway and the Netherlands have achieved the lowest methane intensity metrics with some of the strictest regulations. In many other less-developed economies, regulations are simply not enforced or completely absent, with some governments blocking the influence of environmental advocacy groups. As pointed out in Figs. 6.2 and 6.3. several countries fall significantly behind the good performers, in some cases more than 100 times the emission intensity of the lowest emitters. Creating effective regulations and policies to overcome these challenges requires implementing the necessary institutions (and measurement tools) to ensure accurate monitoring. It also requires balancing policy enforcement against impacts on the cost of producing oil and gas and energy supply security.

Climate policy has often focused on the polluter pays principle to address a range of GHG emissions. Emission pricing is a common approach to regulating GHG emissions, from a carbon tax to Emission Trading System (ETS), where emission allowances are auctioned to industry. While several national, regional, and state-level carbon markets do exist in many countries, few jurisdictions have introduced direct taxation on methane emissions by the oil and gas industry. Norway has set the example around methane regulation, with several policies directly targeting emissions, including the CO_2 Tax Act on Petroleum Activities and the Greenhouse Gas Emission Trading Act. In the U.S., many of the oil and gas production basins are not covered by existing state-level emission trading systems, and no federal carbon pricing mechanism has been introduced by the government.[6] Even the longest-running ETS in the European Union (EU) does not cover methane emissions (Lewis et al. 2022). However, the EU ETS does set requirements for producers to report flaring emissions, which are regulated within the EU.

Other market-based mechanisms have been investigated as potential mechanisms to price methane emissions. One example is severance taxes charged on natural gas extraction, which could be applied to all forms of methane, whether marketed, flared, or vented. While these taxes exist in most U.S. states that produce oil and gas, most allow for exemptions on the release of methane emissions. There have been a few cases where pricing was introduced, with some states exploring the use of severance tax to address rising environmental concerns on flaring and venting practices during the boom in U.S. tight oil production (Rabe et al. 2020). However, the research shows that exemptions on methane emissions have remained the norm with industry opposition, as opposed to reduction.

The U.S., a large methane emitter and Global Methane Pledge participation, has introduced a range of methane regulations for the oil and gas industry. At the Federal level, the Environmental Protection Agency (EPA) proposes rules under the Clean Air Act. These include new initiatives for monitoring oil and gas infrastructure and new low-emission standards to eliminate venting, requirements to capture and sell gas if the infrastructure is available and offering emission compliance options based on the use of advanced LDAR technology (EPA 2021). In addition, the Build Back Better Act of the Biden Administration has introduced a methane fee that could cost the oil and gas industry more than $39 billion (Czapla 2021). The specification of the methane fee were introduced under the Inflation Reduction Act approved in 2022.

The European Commission has also introduced a proposal to regulate methane emissions under the hydrogen and gas market decarbonization package (European Commission 2021). The package sets high standards for the measurement, reporting, and verification (MRV) of methane emissions, mandatory LDAR, and bans on routine venting and flaring. The proposal is believed not to have a significant impact on energy prices; however, it also does not specify any direct pricing mechanism. The proposed gas package focuses on restricting emission quantities with no methane pricing, as in Norway. Even the recently drafted EU carbon border adjustment tax targeting embodied emissions in commodity imports does not target oil and gas industry products (Rioux and Chen 2022).

An important factor in the regulation of the oil and gas industry is verifying current operations and improvements. However, the EPA has faced criticism that the GHG warming impact from methane released from U.S. oil and gas wells was at least 50% higher than its estimates (Penn State 2021). Furthermore, the U.S. government has

[6] Regional markets include the California Air Resources Board (CARB) cap and trade market, and the Regional Greenhouse Gas Initiative (RGGI). The RGGI consists of Connecticut, Delaware, Maine, Maryland, Massachusetts, New Hampshire, New Jersey, New York, Rhode Island, Vermont, and Virginia.

recently made efforts to roll back methane emission rules and statutory authorities, reducing the environmental oversight on producers (Vizcarra 2020).

Russia, one of the world's largest oil and gas producers, has faced some setbacks in achieving its target to utilize 95% of all associated gas to reduce gas flaring (Kobzeva et al. 2021). This official government policy, designed to minimize the waste of natural gas by state-run energy companies, has likely been undermined by other political priorities. While Russia has endorsed the Global Methane Pledge, 2021 flaring data suggest that both Russian total flaring and flaring intensity are on the rise (Davis 2022). In addition, spikes in flaring and large methane leaks are observed across North Africa and South America in countries with significant oil and gas reserves, where political instability can undermine emission governance. While these regions rely on international investors and oil and gas companies from countries with strict environmental regulations, standards do not generally follow them across borders. Improving access to essential emission data is needed to help governments identify lost opportunities from the release of methane during the extraction of fuel by international interests. For example, ensuring leaseholders pay severance tax on extracted natural resources.

The identification of inconsistencies in regulatory oversight is facilitated by increases in independent measurement efforts. Ensuring that governments are sufficiently regulating methane emissions will require close oversight by public initiatives using the range of instruments and services discussed in Sect. 6.2 (e.g., EDF's PermianMAP and Climate TRACE).

Legal requirements on investors are another area where emission governance is gaining traction. Canada has announced that ESG reporting will be mandatory starting in 2024 in line with the guidelines of the Task Force on Climate Related Financial Disclosures (TCFD) (Chell et al. 2022). The Security Exchange Commission (SEC) in the U.S. has also taken action to address increasing investor demand in ESG reporting. The TCFD covers governance, strategy, and risk management, but also metrics and targets. Methane performance data represent a direct, and increasingly available tool that governments and regulators can use to enforce ESG reporting.

6.3.6 Landowners and Royalty Laws

Landowners and royalty obligations have also been used to target oil and gas sector emissions when waste emissions impact their royalty interests. Royalties typically do not cover venting and flaring of methane which are reported as waste products by the industry. However, depending on existing regulations and laws, landowners can attribute lost royalties to poor emission management by the leaseholder.

A study on the taxation of flaring and methane emissions reported that landowners had voiced concerns over costs being deducted before paying out royalties (Rabe et al. 2020). This includes untaxed methane flaring and venting activities, a common practice to approve exemptions on emissions regulation in states with methane legislation that could impact royalties. Cases have been raised where landowners leverage other legal terms enforced on a lease. For example, in Texas, a landowner can take a producer to court for failing to act prudently if venting and flaring gas impact their financial interests (Wells 2016). Independent methane measurements could be used as evidence in a court to defend the interests of the landowners. This includes requiring producers to pay for wasted gas to incentivize better emission management practices.

There are also situations where two producers hold leases to extract oil and gas from the same natural reservoir. Cases have been reported of a producer intentionally pumping (and flaring) associated gas out of a reservoir to create negative pressure that increases the extraction of oil. Such activities can be verified through satellite flare detection. This can reduce extraction rates by other producers operating on the same reservoir, in addition to wasting valuable resources and reducing emission performance. Publicly verified data could be used as evidence in local or possibly international courts to take legal action against competing producers that are intentionally spoiling national resources for economic gain. Such cases would be complicated when activities occur on international borders. Multilateral agreements and commitments by countries that share access to oil and gas reservoirs could facilitate the enforcement of international laws to combat these wasteful practices.

6.3.7 Voluntary Commitments

Several voluntary commitments by oil and gas producers, international organizations and new market initiatives are leading efforts to address methane waste. The Global Methane Pledge was announced at the international climate conference, COP26, in Glasgow in November 2021. Led by the EU and the US, it brings together 111 countries pledging to collectively reduce methane emissions by 30% relative to 2020 levels. A 75% reduction in methane emission by oil, gas, and coal production aligns with IEA's Net Zero by 2050 scenario and would come close to achieving the 30% global methane reduction (IEA 2022e). However, this would require a significant reduction from countries that have not joined the pledge. The pledge is voluntary and provides a leading effort to fill the gap in guidance on reducing methane emissions.

Regarding gas flaring, the World Bank has established the Global Gas Flaring Reduction Partnership (GGFR). It is a multi-donor trust fund bringing together government, energy companies and other multilateral organizations to end routine flaring. It initiated the Zero Routine Flaring (ZRF) initiative, which started in 2015 with the ambition of ending routine flaring by 2030. As a result, many countries have made significant progress in reducing flaring by investing in infrastructure to monetize the otherwise utilized gas. However, non-routine flaring is still a common practice, and many opportunities remain due to a lack of both regulation but also access to funding, especially in under-developed regions where political instability can limit access to capital. Achieving the ZRF's goal will require strong cooperation across all stakeholder groups to fill these gaps. Collective decision-making is needed to coordinate investment by not only increasing regulation but enabling the types of incentives and deterrents discussed above to encourage producers to phase out venting and flaring practices.

The Oil and Gas Climate Initiative (OGCI) is a consortium of national and international oil companies that have issued their own methane emission targets. The OGCI has also made voluntary commitments to help fund decarbonization efforts, including the spread of methane abatement technologies. The actions taken by the group are a move to anticipate future regulation to get ahead of the curve. However, it also allows companies to cooperate in realizing strategic advantages to solve an increasingly difficult problem. One of the internal metrics employed by members is the collective methane intensity of production self-reported by the members.[7] The challenge with reporting data under this voluntary initiative lies in how companies report internal emissions, including the boundaries they apply to methane reporting. This can make independent verification difficult. For example, undetected and unreported leaks from a facility may not be accounted for internally but would be detected through independent monitoring systems. Even when such events are considered, many OGCI members only account for emissions from wholly owned facilities, omitting joint projects with other companies. Such discrepancies can further complicate independent assessments.

Organizations like the OGCI, multinational efforts like the Global Methane Pledge, and multilateral organizations under the World Bank's zero routine flaring initiative represent leading voluntary initiatives to address methane emissions. Shared accounting, verification, and trading systems represent an additional tool these organizations could use to help coordinate voluntary commitments and strengthen market forces within the measurement economy.

6.4 Reducing Methane Emissions

This section details the steps the oil and gas industry can take to reduce methane emissions. In many cases, the marginal cost can outweigh the commercial benefits for the producer. As a result, the various incentives and deterrents, the "carrot" and the "stick", as described in the previous section, are needed to achieve a measurement economy that can achieve emission reductions. The carrot provides positive reinforcement, rewarding producers for lowering emissions, such as through the trade of RSG certificates.

There are two parts to reducing emissions. The first involves installing equipment to prevent and capture methane from being leaked into the atmosphere. A direct solution involves avoiding leaks by replacing old and worn-out equipment, such as out pumps and seals, and installing more efficient electric motors in oil and gas facilities. New devices can also be used to prevent venting operations, including vapor recovery units, and installing new flares, and Leak Detection and Repair (LDAR) ensures

[7] Methane intensity reports the percentage of methane emissions relative to total marketed production.

events are identified early on. The second part involves investing in equipment used to bring captured methane to markets. When re-injecting methane back into the well is not effective or simply not possible, selling the methane for productive industries avoids the need for non-routine flaring. However, such investments add an extra layer of complexity, and financial risk, depending on the location and production cycle of associated gas. For example, how remote the oil and gas fields are and the technical challenge of transporting captured methane to demand centers.

Figure 6.4 shows the global abatement cost curve for preventing and capturing methane leakage and venting using different technologies from the IEA's Global Methane Tracker. The curve is adjusted to account for regional natural gas prices for captured methane that can then be marketed to consumers. The negative costs illustrate where there are real commercial incentives to sell the otherwise unutilized waste rather than sending it to a flare stack. A typical justification for the non-routine flaring is the absence of local gas collecting and transportation infrastructure. Natural gas is commonly distributed in pipelines and if a well pad is not near access a regional pipeline grid flaring (of associated gas) becomes a more economically feasible solution when extracting oil. This is simply because investing in additional pipelines is not always economical. This depends on various techno-economic features of a project. For example, how far it is from a demand center, how complex the terrain is, and the expected utilization of the infrastructure can affect the project's economics.

Companies can also introduce alternative onsite processing options to convert the gas into another commodity that is easier to distribute. One option is onsite electricity production, by combusting the methane in a gas turbine. Other portable onsite solutions involve transforming methane into products that are easier to transport on existing road or rail infrastructure, such as compressed and liquefied natural gas. Methane can also be easily transported by converting methane molecules into other liquid fuels (GTL) or chemicals (GTC), such as methanol and ammonia. GTL uses

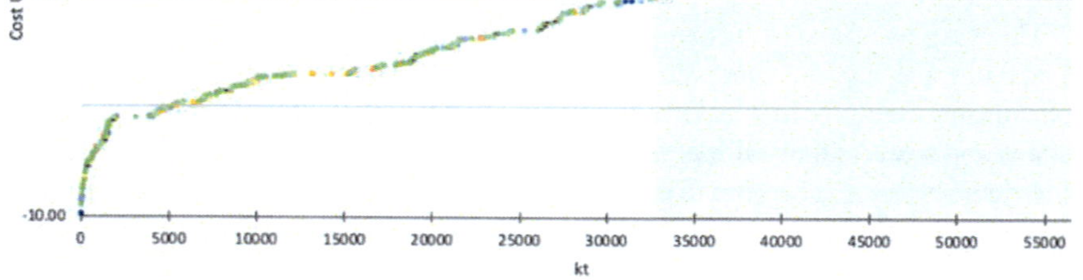

Fig. 6.4 Methane abatement potential. *Source* IEA (2022c)

the "Fischer Tropsch" process with commercialization by oil operators, including Shell and Sasol (Saunier et al. 2019). Recent innovation promises smaller modular plants to allow for mobile deployment at the wellhead. Carbon Limits SA identifies more than 12 companies that improve GTC processing (Pederstad et al. 2015).

Other innovations supporting the utilization of otherwise flared and vented methane target information technology and data processing. Mobile data processing units are being delivered to extraction facilities. They provide gas turbines installed in shipping containers that convert otherwise wasted gas into electricity used to power advanced on-site computation systems and data centers. Crusoe Energy is one company operating these facilities across the US, including portable units that can be used to mine cryptocurrency at the flare stack, which is then sold for a profit. This represents one-way digital money created using blockchain technology as an incentive mechanism to reduce methane flaring and venting. These units access a low cost of fuel to produce electricity below available utility prices and generate returns that can be more than double natural gas commodity prices in some cases, with no investment in distribution infrastructure (Bedolla et al. 2020).

A growing body of research has been introduced on the environmental impact of cryptocurrency mining. This casts some doubt on using this as a methane abatement strategy by the oil and gas sector. The optimal allocation of energy, including how it is used internally by a company or the customers it supplies, is a complex issue. As a profit-maximizing incentive, data processing and crypto mining represent one of many options oil and gas producers can use to generate revenues to finance the abatement of methane waste in remote areas in the short term. These revenues can be used to invest in more difficult methane abatement solutions, direct air capture or other carbon removal solutions. However, cryptocurrency mining is not a sustainable longer-term solution to addressing flaring emissions as a very energy-intensive practice. Delivering the otherwise flared gas to traditional natural gas products that would help displace additional extraction contribute more directly to reducing oil and gas industry emissions.

Sizing of infrastructure to the amount and rate of gas produced over the project's lifetime and mobility are important risk factors to consider when planning to invest in emission abatement. Methane venting and flaring operations can occur in highly variable cycles. In some very short-term tight oil plays, most flaring may occur during the initial 6-month production period. Therefore, processing all the flared gas would require investing in large facilities used for short periods of time at a given production site.

These technical features of oil and gas production are why flexibility and mobility of infrastructure options are crucial. The risks linked to financing such projects are why captured methane is often sent for non-routine flaring or direct venting. Producers must exercise greater agility in planning emission reductions. This requires matching the long run marginal cost of a project against the revenue opportunity of the captured gas.

As illustrated by the negative costs under the methane abatement curve in Fig. 6.4, these opportunities can be highly variable. A significant portion of the abatement solutions creates additional costs for producers even when accounting for additional revenue potential. Investing in the technologies discussed above will require companies to exercise greater capital discipline and agility in planning oil and gas projects. Abatement solutions may still be viable if the producer's profit margins are sufficient to cover the additional cost. However, such efforts introduce financial risks driven by energy price volatility. Price hedging mechanisms can be effective in mitigating this risk, but more is needed to incentivize emissions reduction efforts. This is where the measurement economy can help.

Clearly, oil and gas producers are primary clients of the measurement economy for methane data. First and foremost, producers require this data to optimize internal operations. However, a shared ledger can assist the producer in leveraging data against its interaction with both consumers and investors. One of the challenges faced by the producers is providing a transparent representation of their emissions without undermining the producer's competitive interests and potential trade secrets.

Another risk faced by energy companies is the use of an independent monitoring system by political and environmental advocacy groups to monitor their operations. For example, TROPOMI has been used for the detection of large unidentified methane leaks in major production basins in the United States, Russia, and Central Asia (Fountain 2022). This data can be used with the objective of undermining a producer's ability to operate through the enforcement, or introduction, of new emission regulations and investors that enforce strict climate reporting requirements. The use of the measurement economy as a deterrent, the stick, represents a real challenge that producers must prepare for. This will involve adapting to how new independent methane measurement systems are used. For example, comparing internally reported emission inventories with satellite estimates. These measurements are powerful tools for both the regulator and environmental lobbyists, and producers need to respond to closer monitoring of their operations.

Good actors with a proactive strategy to make use of the increasing amount and sources of methane emission data will be better positioned to identify undetected operational issues and address regulatory and investment risks. Industry support for independent efforts to verify emission data is key to the development of the measurement economy. A distributed system can facilitate this process by balancing the interests of producers and their stakeholders to incentivize emission reductions.

6.5 The Role of Blockchain: Emissions Management and Governance

Producers can leverage emission data to address the financial and regulatory risks associated with reducing emissions discussed in the previous section. The challenge lies in how to manage a growing body of data used across different sets of interests, and opposition, to the oil and gas industry. This is where blockchain can contribute to how methane measurement data is managed. On the producer side, it can help to aggregate and monetize data. On the side of regulators and the public, it can be used for data consolidation to apply pressure on the industry. And, of course, it can facilitate the balance of both the carrot and the stick by connecting the measurement economy to investors, lenders, and the final consumer.

Blockchain networks can help enhance visibility around the verification of emission data and how information is relayed through standard-setting bodies and transferred across relevant counterparties. A shared ledger can also be used to operate emission certificates as digital assets that transfer the cost of emission reduction to consumers as a premium paid for proof of sustainable production. In addition, producers can leverage emission performance metrics with the assistance of blockchain-enabled financial instruments solutions linked to the funding of projects. We need to understand who the stakeholders are and what decisions they are making to understand how and where blockchain can be applied.

Inter-organizational reporting through digital ledgers can create shared advantages for parties on different ends of a supply chain. Individual organizations can share the cost of conducting MRV across the value chain. Upstream suppliers can seek out new opportunities to monetize emission data by increasing visibility on the scope 3 value chain emission impacts for consumers and investors.

Methane venting and flaring represent the very first GHG emissions that occur in the value chain of fuels and derived commodities. The Corporate Value Chain (Scope 3) Standard guides organizations on reporting these emissions. According to the Carbon Disclosure Project, or CDP, which provides one of the largest TCFD-aligned databases in the world, identifying emission reduction opportunities of an organization using Scope 3 reporting has not been very successful (Patchel 2018). Patchel alludes to the role of designed-based approaches in addressing gaps in value chain reporting. This is a distributed problem, and access to information and visibility on the experts that operate verification networks are crucial. Different trusted sources have and will come into conflict. Distributed decision-making coordinated on a blockchain network offers a designed-based approach to facilitate value chain reporting.

6.5.1 *Blockchain for Methane Data Governance*

Leveraging blockchain requires understanding its fundamental design concepts and limitations. Its primary function is to organize consensus in a distributed network. This can be applied to managing operational faults and the occasional unavailability of trusted machines in a network, to more advanced applications that involve preventing malicious attacks by machines that are not inherently trusted and may attempt to undermine the integrity of the network.

The focus on innovation blockchain technology has targeted the latter. This has enabled more decentralized networks capable of protecting transactions and data integrity on a shared ledger without a central counterparty tasked with protecting the network from malicious actors. This provides a form of permissionless (or trustless) consensus, with a variety of applications for financial technology and the enforcement of digital contracts across multi-stakeholders.

The benefits of a blockchain are complicated when decisions, or transactions, between counterparties, are based on information about the real world, such as methane emissions. This requires additional solutions to solve the problem of identifying trusted parties, and sources of data operating within the network. For example, identifying the source of methane measurement, and trusted third parties tasked with verifying it is challenging.

In decision trees used to evaluate if blockchain is a suitable technology to address a problem, a typical question is identifying where there is a conflict of interest or lack of trust. This is clearly an issue for the governance of oil and gas industry emissions, as we have already established. Another question is whether the problem of a lack of trust can be reliably and efficiently solved by a trusted third party. In this case, we need to highlight the role of standard-setting bodies in governing methane emission data. Blockchain applications could be used as part of verification networks that involve the processing of multiple different data sources. Standard-setting bodies and certificate providers would continue to play a role in how the network is coordinated, including the selection of approved auditors. The purpose of the shared ledger is to improve how producers, consumers, investors, and regulators are connected to a larger array of verification solutions.

We identify three areas where blockchain can be applied to the governance of methane emissions:

1. Issuing verifiable credentials for stakeholders and verifiers
2. Aggregating methane data from multiple sources
3. Automation of verification services and monitoring

A Decentralized Autonomous Organization (DAO) represents the formal entity registered within a blockchain network around which these activities are organized. In the case of emission verification for the oil and gas industry, it provides a system for coordinating decisions across producers and verification bodies. The DAO facilitates the creation of certifiable emissions records used by consumers and investors in sustainable commodity markets, but also regulators requiring enforceable data.

6.5.1.1 Verifiable Credentials

At the center of a DAO for emission verification is the identification of professionals that adhere to the principles of methane reporting organized by standard-setting bodies. Verifiable credentials issued by existing trusted agencies allow standard-setting bodies to match industry claims with recognized auditors.

There is a growing interest in blockchain-based identity management systems for digital credentials. One example is the BC Digital Trust technologies developed by the government of British Columbia including a public open-source archive describing a proof-of-concept for issuing verifiable credentials to the BC Oil and Gas Commission (Github 2020). A related project in Canada is led by ATB Ventures, set up by a leading financial institution in Alberta, developing a blockchain identity management solution with the Canadian Government for a National Digital Trust Proof of Concept (ATB Ventures 2022).

These concepts make use of permissioned identity frameworks like Hyperledger Indy for self-sovereign identity, and Hyperledger Aries, a tool kit for managing verifiable credentials. While there is a growing acceptance around the value of digital credentials, getting the oil and gas industry to adopt and recognize them will take time and convincing. An interview with a manager of global talent acquisition for Halliburton Co., a large oil field service company, expressed some reservations about the added value of digital credentials versus traditional credential verification methods. They did recognize that such credentials could help operators verify the competencies of new hires that may not have worked for a major field service operation company, making it more difficult to verify their credentials (Jones 2015).

An area where digital credentials could provide value to oil and gas companies is identifying qualified service providers and auditors for the verification of methane measurements. This is a rapidly growing space with many new entrants. A challenge for standard-setting bodies is ensuring that there are sufficient auditors qualified to independently verify claims by the oil and gas industry. For example, MiQ, a leading standard-setting agency for methane emissions, is constantly hiring and evaluating new qualified auditors to assess claims by its customers in the oil and gas industry. At the start of 2022, MiQ had 10 auditors and was working to increase its network by at least 40% (Borden 2022). Implementing a government-backed digital credential system for institutions that serve oil and gas operators could help scale the onboarding of qualified auditors that server emissions registries like MiQ or performance-based financial instruments.

6.5.1.2 Data Aggregation

In addition to facilitating the identification of specialized roles, permissioned networks can also play a role in how data is aggregated and verified within the measurement economy. The Blockchain Carbon Accounting project, hosted under Hyperledger labs, is developing several tools to support the management of emissions data. This includes work on a distributed data architecture to verify metrics used in the measurement economy depicted in Fig. 6.5 (Chen and Williams Rioux 2022). It can support a variety of GHG management applications such as sustainable aviation fuels, offset credits, and of course, sustainable oil and gas production.

The first component, data aggregation and management, targets the interoperability of emission data with distributed data systems. The data can be supplied from web-based applications such as carbon emission calculators, mobile devices that provide real-time emission data, and the Enterprise Resource Planning (ERP) systems used by the industry. Through various APIs external data providers are connected to a distributed emission verification and monitoring network. It consists of verified credentials, private data channels, and a DAO discussed in Sect. 6.5.1.3. These tools feed into the final stage in Fig. 6.5, the tokenization of emissions. In

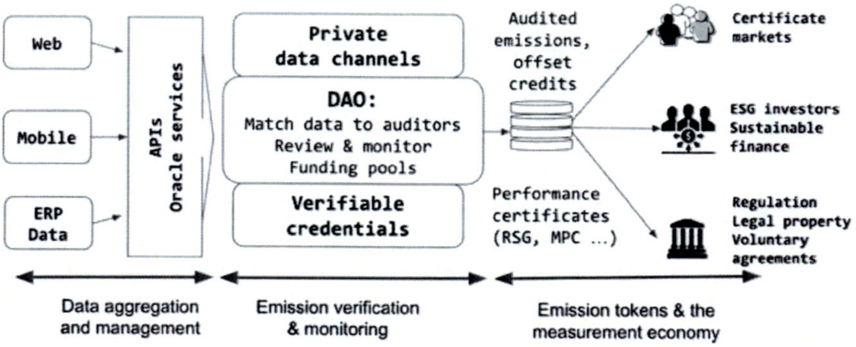

Fig. 6.5 Distributed data architecture for verifying waste emission. *Source* Authors Chen and Williams Rioux (2022)

Sect. 6.4, we discussed several design options for emission tokens and how they can be used within the measurement economy, from voluntary certificate market and sustainable finance to regulation.

Operating the verification network requires connecting trusted information sources that can be used by programmable contracts and the participants of a blockchain network. These are referred to as off-chain solutions, such as Oracle services, that come in different types. Human oracles represent physical organizations or individuals authorized to submit data to the network. Other options relay data from trusted software, such as a web-based service provider, or directly from a physical device installed within a facility, enabled by a hardware Oracle.

There are various options for managing data within the blockchain network. In the Blockchain Carbon Accounting project, Hyperledger Fabric is used to create private emission data channels between verification agencies and providers of data. Fabric is an open-source blockchain framework used for a variety of industrial use cases. Multiple known organizations share a common ledger of data entries and transactions. It can support private data channels between trusted groups to preserve data privacy. Fabric uses a private permissioned consensus mechanism that can support faster transaction speeds and volumes compared to public and permissionless blockchain networks.[8] Fabric is well suited for handling large emission datasets by trusted verification bodies. For example, real-time and high-frequency methane data from onsite and independent emission monitoring systems.

The permissioned configuration of the emission data channel can also be applied to the secure handling of industry data. Emission data provides only part of the information used to issue certificates for energy commodities. For example, Project Canary's TrustWell™ service introduced in Sect. 6.3.2 only provides methane quantities as the numerator in emission factors used to evaluate facility performance. The

[8] A permissionless network allows anyone to participate in the operation of the network (e.g., confirming transactions). This requires decentralized consensus mechanisms that maintain integrity of the network without knowing the other parties in the network, unlike in a permissioned framework.

denominator requires specific data on the quantities and types of fuels produced, i.e., the denominator, to evaluate the underlying emission intensity and determine if the producer is within a certification threshold. Additional providers specializing in production data, such as Validere, are needed to complete the certification process.

6.5.1.3 Automation: Verification and Monitoring

As shown in Fig. 6.5, a DAO acts as a bridge between aggregated data sources and the process of issuing certified emission tokens. The DAO is a smart contract with pre-programmed rules used to facilitate the operation of a verification network. Participants in the DAO include oil and gas data providers, standard-setting agencies, professional auditors, and certificate registries that provide verified metrics and certificates for the oil and gas industry. We define three key functions of the DAO:

1. Match emission claims by industry to auditors with the necessary verified credentials;
2. Review, rate, and monitor services provided by verification agencies;
3. Pool fees from the industry to be paid to auditors for completing verification tasks.

The first point is about setting the rules to identify auditors, for example, by matching verified credentials to verification requests from a specific data source. The DAO can also be used to review, rate, and monitor the performance of participants in a verification network based on services delivered. Voting systems could be set up targeting the reputation of registered auditors and the methodologies they use.

There is a growing body of literature on peer-to-peer self-governance applications using blockchain technology (Wang et al. 2019; Kim et al. 2018). This includes the incorporation of DAOs in the future of regulation (Zwitter and Hazenberg 2020). For example, governments or regulatory bodies could coordinate a DAO for mandatory emission reporting and monitoring programs (Sect. 6.3.5).

Independent observers, including international organizations like the World Bank, that coordinate voluntary commitments could participate in community review. Oversight of emission registries and entities assessing claims is an important challenge as the markets for sustainable commodities, and ESG financing grow. This includes establishing mechanisms to identify underperformers or bad actors in the verification network. A DAO could help standard-setting bodies and regulators identify and minimize greenwashing claims and other efforts to game reporting requirements and certificate registries.

Decentralized governance could also be used to support the onboarding of professionals authorized to submit metrics to a registry. For example, before being authorized to serve data to a registry, specific criteria must be met through participation in the DAO. This could include approval by a recognized standard-setting body or achieving a high reputation by completing verification requests (not serving the registry) for a fee.

Payment to verification agencies is another feature that can be supported by funding pools organized by a DAO. These funds are issued as tokens that participants agree on using to pay each other for services and could include stable coins pegged to the value of fiat currency or the native token of a cryptocurrency. How a fund is accumulated and distributed can be defined by the members. One arrangement is for data providers to post verification tasks by staking a maximum fee threshold. Auditors can be randomly selected based or assigned based on the credentials required. For example, if the data is meant to serve a certificate registry, only auditors qualified to serve it can be selected. The DAO could also be used to increase competition across the fee structure of a verification network.

The features of the DAO help establish credible data points for the measurement economy at scale. This connects the services provided by standard-setting agencies and the registries they operate for emission certificates. Given that a DAO involves a distributed operating model, it may be perceived as a disruptive innovation to current centralized certificate registries and verification practices. Rather than excluding existing agencies, a DAO could be incorporated into their business model; for example, it could reduce the cost and time associated with onboarding and managing members of a verification network and how they are compensated. This can help scale out certificate markets by connecting a growing number of industry (oil and gas) customers to trusted professional auditors in a competitive environment.

The DAO described above offers a mechanism for governing Oracles as sources of trusted inputs. Existing standard-setting agencies could play a role in onboarding these service providers, using a DAO to help scale operations. A more public and decentralized network configuration may be more suitable for the implementation of a DAO than the private data channels used for aggregation. This is because the outputs from the emission verification process apply to a larger number of stakeholders, including participants in sustainable commodity markets and investors (Fig. 6.5). Here accessibility and transparency may be prioritized.

Features supported by blockchain could also help standard-setting agencies navigate commercial privacy concerns and potential trade secrets of energy companies without compromising on transparency. Access to product details provided by a company like Validere, can be restricted only to private channels to protect sensitive commercial data about their products and their customers. Issuing performance certificates requires relating emission data provided by companies like Project Canary, to product quantities. However, this does not necessarily require publishing data about product quantities.

A certificate issued by MiQ for responsibly sourced natural gas verifies that it is produced below a benchmark methane emission factor, 0.05% for stringent quarterly reporting (MiQ 2022). This kind of performance metric can be verified within a blockchain network using a zero-knowledge proof (ZKP). The ZKP uses secure digital signatures to guarantee that a given constraint or threshold is satisfied by data points that are known to have been validated by a trusted auditor. For example, the ratio of publicly disclosed emissions divided by the quantity of fuel falls below the threshold of an RSG certificate known to the public. A ZKP can achieve this without exposing sensitive trade information linked to production volumes. Product quantity data could be securely disclosed to approved counterparties, such as regulators or certificate consumers that need to track the precise embodied emissions linked to the commodity.

6.5.2 Blockchain Law and Regulation

The legal recognition around the use of blockchain-based systems is an important non-technical factor to consider when designing governance solutions. Blockchain law is an emerging concept as the technology is applied to new business and legal applications; however, it is not yet recognized as a field of law. The disruptive nature of blockchain innovation complicates the application of existing legal frameworks to how this technology is used. This has to do with the fact that laws are generally technology agnostics, focusing on the actions of individuals or organizations that use technology to offer products and services.

The agreements established between stakeholders using a smart contract code may be recognized by a legal system. However, the implementation of code does not substitute existing laws, or in other words, code does not supersede the law. In the case of blockchain, there are specific issues related to the location of property and identification of counterparties. Conflict resolution often requires identifying the location of an asset, or the location where an agreement was made. In a distributed network location of digital property, such as a tokenized emission certificate, transactions are confirmed and stored in multiple locations.

Stakeholders will be hesitant to participate in using services based on technology without recognition and definition of terms by legal authorities. Identity solutions based on blockchain infrastructure, such as verified credentials, and administrative tools, could play a role in helping to identify participants in a distributed network for legal compliance purposes. This includes identifying where agreements are signed by counterparties to identify the relevant laws that may be applied. For example, tracking the source of methane emissions data supplied to a data aggregation service or the location of auditors selected to audit the data for a fee.

As discussed earlier, government institutions in Canada are involved in programs using blockchain to create verifiable credentials. Applying these efforts to emission verification services and certificate markets could help overcome legal uncertainties associated with blockchain applications. Representatives of the U.S. Senate have also drafted legislation related to the research and application of blockchain technology (U.S. Congress 2021).

There is also the issue of the legal validity of code and transactions written to a blockchain network. Several U.S. states have introduced laws to support the application of blockchain technology and the enforceability of smart contracts by businesses. The state of Wyoming, a producer of oil and gas, has passed legislation related to the operation of a DAO, and classifying digital assets as legal property (State of Wyoming Senate 2019, 2021).

Given that distributed networks are by design immutable, it may not be possible to change the state of the ledger, such as tokenized methane emissions deemed to be invalid. In this case, other legal resources could be taken if participants operating within the network can be properly identified. For example, flagging, or restricting, the use of transactions and digital assets by registered participants within a regulated certificate market could be a solution.

Another question is how operations coded into a smart contract can be updated to reflect changes in the law. This is an issue of upgradeability of code as opposed to rewriting the state of the ledger. While code deployed to a blockchain network is immutable, smart contracts can be upgraded to implement new or revise existing features. For example, an upgradeable smart contract framework was developed by OpenZepplin using the concept of proxy and logic contracts (Moralis 2022). The proxy is used by clients to access the logic contract. The proxy stores the current address of the logic contract used by clients of the application, which can be updated by a contract administrator. It is designed so that when publishing an update, the state of transactions under previous versions of the contract is preserved while directing users to the new contract logic for future transactions. Such a framework can be used to adapt smart contracts applied to the methane verification and certification landscape. One example is in the implementation of a DAO, where the selection, monitoring, and rating of a verification network evolve over time.

6.6 Emission Tokens and Digital Assets

So far, we have described the role blockchain can play in how emission data is managed. This sets the foundation to use the outputs of the measurement economy for financial transactions, among the most common applications of blockchain technology. The last step, depicted in Fig. 6.5, refers to transforming emission data into tokens and digital assets.

The process of issuing tokens is relatively straightforward with a variety of tools and platforms used today. While the process of tokenizing emissions is significantly enhanced by the distributed data governance systems discussed above, it could be applied directly to the outputs of existing standard-setting agencies, and contents published by certificate registries like MiQ and Xpansiv. The only difference is that these companies issue tokens on centralized databases. The transition to a distributed transaction model is similar to ongoing efforts to digitize financial assets and transactions or for end-to-end tracking for increased transparency in a supply chain (Laaper 2017).

In the following paragraphs, we discuss how tokenization could be applied to performance certificates, tracking emissions in the fuel supply chain, and instruments for sustainable finance.

6.6.1 Performance Certificates

As discussed in Sect. 6.3.2, the markets for methane performance certificates are quickly emerging as part of the measurement economy. These certificates are typically traded in parallel to physical commodity markets. For example, a consumer can buy natural gas on the primary market and RSG certificates from a secondary market provider as a commitment to sustainable production.

This feature of secondary markets for sustainable commodities makes it well suited for operating on a distributed network. Certificates would not have to be integrated directly with existing physical commodity market tools. Buyers could leverage public certificate registries to shop for certificates that satisfy their specific requirements.

There are several emerging solutions targeting the integration of environmental performance certificates within a blockchain network. PureWest Energy, a natural gas production company based in Wyoming has partnered with EarnDLT to issue low methane emissions data verified by Project Canary as digital assets on a blockchain network (PR Newswire 2022). The tokens will be made available for purchase by third-party companies representing their commitment to reduce the embodied emissions associated with natural gas they consume.

The immutability of transactions combined with greater transparency and accessibility provided by a shared ledger creates several benefits by certificate the offtakes. This can accommodate consumers with strict reporting requirements, for example, that require certificates to be bought from the source of production. It also has the flexibility to be used in voluntary markets where a consumer's primary interest is buying certificates to express their commitment to sustainable production in any location.

An ongoing contribution to the Hyperledger Labs Blockchain Carbon Accounting project is working on applying this technology for oil & gas methane certification. It builds on the current Net Emission Token (NET) contract network used by auditors to issue fungible emission tokens to consumers and registered industries (Chen 2022). To apply the NET network to emission performance certification, an emission tracking certificate contract was created. The contract combines multiple NETs with fungible product tokens for the operation of an oil and gas facility. These two tokens reflect the numerator and denominators used when evaluating RSG certificates, as discussed at the end of Sect. 6.5.1.2. They are combined into a unique non-fungible token (NFT) that can be used to describe total emissions and product-specific emission factors (Williams Rioux and Ward 2022).

Figure 6.6 depicts the attributes from an emission certificate issued as an NFT for the Bakken oil and gas production basin in the U.S. mid-western states. The certificate covers production during the year 2020, with total emissions from venting and flaring of methane distributed across oil and gas products measured in tons of oil equivalent (toe) and thousand cubic meters (kcm), respectively.

Emission factors are reported for each product in terms of the total kg CO_2e per unit of fuel. In Fig. 6.7, methane venting and flaring certificates are compared for the

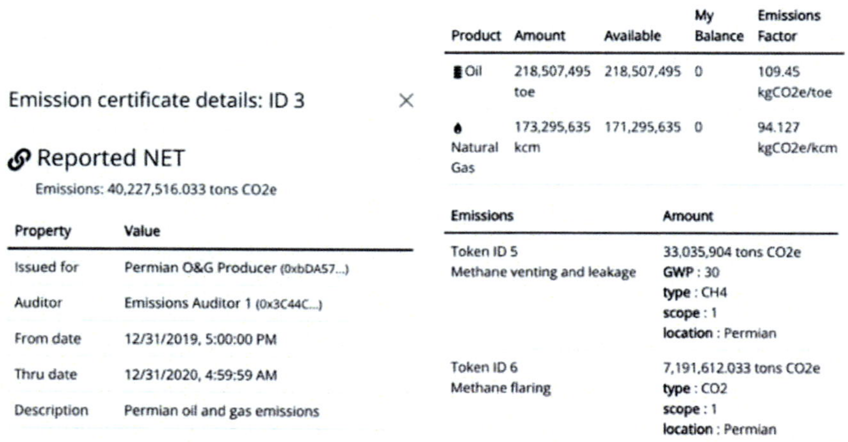

Fig. 6.6 NFT Emission certificate issued for the Bakken oil and gas production basin. *Note* The NET for methane venting (type: CH_4) includes the corresponding global warming potential used, i.e., 30 times that of CO_2 emission. *Source* Authors

aggregate oil and gas production in three basins in the U.S., (Bakken, Niobrara, and Permian), compared to the U.S. national and global average. These figures are from a prototype submission to the 2022 Hyperledger Challenge competition (Williams Rioux 2022).

This simulates a certificate registry where consumers shop for products with explicit metrics. Here emissions are distributed by assigning a proportionate amount of unitless oil and gas product tokens to each certificate. For example, assigning 50 tokens to oil and 50 to gas splits the emissions equally across products. Consumers can then request the holder of the certificate to transfer the product and, by association, the emission tokens.

The units reported in Figs. 6.6 and 6.7. have been converted to physical units. These physical attributes do not need to be posted to the public certificate. For example, an industry can publish unitless token amounts that are distributed to consumers and independently verifiable emission data. However, it may choose not to disclose specific details about the internal production operations, such as physical units of products. Such a certificate could be simplified even further by issuing the certificate using a ZKP discussed at the end of Sect. 6.5.1.3. In this case, the certificate only checks that a ratio of emission and product tokens verified by an auditor, and shared on a private data channel, satisfies the required emission threshold.

These certificates can be beneficial to projects in remote regions where waste emissions are a significant problem and regulation is lacking. Take, for example, regions in North Africa that routinely flare natural gas during oil production. Importers of oil from such regions typically do not take into consideration the embodied emissions from methane emissions and flaring.

ID	Net Emissions (tons CO2e)	Products	Description
5	2,462,743,508 314.526 kgCO2e/toe	4,378,031,482 toe 4,013,926,980 kcm	World Average oil and gas emissions
✓3	40,227,516.033 109.45 kgCO2e/toe	218,507,495 toe 173,295,635 kcm	Permian oil and gas emissions
1	45,532,021.213 551.008 kgCO2e/toe	60,597,652 toe 25,623,756 kcm	Bakken oil and gas emissions
4	430,068,016 262.513 kgCO2e/toe	822,827,823 toe 948,195,526 kcm	U.S. Average oil and gas emissions
2	5,043,360.832 53.371 kgCO2e/kcm	56,765,356 kcm 32,449,236 toe	Niobrara oil and gas emissions

Fig. 6.7 NFT Emission certificates issued for three different oil and gas production basins. Note: Certificates are color shaded to highlight the difference in emission factors with respect to the U.S. average (green for low and red for higher). *Source* Authors

6.6.2 Tracking Emission Reductions

Tracking emission reductions can be applied to a range of stakeholders in the oil and gas industry. The emission certificates discussed above can directly support Scope 3 value chain reporting by the consumers of sustainable commodities. This includes participation in low-carbon fuel standards and certifying the supply of low-carbon LNG and hydrogen fuel.

Suppliers could use emission certification in different ways than consumers with low-carbon footprints. Producers could consider trading emission certificates to reduce the verified methane or flaring intensity. Oil and gas producers could also consider the integration of offset credits in their emission certificates to appeal to markets for carbon-neutral products.

Certificates can also be linked to international efforts to track emission reduction, such as the World Bank's GGFR partnership targeting gas flaring by oil producers (Sect. 6.3.7). The World Bank has set up the Imported Flared Gas (IFG) Index as a measure of a country's exposure to gas flaring through international trade (Bamji and Agrawal 2021). The index is not designed for performance certification or to market any product but is intended to help inform and influence policymakers. It highlights the need for shared responsibility by both the producer and consumer.

The index offers an opportunity for producers with a significant methane problem to identify importers with a shared interest in advancing their abatement technology. Consider an importer exposed to high IFG index values. Rather than searching for alternative supplies, which would only displace the problem to another importer, it could incentivize a producer from a country with poor flaring performance to invest in reducing its flaring volumes.

Emission tracking certificates could be linked directly to the importers corresponding IFG index. This is like RSG certificates sold for a premium to importers committed to reducing their IFG index exposure. Policymakers could also incentivize importers to participate in these markets as a national policy to encourage waste reduction. If the reduction in gas flaring increases potential supplies of natural gas for the importer, it would also support energy security objectives. For example, transitioning methane flaring in North Africa to natural gas imports for Europe.

Tracking reductions using emission token systems can integrate directly into the regulation of oil and gas production. For example, identifying severance tax on waste resource extraction or even legal enforcement of royalty obligations to landowners (Sect. 6.3.5). However, incentivizing or enforcing such applications would not necessarily be straightforward. This requires laws and regulations to be aligned with emission reduction goals, also shifts in public policies and committing to technology to enforce them.

6.6.3 Financing Emission Reductions

Investing in projects to reduce emissions can be faced with significant risk, negative returns, and other financial barriers. Oil and gas producers will need tools to overcome these challenges. When innovative solutions are needed to make use of methane, such as in remote regions, financial incentives may be needed to justify investments. In developing markets, methane abatement may have viable monetization opportunities. However, capital may be difficult to acquire or simply not accessible.

In Sect. 6.3.3, we discussed the role of ESG investors and sustainable financing could help overcome some of the investment barriers. We look specifically at integrating emission performance metrics into KPI-based financial instruments. Digital emission tokens represent suitable candidates to operate SLLs and SLBs. These instruments do not restrict how funds are spent, as in the use of proceeds, which would be hard to enforce with a digital emissions certificate.

What benefits would a blockchain-enabled emission certification platform bring to KPI-based financing? A major issue with existing ESG loans is that they historically have not compensated or charged companies enough for succeeding or failing to achieve their goals, respectively (Tobin and Scigliuzzo). Deploying emission certificates as digital assets offers a new pathway to operationalize KPI-based financing as an incentive for sustainable oil and gas production.

A smart contract linked to the environmental performance of the beneficiary of an investment could be deployed by financial institutions. The contract receives certificates from the defined borrower, comparing the emission factors to predefined thresholds set by the lender to deterministically adjust interest rates. The contract legally binds the lender and borrower to rates based on the surrender of these digital emission assets.

Enforcing these contracts faces the challenge of converting independently verified emission metrics into digital assets for KPI-based financing. Progress is being made towards this goal with S&P Commodity Insights introducing an additional 19 satellite-based methane intensity calculations (S&P Global Platts 2022) in the first half of 2022. While this serves as inputs to its voluntary markets for methane performance certificates, these types of initiatives could provide a good launch point for issuing KPI-based instruments linked to new projects targeting methane reduction. Section 6.5 discussed the role of distributed data governance tools to bridge these different initiatives with common data requirements.

KPI-based financial instruments will require surrendering frequent emission intensity reports with reduced interest payments under an SLL agreement requiring several reporting cycles. A potential issue is that producers must wait for performance improvements to be reported before unlocking benefits. This may not provide sufficient incentives for a producer to initiate an emission abatement solution for an oil and gas asset operating at low marginal profits with high investment risk.

Lenders could design contracts to gradually release additional financial support to a project developer as they surrender certificates and report performance gains. This option can introduce early funding advantages before the interest rate adjustment of an SLL takes effect. Applying emission certificates to the implementation of SLB can also help realize projects with significant financial barriers. Preferential rates are provided at the start of the project, only adjusting rates up if the target performance improvements are not met. This can introduce higher rate adjustment risk towards the end of a project and should prioritize abatement efforts with a higher probability of success.

6.7 Summary and Conclusion

Methane emissions representing a significant portion of the oil & gas industry's direct impact on GHG emissions globally. When accounting for other downstream scope 3 emissions resulting from the consumption of energy commodities supplied by this industry, this represents a smaller share of its total climate impact. However, these commodities represent essential inputs for the global economy requiring decades to phased down.

One of most effective ways to mitigate the impact in the short term will involve investing in more efficient production methods that prioritize a reduction in wasted methane. This waste represents emissions equivalent of about 1 billion cars. By preventing this waste, the oil & gas industry can support international climate commitments while also covering the cost of mitigation by selling responsibly sourced natural gas.

A variety of solutions using proven technologies to reduce GHG emissions linked to methane venting and flaring operations, or infrastructure leaks. The reality is most of them can be achieved at no net cost to industry. However, implementing them requires a delicate balance of incentives. These are needed to overcome the complexities linked to investing in and operating abatement solutions, and to achieve the goals of the Global Methane Pledge by 2030.

Among the most impactful incentives are government regulation and legal frameworks that set the requirements and establish precedents to drive change among oil and gas operators. These efforts are also complemented by new financial tools and voluntary markets. They provide economic incentives to help producers manage the costs associated with sustainable production, including access to capital. The also allow companies to market to consumers who help drive the demand for responsible production and absorb the cost of mitigation.

All these incentives are linked to the development of a measurement economy based on secondary environment attributes. We discuss how innovation in blockchain technology and the broader DLT space can help establish the measurement economy. This includes applications related to identity management and managing verifiable credentials, to smart programmable contracts. We build on the topic of smart contracts, addressed in other chapters of this book, as a tool to coordinate decisions and improve the flow of information related to methane emissions.

We discuss how blockchain can be applied to the aggregation of emission data and automating the verification and monitoring of performance data. These solutions focus on how to address existing bottlenecks and bridge silos in existing MRV frameworks. Finally, we highlight how blockchain can be applied to the tokenization of emission data. This area of innovation looks at ways to improve secondary markets for environmental attributes. It includes the minting and trade of performance certificates for sustainable commodity markets. We also explore how tokenized emission data can be applied to the financing emission reductions and tracking scope 3 value chain impacts.

Achieving methane emission reduction requires a complex ecosystem of stakeholders, institutions, standards and enabling market forces. Applying blockchain based solutions to this issue must be done with targeted precision as the benefits of innovation can get overshadowed by day-to-day real-world operations. Verification and audits of methane reduction projects requires onsite visits and trust across organizations. The benefits of digital identify and credential systems will require integration with existing intuitions that will require education programs and adopting new technology standards, all of which will take time.

Some of the more direct benefits linked to blockchain involve improving the transparency and interoperability of environmental attributes across the measurement economy. Scope 3 value chain reporting, and tracking of embodied emission, is a broader issue that cuts across the energy industry. Communicating efforts to reduce methane emissions sits at the very foot of the oil & gas industry's efforts to resolve the scope 3 impacts of its products. Therefore, this is a natural starting point where the industry can apply blockchain based solutions, whether to establish new financial instruments or enable more impactful sustainable commodity markets.

References

ATB Ventures (2022) ATB Ventures is working with the government of Canada to power its national digital trust proof of concept. Newswire.ca, February 15, 2022. https://www.newswire.ca/news-releases/atb-ventures-is-working-with-the-government-of-canada-to-power-its-national-digital-trust-proof-of-concept-836768875.html

Bedolla L, Cai W, Martin Z, Yu F (2020) Policy and Technology Solutions to Reduce Harmful Gas Flaring | Columbia SIPA. Columbia SIPA. https://www.sipa.columbia.edu/policy-and-technology-solutions-reduce-harmful-gas-flaring. Date accessed May 27, 2022

Bamji Z, Agrawal H (2021) A new index highlights the need for shared responsibility to end gas flaring. World Bank Blogs, June 25, 2021. https://blogs.worldbank.org/opendata/new-index-highlights-need-shared-responsibility-end-gas-flaring

Berg RC (2021) The role of the oil sector in Venezuela's environmental degradation and economic rebuilding. Center for Strategic and International Studies, October 12, 2021. https://www.csis.org/analysis/role-oil-sector-venezuelas-environmental-degradation-and-economic-rebuilding

Borden K (2022) Methane certification designed to differentiate production—compressorTECH2. CompressorTECH2, January 25, 2022. https://www.compressortech2.com/news/methane-certification-designed-to-differentiate-production/8017682.article

Chell C, Ullah S, Roberts L (2022) It's official: mandatory ESG disclosure is coming to Canada. MLT Aikins, April 8, 2022. https://www.mltaikins.com/esg/its-official-mandatory-esg-disclosure-is-coming-to-canada/

Chen S (2022) Emissions tokens network. Hyperledger Foundation. Last modified June 8, 2022. https://wiki.hyperledger.org/display/CASIG/Emissions+Tokens+Network

Chen S, Williams Rioux B (2022) Supply chain tokens for oil & gas emissions reductions. Confluence Mobile—Hyperledger Foundation. Last modified December 5, 2022. https://wiki.hyperledger.org/pages/viewpage.action?pageId=62241904

Climate Trace (2021) Bringing radical transparency to global emissions bringing radical transparency to global emissions. Climate TRACE, November 1, 2021. https://www.climatetrace.org/public/papers/2021/ClimateTRACE_BringingRadicalTransparencytoGlobalEmissions_FullReport.pdf

Credit Agricole (2021) Eni issues the world's first sustainability linked bond in the oil & gas sector. Crédit Agricole CIB, June 7, 2021. https://www.ca-cib.com/pressroom/news/eni-issues-worlds-first-sustainability-linked-bond-oil-gas-sector

Czapla E (2021) Methane fees for petroleum and natural gas systems—AAF. The American Action Forum, November 9, 2021. https://www.americanactionforum.org/insight/methane-fees-for-petroleum-and-natural-gas-systems/

Davis M (2022) New flaring data shows unacceptable flatlining and boldens the imperative to act. FlareIntel, May 6, 2022. https://flareintel.com/insights/new-flaring-data-shows-unacceptable-flatlining-and-boldens-the-imperative-to-act

Davis M, Toledano P, Schorr T (2022) North Africa can reduce Europe's dependence on Russian gas by transporting wasted gas through existing infrastructure. FlareIntel, March 29, 2022. https://flareintel.com/insights/north-africa-can-reduce-europes-dependence-on-russian-gas-by-transporting-wasted-gas-through-existing-infrastructure

EDF (2017) Vast energy resources wasting away in the Texas Permian Basin. EDF News and blogs, March. https://blogs.edf.org/energyexchange/files/2018/06/Permian-Flaring-Report-2017-3.pdf

EDF (2021) The burning question: how to fix flaring. ESG by EDF, October 21, 2021. https://business.edf.org/files/ESG-by-EDF-Flaring-Report-Book-V2-Reduced.pdf

EDF (2022) Methodology for EDF's Permian Methane Analysis Project (PermianMAP). Environmental Defense Fund, March 25, 2022. https://www.edf.org/sites/default/files/documents/PermianMapMethodology_1.pdf

EIA (2022) Annual Energy Outlook 2022. U.S. Energy Information Administration, March 3, 2022. https://www.eia.gov/outlooks/aeo/pdf/AEO2022_ChartLibrary_full.pdf

EPA (2021) U.S. to sharply cut methane pollution that threatens the climate and public health | US EPA. EPA, November, 2021. http://epa.gov/newsreleases/us-sharply-cut-methane-pollution-threatens-climate-and-public-health

European Commission (2021) Reducing methane emissions in the energy sector. European Commission, December 15, 2021. https://ec.europa.eu/commission/presscorner/detail/en/QANDA_21_6684

Fisher J (2022) EU reveals its plans to stop using Russian gas. BBC, May 18, 2022. https://www.bbc.com/news/science-environment-61497315

Fountain H (2022) Seen from space: huge methane leaks. The New York Times, February 4, 2022. https://www.nytimes.com/2022/02/04/climate/methane-leaks-satellites.html

Freitas G (2022) 'Responsibly sourced' gas finds a niche, but some cry greenwashing. BNN Bloomberg, January 19, 2022. https://www.bnnbloomberg.ca/responsibly-sourced-gas-finds-a-niche-but-some-cry-greenwashing-1.1709728

Github (2020) bcgov/von-oil-gas-poc: proof-of-concept for oil & gas credential issuing. GitHub. Last updated December 16. https://github.com/bcgov/von-oil-gas-poc

Hallahan K, Corral E (2021) Certified natural gas: midstream sector begins embracing concept, standards. S&P Global, October 14, 2021. https://www.spglobal.com/commodityinsights/en/market-insights/latest-news/natural-gas/101421-certified-natural-gas-midstream-sector-begins-embracing-concept-standards

IEA (2021) Methane Tracker 2021—Analysis—IEA. International Energy Agency, January 2021. https://www.iea.org/reports/methane-tracker-2021.

IEA (2022a) Overview—Global Methane Tracker 2022—analysis—IEA. International Energy Agency, February. https://www.iea.org/reports/global-methane-tracker-2022/overview

IEA (2022b) Flaring emissions—analysis—IEA. International Energy Agency, September. https://www.iea.org/reports/flaring-emissions

IEA (2022c) Methane tracker data explorer—data tools—IEA. International Energy Agency. Last updated February 23, 2022. https://www.iea.org/articles/methane-tracker-data-explorer

IEA (2022d) Total methane emissions and methane intensity of production in selected oil and gas producers, 2021—charts—data & statistics—IEA. International Energy Agency. Last updated October 26, 2022. https://www.iea.org/data-and-statistics/charts/total-methane-emissions-and-methane-intensity-of-production-in-selected-oil-and-gas-producers-2021

IEA (2022e) Strategies to reduce emissions from fossil fuel operations—Global Methane Tracker 2022—analysis—IEA. International Energy Agency, February 2022. https://www.iea.org/reports/global-methane-tracker-2022/strategies-to-reduce-emissions-from-fossil-fuel-operations#abstract.

International Monetary Fund (2022) Global price of LNG, Asia [PNGASJPUSDM]. Retrieved from FRED, Federal Reserve Bank of St. Louis, December 11, 2022. https://fred.stlouisfed.org/series/PNGASJPUSDM

Jones V (2015) Outlook: is there a role for digital credentials in oil, gas. RigZone, June 9, 2015. https://www.rigzone.com/news/oil_gas/a/138957/outlook_is_there_a_role_for_digital_credentials_in_oil_gas/?all=HG2

Kim HM, Laskowski M, Nan N (2018). A first step in the co-evolution of blockchain and ontologies: towards engineering an ontology of governance at the blockchain protocol level. arXiv, January 6, 2018. https://arxiv.org/abs/1801.02027

Kobzeva O, Astakhova O, Soldatkin V, Donovan K (2021) Russian oil producers struggle to contain gas flaring—document. Reuters, April 9, 2021. https://www.reuters.com/article/russia-flaring-idUSL8N2M23D0

Laaper S (2017) Using blockchain to drive supply chain transparency and innovation. Deloitte. https://www2.deloitte.com/us/en/pages/operations/articles/blockchain-supply-chain-innovation.html. Date accessed 2 June 2022

Lewis KK, Kryczka D, Waslicka E (2022) Methane: the world's overlooked greenhouse gas comes to the spotlight. EY, February 17, 2022. https://www.ey.com/en_pl/law/methane-the-worlds-overlooked-greenhouse-gas-comes-to-the-spotlight

MiQ (2022) The MiQ Standard. MiQ. https://miq.org/the-technical-standard/. Date accessed 18 Aug 2022
Moralis (2022) What are upgradable smart contracts? Full guide. Moralis. https://moralis.io/what-are-upgradable-smart-contracts-full-guide/. Date accessed 2 Sept 2022
Patchel J (2018) Can the implications of the GHG protocol's Scope 3 standard be realized? J Clean Prod 185:941–958. https://doi.org/10.1016/j.jclepro.2018.03.003
Pederstad A, Gallardo M, Saunier S (2015) Improving utilization of associated gas in US tight oil fields. Clean Air Task Force, October 2015. https://www.catf.us/wp-content/uploads/2015/04/CATF_Pub_PuttingOuttheFire.pdf
Penn State (2021) Ethane proxies for methane in oil and gas emissions. ScienceDaily, June 24, 2021. https://www.sciencedaily.com/releases/2021/06/210624135546.htm
Plevin R, Delucchi MA, O'Hare M (2017) Fuel carbon intensity standards may not mitigate climate change. Energy Policy 105:93–97. https://doi.org/10.1016/j.enpol.2017.02.037
Project Canary (2020) TrustWell^{TM} standard definitional document. Project Canary, September 1, 2020. https://www.projectcanary.com/wp-content/uploads/2021/01/IES-TrustWell-Ratings-Definition-Doc.pdf
Project Canary (2022) Project Canary sets its RSG trademark free. Project Canary Blog. https://www.projectcanary.com/blog/project-canary-sets-its-rsg-trademark-free/. Date accessed 4 Jan 2023
PR Newswire (2022) PureWest grow market for certified gas through partnership with EarnDLT and Project Canary to tokenize verifiable environmental attributes. November 15, 2022. https://www.prnewswire.com/news-releases/purewest-grows-market-for-certified-gas-through-partnership-with-earndlt-and-project-canary-to-tokenize-verifiable-environmental-attributes-301678365.html
Rabe B, Kaliban C, Englehart I (2020) Taxing flaring and the politics of state methane release policy. Wiley Online Library, January 13, 2020. https://doi.org/10.1111/ropr.12369
Rioux B, Chen's D (2022) Potential implications of the EU carbon border adjustment mechanism. KAPSARC, March 22, 2022. https://www.kapsarc.org/research/publications/potential-implications-of-the-eu-carbon-border-adjustment-mechanism/
Saunier S, Bergauer M-A, Isakova I (2019) Black carbon in the Arctic—best available technologies for flaring—carbon Limits. Carbon Limits, December 1, 2019. https://www.carbonlimits.no/project/black-carbon-in-the-arctic-best-available-technologies-for-flaring/
S&P Global (2021) Methane performance certificate assessments. S&P Global Commodity Insights. https://www.spglobal.com/commodityinsights/en/products-services/energy-transition/methane-performance-certificates. Date accessed 8 July 2022
S&P Global (2022a) Specifications Guide Platts Methane Performance (MPC). S&P Global Commodity Insights, November 2022. https://www.spglobal.com/commodityinsights/PlattsContent/_assets/_files/en/our-methodology/methodology-specifications/methane_performance.pdf
S&P Global (2022b) Methane intensity premiums. S&P Global. http://spglobal.com/commodityinsights/en/products-services/natural-gas/methane-intensity-premiums. Date accessed September 2, 2022
S&P Global Platts (2022) S&P commodity insights launches 19 new satellite-driven methane intensity calculations and corresponding premiums for U.S. natural gas production basins. PR Newswire, April 5, 2022. https://www.prnewswire.com/news-releases/sp-commodity-insights-launches-19-new-satellite-driven-methane-intensity-calculations-and-corresponding-premiums-for-us-natural-gas-production-basins-301517605.html
State of Wyoming Senate (2019) SF0125—digital assets-existing law. State of Wyoming Legislator, February 26, 2019. https://www.wyoleg.gov/Legislation/2019/sf0125
State of Wyoming Senate (2021) SF0038—decentralized autonomous organizations. State of Wyoming Legislator, April 21, 2021. http://www.wyoleg.gov/Legislation/2021/SF0038

Tobin M, Scigliuzzo D (2021) Wall Street's ESG loans charge corporate America little for missed goals. Bloomberg, September 7, 2021. https://www.bloomberg.com/news/articles/2021-09-08/esg-financing-comes-with-few-penalties-for-missing-goals?leadSource=uverify%20wall.

U.S. Congress (2021) H.R.3639—117th Congress (2021–2022): Blockchain Innovation Act. Congress.gov, May 31, 2021. https://www.congress.gov/bill/117th-congress/house-bill/3639/text?r=16&s=1

Vizcarra H (2020) EPA's final methane emissions rules roll back standards and statutory authority. Environmental & Energy Law Program, September 9, 2020. https://eelp.law.harvard.edu/2020/09/epas-final-methane-emissions-rule-rolls-back-standards-and-statutory-authority/

Wang S, Ding W, Li J, Yuan Y, Ouyang L, Wang F-Y (2019) Decentralized autonomous organizations: concept, model, and applications. IEEE Trans Comput Soc Syst 6(5):870–878. https://doi.org/10.1109/TCSS.2019.2938190

Wells B (2016) Is flaring just bad for business, or is it a violation of the landowner's contract? Forbes, October 3, 2016. https://www.forbes.com/sites/uhenergy/2016/10/03/is-flaring-just-bad-for-business-or-is-it-a-violation-of-the-landowners-contract/?sh=6163cf9649c5

Williams Rioux B (2022). Tracking GHG emissions from methane leakage and flaring with supply chain tokens. Last updated November 5, 2022. Hyperledger Foundation. https://wiki.hyperledger.org/display/events/Tracking+GHG+emissions+from+methane+leakage+and+flaring+with+supply+chain+tokens

Williams Rioux B, Ward C (2022) A non-fungible token model for tracking emissions in the fuel value chain. Soc Sci Res Netw, April. https://doi.org/10.2139/ssrn.4081426

World Bank (2021) Global gas flaring tracker report. World Bank, April 28, 2021. https://www.worldbank.org/en/topic/extractiveindustries/publication/global-gas-flaring-tracker-report

World Bank (2022a) Global flaring data. World Bank. https://www.worldbank.org/en/programs/gasflaringreduction/global-flaring-data. Date accessed 8 May 2022

World Bank (2022b) ZRF FAQ. World Bank. https://www.worldbank.org/en/programs/zero-routine-flaring-by-2030/qna#4

YCharts (2022) Henry hub natural gas spot price. YCharts. https://ycharts.com/indicators/henry_hub_natural_gas_spot_price. Date accessed 7 Oct 2022

Yergin D (2021) Daniel Yergin: why the energy transition will be so complicated. Financial Post, December 14, 2021. https://financialpost.com/commodities/energy/oil-gas/daniel-yergin-why-the-energy-transition-will-be-so-complicated

Zwitter A, Hazenberg J (2020). Decentralized network governance: blockchain technology and the future of regulation. Front. Blockchain, March. https://doi.org/10.3389/fbloc.2020.00012

Bertrand Williams Rioux is the director of Two Ravens Energy & Climate Consulting and a former Research Fellow at the King Abdullah Petroleum Studies And Research Center (KAPSARC), an independent think tank in Saudi Arabia. He has more than 10 years of experience in energy economics & policy research. Bertrand is an active member of the Hyperledger Climate Action Special Interest Group and contributor to the open source hyperledger-labs blockchain-carbon-accouting project. He holds graduate degrees in Atmospheric Physics, Marine Engineering, and Blockchain & Digital Currency.

Chapter 7
Carbon Capture and Storage

Si Chen, Soheil Saraji, and Fred J. McLaughlin

7.1 Introduction

Can we burn fossil fuels without causing climate change?

Carbon capture and storage (CCS), which captures CO_2 and stores it underground, promises to do just that. Around since the early 1970s, CCS has been already proven in a variety of projects around the world. Now it's gaining prominence as a credible way for the existing oil and gas industry to solve the climate problem. At a time when political consensus on climate is still rare, it enjoys bipartisan political support in the U.S. In 2018, the U.S. Congress increased the section 45Q tax credits while removing the original program limits (Rogers and Dubov 2021). In 2021, the infrastructure bill allocated over $10 billion for carbon capture, storage, and direct air capture projects, as well as low-interest loans for pipelines (Johnson et al. 2021).

The industry is responding enthusiastically as well. In the two years ending 2021, the number of CCS facilities grew from 9 to 29 worldwide, with additional facilities planned to triple the global capacity to 111 million tons per year. And that's just the beginning. ExxonMobil, the current world leader in CCS, expects it to become a $4 trillion market by 2050 (Gordon 2022). This would imply 40 billion tons of CO_2 per year at $100 per ton (Vikas and Aiyer 2021), in line with IEA estimates for CCS to grow 40-fold by 2030.

Making that happen will require a mirror opposite of the natural gas supply chain to be built—wells must be drilled, pipelines built, and gasses shipped over long distances. The oil and gas industry knows how to do it, and it also knows how hard

S. Chen (✉)
Open Source Strategies, Inc., Los Angeles, CA, USA
e-mail: sichen@opensourcestrategies.com

S. Saraji · F. J. McLaughlin
Energy and Petroleum Engineering, University of Wyoming, Laramie, WY, USA
e-mail: ssaraji@uwyo.edu

S. Saraji and S. Chen, *Sustainable Oil and Gas Using Blockchain*,
Lecture Notes in Energy 98, https://doi.org/10.1007/978-3-031-30697-6_7

it is to do it. In this chapter, we'll take a look at CCS technology, the challenges to scaling it, and how the blockchain could help address those challenges.

7.2 The Technology Behind Carbon Capture and Storage

Today, CCS projects are generally deployed for two reasons. The most widespread CCS projects utilize CO_2 for enhanced oil recovery (EOR, or sometimes referred to as CO_2-EOR). EOR involves injecting captured CO_2 into a hydrocarbon field, which is commonly mature or depleted, to increase production. The injected CO_2 increases reservoir pressure and lowers the viscosity of hydrocarbons, allowing more hydrocarbons to be produced. Utilization projects typically recycle the CO_2 as it is co-produced with the hydrocarbons, which can then be reused in the same field, sold to another project, or reinjected into the field for long-term storage. EOR projects require a relatively large volume of CO_2, with upwards of 1 million metric tonnes per annum consumed by a typical field (Verma 2015). To provide an example, the largest CO_2-EOR project in the Rocky Mountains has utilized over 130 million metric tonnes of CO_2 since project inception (Bebo et al. 2016).

As an alternative, saline CO_2 storage (sometimes referred to as geologic storage) involves injecting captured CO_2 into a deep geologic reservoir. Saline storage projects aim to sequester CO_2 indefinitely as a means to mitigate greenhouse gas impacts on the world's climate.

Successful CCS projects have four components that must be met: proper geologic conditions, an emissions source with capture technology, existing regulatory frameworks relevant to the site, and an economic/business case. Additionally, successful CCS projects will need to consider public acceptance, environmental impacts, infrastructure, workforce, and numerous other auxiliary factors. However, there have been no CCS projects that have proceeded to date without meeting the four primary components.

The physical conditions for successful geologic storage and EOR are well-known and have been proven in-field (Michael et al. 2010; Eiken et al. 2011). For CCS, the geologic system must include a reservoir for CO_2 injection and an overlying caprock that is capable of retaining the injected CO_2 (Fig. 7.1). It is preferred for target reservoirs to be deep enough for CO_2 to remain at a supercritical state. This depth can vary relative to different factors but is generally deeper than 3,000 feet (approximately 915 m). Reservoirs that meet proper CCS geologic conditions must also have proper pore space resources and permeability capable of meeting injectivity goals (rate and volume) and be able to withstand the increases in pore pressure associated with the injection. To meet permitting requirements, fluids within the reservoir need to be >10,000 TDS (DOE 2017). Caprock(s) that overlie the reservoir must be able to retain injected CO_2. Preferred caprock characteristics include very low permeability, a wide areal extent, and lateral and vertical consistency with respect to geologic character. Other considerations of the geologic system include risk, such as faults or fractures that allow for fluid migration or that are susceptible to induced seismicity.

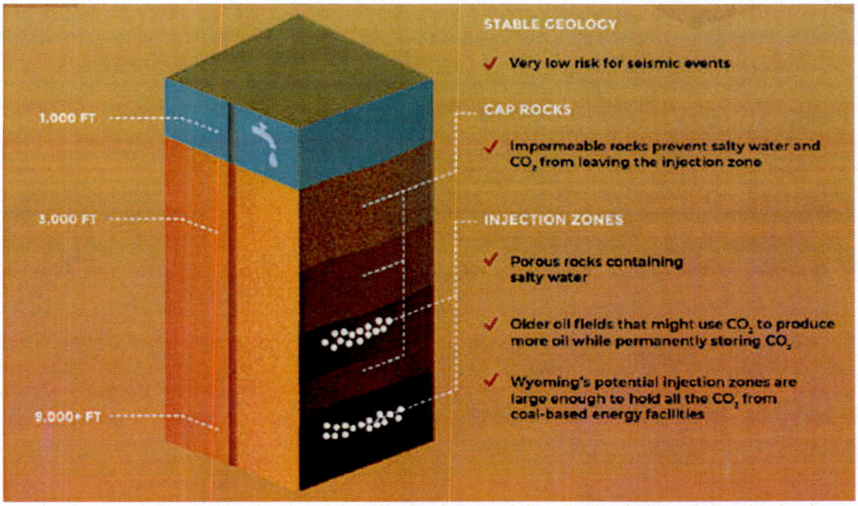

Fig. 7.1 A cartoon schematic of a generalized CCS geologic system in Wyoming (courtesy of the University of Wyoming School of Energy Resources)

The CCS project must also prove the ability to operate an injection site without imposing risk to regional freshwater reservoirs. Successfully permitting injection wells requires meeting proper geologic conditions.

If a CCS project has proper geologic conditions, it may consider installing a capture facility to capture and purify a commercial-scale CO_2 source. Many CCS projects aim to capture CO_2 from emissions from large point sources, either post-combustion, after fuel is burned, or pre-combustion, when fuel is burned to create a carbon-free fuel such as hydrogen for later use. Target point sources for post-combustion CO_2 capture include power facilities such as coal- and natural-gas-fired power plants, and industrial facilities such as cement, fertilizer, gas processing, and steel manufacturing plants.

CO_2 capture technology has long been used for different commercial applications, such as gas processing. Amine-based technologies are most commonly used for point source emissions capture, though CCS-focused research continues to explore new technologies such as membranes and distillation through cryogenic processes. Other new technologies seek to capture CO_2 directly from the air (referred to as direct air capture [DAC]). Though amine capture technologies have been commercially proven, the costs associated with retrofitting these technologies to a point source have been prohibitive to the CCS industry (Fig. 7.2). However, recent support for CCS is bolstering new projects.

Once captured, the gasses could then be transported via pipeline and stored, usually in an underground structure such as a cavern, deep saline aquifer, or depleted oil field. Alternatively, they could be used in a variety of products ranging from cement to plastics. Finally, they could be used to produce more oil through EOR.

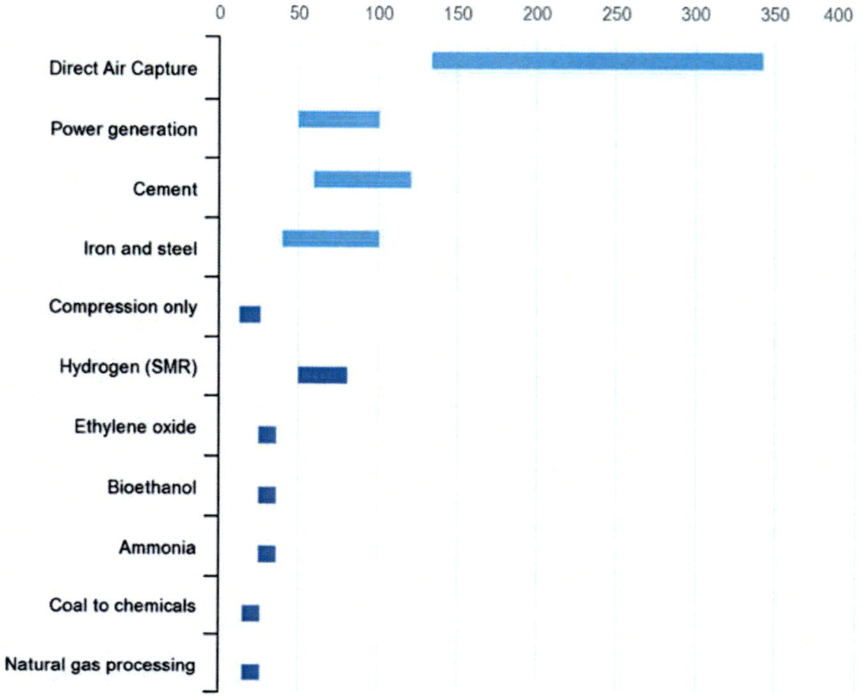

Fig. 7.2 Levelized cost of CO_2 capture by industry and initial concentrations in emissions (IEA 2022). *Source* IEA, Levelised cost of CO_2 capture by sector and initial CO_2 concentration, 2019, IEA, Paris https://www.iea.org/data-and-statistics/charts/levelised-cost-of-co2-capture-by-sector-and-initial-co2-concentration-2019, IEA. License: CC BY 4.0

In each use case, the climate benefit is different. CCS at power plants or industrial sites offsets the emissions that would have been released from combustion. Direct air capture (DAC) could remove emissions from any source and years after its emission. Underground storage counts as permanent removal of GHG emissions, while use in industrial materials may cause the emissions to be released within a relatively short period of time, thus providing no real climate benefit. Finally, while EOR permanently injects CO_2 underground, it also produces more oil which could release additional emissions down the road. Furthermore, trying to store too much CO_2 degrades the economics of EOR, so the combination of CCS and EOR needs to be optimized based on the value of CO_2 stored versus the value of oil produced (Li et al. 2021).

7.3 Policy, Investment, and Economics

Now comes the bill.

According to the World Economic Forum, \$5 trillion in CCS investments will be needed by 2050 (Vikas and Aiyer 2021). A Global CCS Institute study suggested that between \$650 billion to \$1.28 trillion would be needed to build capacity for 5.6 gigatons of CCS per year (Rassool 2021). If expanded to the scale of the IEA and ExxonMobil estimates, this would imply up to \$10 trillion may be needed. In addition, further capital is needed for pipelines. A Harvard Belfer Center study calculates that a 110,000 km network would be required for CCS at the cost of \$170–\$230 billion (Moch et al. 2022). ExxonMobil is already looking for industry and government to invest \$100 billion to turn the Houston Ship Channel into a global hub for CCS (Bussewitz 2021).

A recent Global CCS Institute study highlighted three fundamental issues for funding CCS. First, there is no economic value to the output other than abating CO_2 emissions from another party. As such, CCS depends on a functioning carbon market. Second, like other energy systems, CCS requires shared infrastructure such as transmission pipelines. Therefore, properly crediting each member of the CCS network is critical to attracting financing. Finally, CCS requires permanent storage of the captured CO_2. This creates liabilities to potential investors, which must be addressed (Rassool 2021).

Furthermore, while the economics of CCS continues to improve, it's still an expensive technology. The IEA estimates that carbon capture could cost between \$50 and \$100 per ton of CO_2 at power plants, \$40–\$100 per ton at steel mills, and \$50–\$80 per ton during hydrogen manufacturing with natural gas. Transporting CO_2 via an onshore pipeline is between \$2 and \$14 per ton, while storing underground is generally under \$10 per ton. If paired with EOR, the cost of storage could even be negative, thanks to revenues from oil production (Baylin-Stern and Berghout 2021). The Harvard Belfer Center study pegged the numbers at \$49–\$150 per ton for natural gas plants, \$8–\$133 per ton at steel mills, and \$65–\$136 per ton for hydrogen. Transport and storage costs would come in at \$17–\$23 per ton (Moch et al. 2022). Thus, even at the more generous amounts of up to \$50 per ton, IRS 45Q tax credits alone would not be enough.

7.4 Safety, Risks, and Regulations

More importantly, will we be allowed to do all this?

Although there are technical challenges to CCS, they are not a big concern for those in the oil and gas industry. Far more complicated are the challenges of assuaging the general public to obtain the necessary permits. Ensuring the safe and permanent storage of CO_2 for extended periods of time (i.e., decades or longer) is an important challenge for underground carbon sequestration. The public is concerned about

CO_2 leaking into drinking and agricultural water supplies and the build-up of pressure underground causing earthquakes. Thankfully, there are technical solutions to significantly reduce these risks through extensive study of the underground storage sites, meticulous design and operation of the injection well, and comprehensive long-term monitoring of the stored CO_2. The public, however, does not seem to recognize or trust these technical solutions. This may partly be due to the lack of public engagement and education in this area and partly because of the lack of long-term national and international experiences with underground carbon storage projects.

These concerns have led to stringent governmental requirements and regulations regarding carbon geo-storage. For example, the United States Environmental Protection Agency (EPA) has put forward an extensive requirement list for permitting a CO_2 injection well (called Class VI wells) (Environmental Protection Agency 2022):

- Extensive site characterization requirements
- Injection well construction requirements for materials that are compatible with and can withstand contact with CO_2 over the life of a GS project
- Injection well operation requirements
- Comprehensive monitoring requirements that address all aspects of well integrity, CO_2 injection and storage, and groundwater quality during the injection operation and the post-injection site care period
- Financial responsibility requirements assuring the availability of funds for the life of a GS project (including post-injection site care and emergency response)
- Reporting and recordkeeping requirements that provide project-specific information to continually evaluate Class VI operations and confirm USDW protection

The combination of permitting and financial responsibility requirements has made it difficult to get actual CCS projects off the ground. Mr. Steve Swanson, CEO of North Shore Energy LLC, an independent oil and gas producer in southwestern Wyoming, described the challenges in an interview with us regarding developing CCS projects:

> We have the willpower. We have the start of some economic incentives. But the big mile, the big roadblock, and the big hurdle to get over are regulatory. And so, as an independent producer, or really anybody who is interested in sequestration, if you're going to use oil and gas, and depleted reservoirs to store the CO_2, you must work with the existing regulatory framework. And so, there are two different types of wells that are permitted for sequestration; a class two well, which is ubiquitous. They're used for enhanced oil recovery in the oil and gas business. And it's generally run through the state's oil and gas commission. That well-understood routine regulatory environment that we're comfortable operating in. The other option for permanent sequestration is through what's called a class six injector Well, that's used for permanent storage. And the Class VI wells are not well understood. There are only really two functioning class six wells in the country right now. And the reason why that's the case is there are two hurdles that you have to get over to get a Class VI permit. The first is a technical analysis that's related to the suitability of the reservoirs and basically all the things that are below the surface, and whether it's suitable for permanent sequestration. And that's been defined generally in several different ways. But you're basically looking at 100 years to make sure that the CO_2 that you inject underground can be stored in a way that it doesn't migrate or leak for a long period of time.
>
> Unfortunately, this technical analysis and suitability of the downhole environment for CO_2 sequestration has been handled by the EPA on a federal level. It is now starting to move

> to the state level; each state can apply for primacy. Wyoming and North Dakota as we speak, are the only two that have been granted primacy and North Dakota is the only one that's issued Class VI permits to date. But we, for example, are the first ones to apply for a Class VI permit in Wyoming. So, we're kind of on the front edge of that and I think we're working our way through the technical review period of that. Wyoming has delegated the regulatory authority for the Class VI well to the DEQ, Department of Environmental Quality, instead of the oil and gas commission, who regulates the class of wells that I referred to earlier. Okay, so you must stand up an entirely new, regulatory body to deal with this, and then DEQ is an existing body, but they deal mostly with air and water pollution, not CO_2 sequestration. And so over time, hopefully, they'll be experienced enough to review these but it's taking a while for them to get up to speed like the oil and gas commission would.

The industry feels confident that working with the regulatory bodies, implementing and learning from new projects will eventually bring everyone up to speed in a reasonable time frame. The second hurdle, however, seems to be a much bigger concern for the industry, as Mr. Swanson continues:

> The other hurdle is much larger and potentially fatal to CCS. And that is the financial assurance hurdle. So, there's a two-prong test for these classes. The first is the technical review and I think we can say that that's in hand. The financial assurance piece is not in hand. It is a serious issue as we speak, and I think the resolution of it will hold the balance and the success or failure of CCS as an industry. The regulator wants to make sure that if there is damage of whatever nature, but mostly environmental damage under that general definition, there are funds available to remediate it. And so, that's no different really than the regulatory environment that we operate in 24/7. So that alone is not the problem. The problem is that when you run your models, and you're looking at permanent sequestration over a long period of time or decades, up to 100 years, and when you start stress testing those models by assuming the worst event possible at the beginning period and extrapolate that out throughout the time of the model. You create an environment where your financial assurance is well beyond the capability of anybody to satisfy. So, for example, the EPA when they granted two Class VI permits to Archer Daniels Midland back 10-15 years ago, their model generated a financial requirement of $40 million per well. ... You know, even ADM, one of the largest publicly listed industrial commodity companies in the world still couldn't satisfy the requirements of the EPA, which had to be modified to accommodate them so they could get approval for that Class VI permit. The point is that if the states, when they received primacy, follow the EPA model for identifying and quantifying financial assurance only major oil companies would be able to qualify. And so, the number of successful permanent storage ... will be extremely limited. And our experience so far is not promising. So, as I said, North Dakota has issued one Class VI permit, and the financial assurance obligation, although not $40 million from EPA, is $20 million. And that's what had to be posted in terms of insurance or bonding or surety. Those typical types of things that we use to cover financial obligations into the future had to be met for $20 million as well. And, you know, speaking from our own perspective, as an independent, I can say that that's a chilling effect on our being able to pursue CCS. So, getting back to your original question about why it's not more popular or more ubiquitous in the world, that's the single reason that has kept the industry from sequestering more CO_2.

The combination of complicated and evolving regulatory permits, high financial assurance requirements, the need for massive global investments, and the continuing lack of sufficient economic incentives (beyond 45Q tax credits) are the main factors holding back CCS from playing a significant role in mitigating carbon emissions. However, the recent support from large investment firms and national governments towards increasing investment in low-carbon technologies, development of real-time carbon monitoring technology, and standardization of carbon emission/storage data are positive moves that could help the CCS industry in the coming years. These positive changes, along with more demonstration projects, and ongoing discussions with the state and federal governmental bodies, could help to relax the stringent regulatory requirements for CCS. This will allow small and medium-sized independent companies with experience in oil and gas operations to move forward with CCS and play a bigger role in the global energy transition.

7.5 Accounting for CCS

Once we move forward with CCS, it will need to be supported by a framework that properly accounts for all the activities and pays all the participants. Accounting for CCS must track and verify the emissions throughout the process, from capture to transport to storage to post-storage monitoring.

The entire operation in CCS could be thought of as a value chain, similar to oil and gas, but in reverse: the commodity, in this case CO_2 and other GHG emissions, is captured, transported, and stored back in the ground, sometimes in the same geologic formations where the original oil and gas came from. It involves several separate participants and occurs over medium to long distances—transporting GHG over 250 km of pipeline is easily feasible, and longer distances are possible with additional processing in between. Therefore, CCS has many of the same issues of the oil and gas supply chain, such as:

- tracking the commodity through the supply chain,
- properly crediting the participants based on the value of CCS,
- monitoring the healthy functioning of the supply chain, such as equipment condition and leakage,
- validating the final product. In this case, it is the net emissions removals for the energy produced, be it electricity from a power plant, low-carbon product (e.g., ethanol, fertilizers), or zero-carbon fuel such as hydrogen.

The key steps in CCS supply chain emissions can be summarized as:

- *Capture*—Emissions are captured at the point of combustion, whether to produce electricity, hydrogen fuel, or other products such as ethanol, or fertilizers.
- *Transport*—Gasses are transported over pipelines. Pipelines could aggregate GHG from multiple capture sites, such as different power plants.
- *Storage*—Gasses are injected and stored in underground structures.
- *Insurance*—While underground storage is reliable over time periods of 100 years or more, there is a small risk of leakage, which could cause the stored gasses to be released back into the atmosphere. This would pose both immediate risks and

long-term negative climate impacts. Government agencies such as the EPA in the U.S. would be expected to monitor stored gasses and require either the site owner, the party who injected the gasses underground, or most likely both to remediate such leakage. Third-party insurance should be developed to protect against such risks.

- *Finished product*—In the case of pre-combustion CCS, the resulting hydrogen would need to be sent to customers while retaining its carbon-free attribute. In the case of post-combustion CCS, the finished product, be it electricity, steel, or some other industrial product, would also need to have a carbon-free attribute attached to it. The finished product also could be the stored carbon in the form of offsets or credits.

For example, a CCS project could be developed by these participants in the following sequence (see Fig. 7.3):

a. A Site Owner invests in obtaining the permits, such as those for Class VI wells, to store a quantity of emissions on their site. As part of obtaining the permit, the Site Owner purchases a surety bond or financial guarantee for the stored emissions from an Insurance Provider.
b. The Site Owner then looks for buyers for the permitted storage capacity.
c. Separately, the Project Developer contracts with an End User, such as an electric utility with a natural gas plant, to capture and store emissions so that they could meet their emissions targets.
d. With this contract, the Project Developer contracts with a Capture Operator to install carbon capture equipment at the End User's power plants.

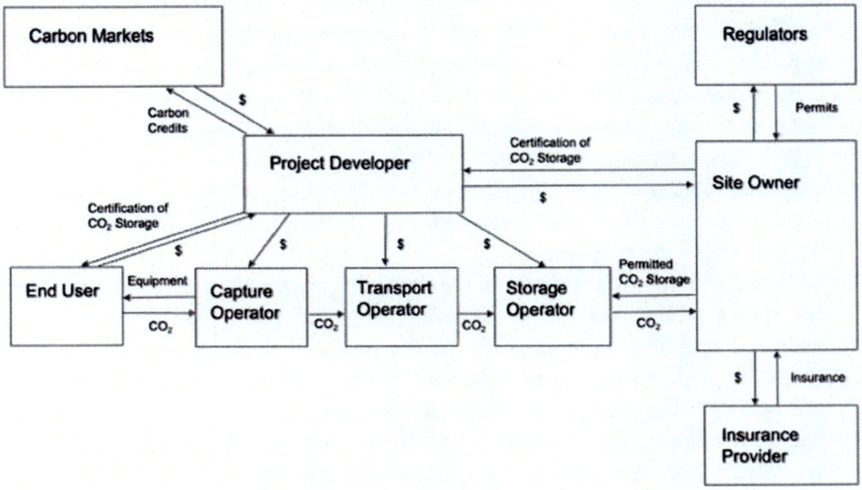

Fig. 7.3 The CCS value chain

e. The Project Developer also contracts with a Transport Operator, such as a pipeline, to transport the captured emissions and a Storage Operator to inject the captured emissions.
f. The Project Developer then purchases the rights to permitted storage capacity from the Site Owner.
g. Once the project is operational, the End User pays the Project Developer based on the volume of emissions captured and stored. The Project Developer then pays the Capture Operator, Transport Operator, Storage Operator, and Site Owner.
h. In return, the Project Developer furnishes the End User with proof of the amounts of emissions captured, transported, and finally stored.
i. Alternatively, the Project Developer could try to monetize the CO_2 captured and stored through Carbon Markets.
j. Finally, equipment at the storage site provides ongoing monitoring. If there is any leakage, the Insurance Provider will pay for the remediation.

For CCS to become a real force for climate change and the oil and gas industry, we must solve several problems:

- raising the capital needed to deploy it at scale,
- pricing the different use cases for CCS based on their climate benefits,
- distributing the economic value of CCS through its supply chain,
- validating the climate benefits of CCS to different industries, regulators, investors, and the general public.

Therefore, a carbon accounting scheme for CCS should meet these needs for all the different participants in the CCS value chain, which include:

- Site owner and permit holder—The owner of the site which has been permitted for CCS. Having made the investment to secure the permits, they should be paid a premium for the use of their site when captured emissions are stored there.
- Capture operator—The operator which installs and operates the carbon capture equipment at the point of emission. They should be paid based on the amount of GHG which are captured.
- Transport operator—The operator of the transport infrastructure, including pipelines and any rail or truck transport. They should be paid based on the amount of GHG which are transferred.
- Storage operator—The operator of the storage equipment at the storage site. They should be paid based on the amount which is stored into the site.
- Insurance provider—The provider of the financial guarantee or surety. They should be paid an ongoing premium based on the cost of the guarantee, which may be tied to the amount that is stored at the site.
- End user—The utility or buyer of the lower carbon fuels or electricity. They should pay a premium for the product, which could then be transferred to the other participants.

7.6 Blockchain for CCS

If this sounds like a lot of work, it is.

Doing all this accounting won't be easy, given the number of participants spread out across different regions, industries, and even jurisdictions. We believe blockchain technology is an ideal tool for carbon accounting for CCS operations and could provide further economic incentives by converting carbon as a commodity to digital assets that could be traded in the market.

Blockchain technology is a distributed and decentralized ledger that provides immutable data records and transparency. It could be used to allow multiple parties to transact with each other by giving all the parties identical records of the transactions. Thus, parties could trade digital tokens, sign off on documents, vouch for the validity of data, or vote on different proposal claims on a blockchain. Because all the parties have access to the same data, they could all verify that the transactions are true and correct, without relying on an external party such as a government agency. Thus, it is an ideal tool for cases such as carbon accounting in CCS, where there may be multiple participants across industries and jurisdictions, and no government agency is available to coordinate their interactions.

Blockchain could play a significant role in promoting CCS as a key element of the global carbon economy by

- credibly proving GHG emissions captured to customers, and their constituents (consumers, regulators, general public),
- creating revenue streams to help offset the cost of CCS,
- certifying end products such as energy produced or hydrogen fuel as "zero carbon",
- creating a digital market for credits or offsets related to the CCS process,
- creating value for CCS site permits, and
- structuring insurance or surety required for site permits.

Two separate blockchain technologies could be used to achieve these goals. First, a data ledger could be used to record data from critical supply chain events. The ledger essentially stores perfect bookmarks into securely stored/accessed historical event records—proving authenticity—and any dependent contractual claims between two parties. This ledger would work like this:

- All parties on the supply chain (capture, transport, storage, monitoring, auditing, regulating, and insurance) are vetted participants—having various access and reporting obligations to fulfill the integrity of measured dependencies.
- At the point of capture, both pre-combustion and post-combustion, instruments could record the amount of GHG emissions captured.
- In the pipeline, the amount of GHG received from the capture sites and the amount of GHG sent to the final storage site could both be recorded by instruments (e.g., IoT devices) and stored on a data ledger. The difference is the loss in transit which could be counted as the emissions footprint of the pipeline, along with other inputs

such as energy, vehicles, etc. This data could be shared with government agencies for monitoring and universities and research laboratories for study and analysis.

- At the storage site, the amount of GHG sent into storage could be stored on the ledger.
- If any oil is produced using the Enhanced Oil Recovery from the stored GHG, it must also be recorded on the ledger.
- Regular monitoring instruments could capture seismic readings of the storage site to determine the amount of GHG still stored there.
- Other monitoring instruments could be integrated into this framework to measure (if any) escaped gas in the long-term. They may include soil monitoring devices, water well monitoring devices, and air monitoring devices.

Recording of data on the ledger would require validation by either a trusted auditor or a distributed oracle. Under the auditor model, one party would be entrusted with verifying that the data entered into the ledger, whether captured directly from instruments or entered manually, is correct. Implicitly, we're trusting that the reputation of the auditor is sufficiently important that they would act honestly in certifying the data. Alternatively, a blockchain technology known as an oracle could validate the data. The oracle would involve multiple parties who would vouch for the correctness of the data and stake tokens. Later, another party could challenge the validity of the data, and a supervisory group would decide whether the data is valid. If the data is determined not to be valid, then the parties who originally vouched for it would lose their reputation tokens (Breidenbach et al. 2021). Such an oracle formalizes the reputation and trust of traditional auditors and opens up the auditing of data to more parties. If the data being audited is sensitive, then the oracle could be limited to vetted members.

Once the data is recorded on the ledger, smart contracts, or programs running on the blockchain, could then trigger payments to the different parties in the carbon capture supply chain. For example, when data of the amount of GHG transported by the pipeline operator is validated on the ledger, a payment could be automatically triggered to them. Similarly, when data for the amount of GHG stored into the site, payments to both storage operator and the insurance provider could be automatically triggered. These payments could be made with digital currencies such as stablecoins, which would be faster and less costly than traditional bank transfers. They would also enable more frequent payments to be made.

A digital token representing the amount of GHG stored could be used to create economic value for CCS. For example, 1 token could represent 1 ton of CO_2 equivalent stored gasses. The tokens could then be used to exchange the value of the captured emissions in the value chain. During the carbon capture cycle, all the data on the data ledger would be used to validate that the emissions are properly captured and stored. When that has been done, the project developer could be issued tokens for the captured emissions. These carbon tokens of CCS create immutable records of the CO_2 captured, transported, and stored. They could be traded between parties and settled in real-time on the blockchain. They could also be listed on online exchanges for individuals or businesses to offset their carbon footprints. Finally, they could be sold to the end users of the energy supply chain, such as transportation companies and power-generating utilities, to meet their carbon emissions targets. The combination of open market sales and transactions with end users could help establish the value of carbon capture. Large sales of the tokens to customers could be used as a way to pre-sell emissions removed by CCS to finance projects, and later sales after storage could help provide additional income.

Smart contracts could also help provide the insurance needed for the safe storage of the emissions, thereby meeting the financial requirements for CCS permits, but without the high overhead of traditional insurance. The smart contract would define an agreed-upon reserve amount relative to the amount of GHG stored underground. The insurance providers would contribute the initial reserve amount by staking digital currencies in the contract, which means the funds would be kept in the contract rather than made available for withdrawal. Each time emissions are stored underground, some payments from the storage operator are transferred to the insurance provider as the purchase of the insurance. These payments are also staked in the contract and counted as additional reserves. All the staked funds could be invested. Thus, the reserve amount would increase over time until a targeted reserve amount is reached. When the amount of staked funds exceeds predefined reserve requirements, the additional amounts could be released to the insurance providers. Finally, when the data ledger shows that stored emissions have been released, the predefined insurance contract could trigger payments from the insurance fund to remedy the released emissions.

When the site owner obtains the permits to store captured emissions, they could issue a "permitted storage token" equal to the total amount of emissions that could be stored on their site. This token would be validated against the actual permit issued to the site owner. It would then be marketed to any party that needs to store emissions and could get them to the site. When the emissions are stored at the site, the storage operator must purchase permitted storage tokens from the site owner. This token then becomes a record of how much emissions could be stored at the site and how much is still available, and it could be used to finance the permitted storage site. The use of a token which could be marketed on a blockchain to vetted buyers improves the transparency of the market, making it easier for project developers and site owners to enter into transactions. It also reduces the risk of overselling by site owners.

Going a step further, CCS projects could raise funding with long-term offtake agreements based on smart contracts and digital carbon tokens. For example, the project developer could enter into a long-term agreement with the site owner for the permitted storage capacity. Similarly, the project developer could enter into a long-term agreement with a utility or carbon market buyer to sell the CO_2 captured and stored. These agreements could be structured as the exchange of tokens for digital currencies, such as stablecoins, over a long period of time. If collateral is needed, the smart contract could require staking of digital currencies. With such offtake agreements in place, both site owners and CCS project developers should be able to obtain long-term financing at attractive rates from banks or securitization in the capital markets.

Last but not least, blockchain could significantly improve the efficiency of invoicing and payments in the CCS process. Imagine the parties in the CCS supply chain all having to maintain their own databases of the amount of CO_2 emitted, captured, transported, and stored. All the parties would need to reconcile their data with their counterparties. They would then send invoices to each other, reconcile the invoices they receive, and then pay them with wire transfers or even paper checks. Done with traditional paper and email processes, this will be tedious, inefficient, and expensive. Fortunately, the blockchain offers a more elegant, faster, and cheaper alternative. It is a common platform for storing and accessing immutable data records such as the amount of verified CO_2 capture and storage, as well as an electronic payment system where transactions could be instantaneous and nearly costless. The two could be linked together to automate all these back office processes, with significant gains in efficiency. One study, for example, suggests that blockchain-based mechanisms could reduce the costs of sustainability-linked bond issuance by as much as 90% (Haahr et al. 2019).

7.7 Valuing Carbon Capture with Emissions Tokens

So how do we get paid for doing all this work?

Our carbon emissions tokens would be linked to a data ledger that shows exactly how the emissions were captured, transported, and stored. They would be linked to insurance for the long-term storage of emissions and to the purchase of permitted storage tokens to show that the storage is at a permitted site. The total emissions footprint of CCS is the net amount stored, minus operational GHG emissions, for example from buildings, vehicles, maintenance, and travel, based on the GHG Product Protocol. This should be a negative number representing emissions removed but minus some for operating emissions costs.

While these negative emissions could be marketed as carbon offsets, the carbon offsets market today suffers from a combination of high costs, low prices, and, ultimately uncertainty over their effectiveness. This is discussed further in the chapter on carbon credits. Meanwhile, the emissions removed by CCS are highly certain. This means that it could be used to label commodities as low or zero-carbon products to meet specific customer or regulatory requirements. These use cases would probably generate higher value for CCS emissions removals than offsets.

For example, when tokens are transferred to the original emitting party, such as a natural gas power plant or steel mill, they represent the total emissions removed by CCS. They could offset that party's emissions footprint, so that their product is "carbon neutral." Alternatively, the power plant or mill where the CCS originated could sell its tokens and give up any claims to the benefits of CCS. For example, an oil and gas company could buy them back from the originating site to market "carbon-neutral LNG" to the next buyer. A transportation company could buy them for its fleet of natural gas-powered vehicles. A hydrogen producer could buy the tokens to convert hydrogen produced with natural gas from "gray" to "blue hydrogen," as discussed in the chapter on sustainable aviation and transportation fuels. Finally, even end users such as small businesses and individual consumers could buy them to offset the emissions from their natural gas use or products which were made from natural gas.

A token of the emissions removed from CCS detaches the physical commodity from its climate benefits, so that the physical commodity could then be transported through the regular supply chain while the climate benefits could be sold to the highest bidder. This helps provide funding for CCS without building out a complete supply chain first (Chen 2022). For example, the optimal sites for CCS may be in different regions or even countries than the ones with the highest carbon prices. Some may object to such transfers as delaying decarbonization in the original sites, but they would bring much-needed capital to deploy and scale up CCS during the initial rollout period, which could be a decade or more.

7.8 Accounting for Enhanced Oil Recovery with Carbon Storage

Does Enhanced Oil Recovery (EOR) with CCS increase or decrease overall emissions?

The production of additional oil under EOR with CCS makes it more economical to capture and store the CO_2 emissions than CCS alone. At the same time, the additional oil produced with the help of CCS would release additional GHG emissions if it's burned. So, we should be asking whether EOR with CCS reduces overall emissions, since we're storing at least some emissions from natural gas power plants underground? Or would producing more oil cause more emissions down the road? But then, if EOR doesn't use CO_2 from CCS, it could use other means to produce CO_2 at the oil field. Wouldn't that produce even more emissions? So, isn't EOR with CCS better than EOR without CCS, and CCS with EOR better than no CCS at all?

Whatever your feelings, it's indisputable that EOR changes the economics of CCS, and the climate benefits of CCS are controversial when it's paired with EOR. Even the IRS 45Q tax credit for CCS assigns a lower credit amount if the CCS is used for EOR. In 2021, a bill was introduced in the U.S. Congress to remove tax credits altogether for CCS if they were used for EOR. It gained the support of over twenty environmentalist groups but did not pass (Douglas 2021).

Over time, we could expect these issues to create a tiering in the carbon market of CCS with or without EOR. This would make the provenance and use of CCS important in addition to the quantity of GHG emissions captured and stored. CCS with EOR may qualify for compliance emissions targets in some jurisdictions and markets, but not others. Alternatively, oil extracted with CCS could be marketed as having lower carbon intensity because some emissions were captured and stored, but only if it could be proven that it was truly captured with CO_2 emissions from natural gas power plants or industrial sources.

With our proposed blockchain mechanism, both the origin and the use of the CO_2 emissions would be stored on the data ledger and hence available for review. The CO_2 capture equipment could be geo-tagged and independently verified to be connected to a natural gas power plant or industrial emissions source. The amount of CO_2 captured at the original power plant could be compared to the amount used for any EOR at the storage site. If some amount of the stored CO_2 is used for EOR, the stored amount will first need to be verified against the captured amount to make sure that it really did come from a power plant source. If so, the carbon tokens issued from the data ledger could be tagged as being used for EOR. The carbon tokens would probably then have different values if sold in the marketplace, as they would be treated differently for tax credits and emissions accounting purposes (Table 7.1).

Table 7.1 Summary of blockchain technologies for carbon capture and storage

Technology	Application	Benefits
Data Layer	Permissioned or public blockchain of supply chains	Record amount of CO_2 captured, transported, and stored Can be linked directly to IOT devices for real-time data
DAOs and Oracles	Decentralized voting based on reputation	Validate data which will be stored in the Data Layer Certify whether CCS was used for EOR or not
Non-Fungible Tokens (NFT's)	Tokens on public blockchains linked to specific lots of commodities	Store net embedded emissions of commodities such as electricity, hydrogen, or steel Account for benefits of CCS Book and claim transfers to fund green premium
CO_2 Tokens	Tokens linked to the amount of CO_2	Raise funding based on CO_2 captured and stored
CO_2 Permitted Storage Tokens	Tokens linked to permitted CO_2 storage capacity	Raise funding for site owners based on permitted CO_2 storage capacity
Smart Contracts	Executable programs on blockchain	Offtake commitments for CO_2 captured and stored and low-carbon commodities based on embedded emissions Structure insurance for CO_2 storage

7.9 Summary and Conclusion

CCS offers a promising solution for reducing or even removing the climate impact of the oil and gas industry's products. There are, however, many regulatory, economic, and technical challenges that need to be addressed. Overcoming these challenges would require a sophisticated carbon accounting mechanism that meets the needs of all the participants in the CCS value chain. The blockchain could be an ideal tool to support all the participants in the CCS supply chain. It provides a chain of custody to verify and validate stored carbon and provide long-term credibility to the data. It can also provide further economic incentives by digitizing the stored carbon and creating mechanisms for trading and insuring stored emissions as well as creating value for the site owners who obtain the permits for CCS. Finally, it could significantly streamline the business processes for all the participants in CCS.

References

Baylin-Stern A, Berghout N (2021) Is carbon capture too expensive?—Analysis. IEA, February 17. Retrieved January 12, 2023, from https://www.iea.org/commentaries/is-carbon-capture-too-expensive

Bebo D, Copithorne L, Nathanail AN, Robinette E, Yarger R (2016) Enhanced oil recovery screening. Doctoral dissertation, University of Wyoming. Libraries

Breidenbach L, Cachin C, Chan B, Coventry A, Ellis S, Juels A, Koushanfar F, Miller A, Magauran B, Magauran D, Nazarov S, Topliceanu A, Tramèr F, Zhang F (2021) Chainlink 2.0: next steps in the evolution of decentralized oracle networks. Chainlink, April 15. Retrieved January 12, 2023, from https://research.chain.link/whitepaper-v2.pdf

Bussewitz, C (2021) Exxon seeks $100 billion for Houston carbon capture plan. AP News, November 1. Retrieved January 12, 2023, from https://apnews.com/article/climate-technology-business-paris-f76df7ee4e6a8a4b6bab96badb2eb41a

Chen S (2022) Blockchain mechanism for tracking GHG emissions through supply chain. SSRN, April 27. https://ssrn.com/abstract=4082449

Douglas L (2021) U.S. lawmaker introduces bill to eliminate carbon credits for oil recovery. Reuters, December 13. Retrieved January 12, 2023, from https://www.reuters.com/markets/commodities/us-lawmaker-introduces-bill-eliminate-carbon-credits-oil-recovery-2021-12-13/

Eiken O, Ringrose P, Nazarian C, Torp TA, Høier L (2011) Lessons learned from 14 years of CCS operations: Sleipner, In Salah and Snøhvit. Energy Procedia 4:5541–5548

Environmental Protection Agency (2022) Class VI—Wells used for geologic sequestration of carbon dioxide | US EPA. Environmental Protection Agency, December 9. Retrieved January 12, 2023, from https://www.epa.gov/uic/class-vi-wells-used-geologic-sequestration-carbon-dioxide

Gordon O (2022). CCUS: where is carbon capture working?—Energy Monitor. Energy Monitor, April 28. Retrieved January 12, 2023, from https://www.energymonitor.ai/tech/carbon-removal/carbon-capture-where-is-it-working

Haahr M, Foster K, Blakstad S, Blakstad B, Suratpipit Y, Allen R, Haglund Lang L (2019) Blockchain. Gateway for sustainability linked bonds. HSBC Centre of Sustainable Finance, September 24. Retrieved January 12, 2023, from https://www.sustainablefinance.hsbc.com/mobilising-finance/blockchain-gateway-for-sustainability-linked-bonds

IEA (2022) Levelised cost of CO_2 capture by sector and initial CO_2 concentration, 2019—charts—data & statistics. IEA, October 26. Retrieved January 12, 2023, from https://www.iea.org/data-and-statistics/charts/levelised-cost-of-co2-capture-by-sector-and-initial-co2-concentration-2019

Johnson K, Schilling E, Tucker A (2021) Bipartisan infrastructure bill invests billions in CCUS | Holland & Hart LLP—JD Supra. JD Supra, November 22. Retrieved January 12, 2023, from https://www.jdsupra.com/legalnews/bipartisan-infrastructure-bill-invests-9111801/

Li D, Saraji S, Jiao Z, Zhang Y (2021). CO_2 injection strategies for enhanced oil recovery and geological sequestration in a tight reservoir: an experimental study. Fuel 284:119013. https://doi.org/10.1016/j.fuel.2020.119013

Michael K, Golab A, Shulakova V, Ennis-King J, Allinson G, Sharma S, Aiken T (2010) Geological storage of CO_2 in saline aquifers—a review of the experience from existing storage operations. Int J Greenhouse Gas Control 4(4):659–667

Moch JM, Xue W, Holdren JP, Braunstein J, O'Sullivan ML, Lee H, Schrag D, Bunn M, Davidson M, Peng W, Wang P, Mao Z, Allison G, Dunford JF, Glick J, Nye JS (2022) Carbon capture, utilization, and storage: technologies and costs in the U.S. context. Belfer Center, January 27. Retrieved January 12, 2023, from https://www.belfercenter.org/publication/carbon-capture-utilization-and-storage-technologies-and-costs-us-context

Operations for Geologic Storage Projects (revised edition ed.) (2017) DOE/NETL-2017/1848

Rassool D (2021). Unlocking private finance to support CCS investments. Global CCS Institute. Retrieved January 12, 2023, from https://www.globalccsinstitute.com/wp-content/uploads/2021/06/Unlocking-Private-Finance-for-CCS-Thought-Leadership-Report-1.pdf

Rogers M, Dubov B (2021) US tax credit encourages investment in carbon capture and storage. White & Case LLP, January 29. Retrieved January 12, 2023, from https://www.whitecase.com/publications/insight/carbon-capture/us-tax-credit-encourages-investment

Verma MK (2015) Fundamentals of carbon dioxide-enhanced oil recovery (CO_2-EOR): a supporting document of the assessment methodology for hydrocarbon recovery using CO_2-EOR associated with carbon sequestration. US Department of the Interior, US Geological Survey. Washington, DC, p 19

Vikas N, Aiyer S (2021) Why private capital is the key to unleashing carbon capture. The World Economic Forum, March 3. Retrieved January 12, 2023, from https://www.weforum.org/agenda/2021/03/financial-disclosures-on-climate-can-help-scale-up-carbon-capture/

Si Chen is the president of Open Source Strategies, Inc. in Los Angeles, CA, which specializes in open-source software for climate finance and investing. He leads the development of open-source blockchain carbon accounting software at Hyperledger Labs. Previously, he has managed investment portfolios for institutional pension funds, central banks, and hedge funds and has been published in *The Journal of Portfolio Management*. He is also the co-founder and CTO of GraciousStyle.com, an online retailer.

Dr. Soheil Saraji is an associate professor of Energy and Petroleum Engineering, an adjust professor at the School of Energy Resources, and co-director of the Hydrocarbons Research Laboratory at the University of Wyoming. He has eighteen years of research experience and more than 35 peer-reviewed journal publications in subsurface energy extraction, storage, and carbon geo-sequestration. Furthermore, Dr. Saraji is a pioneer in applied blockchain research for the oil and gas industry. He has developed new courses and research initiatives on this topic at the University of Wyoming.

Dr. Fred J. McLaughlin Ph.D., PG is a geologist and currently serves as the Director for the Center of Economic Geology (CEGR) at the University of Wyoming. Dr. McLaughlin joined CEGR at its founding and has been involved in and/or led all of its CCUS projects to-date. Dr.

McLaughlin has close to 20 years of experience working on the geology of the Rocky Mountains, including petrology and energy geology, and is a recognized expert in carbon storage.

Chapter 8
Sustainable Aviation and Transportation Fuels

Si Chen

8.1 Introduction

Could there be an oil and gas industry beyond petroleum or natural gas?

Both the oil and gas industry and governments around the world are getting ready to make a trillion dollar investment for just that eventuality. They are planning for a world where all the major assets of the industry, including refineries, tankers, pipelines, terminals, and fueling stations, are used to support fuels made from food waste, algae, and zero emission hydrogen. They would power the traditional users of petroleum and natural gas in heavy transportation, aviation, and electricity generation, turning the oil and gas companies of today into clean energy providers of tomorrow.

This is the next gold rush.

The aviation industry is already fully on board with sustainable fuels. IATA expects Sustainable Aviation Fuels (SAF) to contribute 65% of the emissions reductions needed to achieve its net zero emissions by 2050 goal (IATA 2022). The Clean Skies for Tomorrow Coalition, formed in September 2021 by 60 major airlines, aircraft manufacturers, banks, and corporations, committed to powering global aviation with 10% sustainable aviation fuels (SAF) by 2030 (Hillyer 2021b). Coming from less than 0.1% today, this amounts to a 65% annual increase over the next nine years (Wolff and Riefer 2020). American Airlines took its first delivery of SA in mid-2020, expects to use 9 million gallons by 2023, and has already entered into agreements for more than 120 million gallons of SAF (American Airlines 2021). United Airlines has been using SAF since 2016, for a total of 5 million gallons to date, with further commitments to purchase over 1.5 billion gallons (United Airlines 2022). Altogether, this means that the global supply of SAF is expected to rise to 3 million metric tons (1 billion gallons) by 2025, from less than 200,000 tons in 2019 (Fig. 8.1).

S. Chen (✉)
Open Source Strategies, Inc., Los Angeles, CA, USA
e-mail: sichen@opensourcestrategies.com

S. Saraji and S. Chen, *Sustainable Oil and Gas Using Blockchain*,
Lecture Notes in Energy 98, https://doi.org/10.1007/978-3-031-30697-6_8

Net Zero Commitments		
Carrier	**Net zero by 2050**	**SAF blending by 2030 (%)**
Air France-KLM	Y	5
American Airlines, Inc. (B-/Stable)	Y	120m gallons over 10 years
British Airways Plc. (BB/Negative)	Y	10
Delta Air Lines (BB+/Negative)	Y	10
Easy Jet	Y	-
Etihad Airways PJSC (A/Stable)	Y	-
Iberia	Y	10
JetBlue Airways Corporation (BB-/Neg)	Y (by 2040)	10
Lufthansa	Y	-
Qantas	Y	-
Ryanair Holdings plc (BBB/Stable)	Y	12.5
United Airlines, Inc. (B+/Negative)	Y	1.5bn gallons over 20 years
Source: Fitch Ratings, Company Sustainability Reports		

Fig. 8.1 Airline net zero and sustainable aviation fuel commitments. *Source* Fitch Ratings (2022)

Governments are lining up to support the airlines. Britain is calling for 10% SAF by 2030 (Slavin 2022). The EU has introduced a ReFuelEU mandate that requires a minimum of 2% of SAF by 2025, rising to 5% by 2030 and 63% by 2050 (IATA, n.d.). Finally, the US has introduced a Sustainable Aviation Fuel Grand Challenge to scale SAF production to 3 billion gallons by 2030, while introducing a new tax credit of $1.25 per gallon for SAF with at least a 50% reduction in lifecycle GHG emissions, rising to $1.75 per gallon if it achieves 100% emissions reduction (Hebert 2022). This Grand Challenge may end up being just enough to meet the needs of U.S. airlines, who are expected to consume 9 million metric tons, or 3 billion gallons, by 2030 per year (McGarrity 2021).

The rest of the transportation industry is not sitting still either. Fedex plans to "*obtain 30% of our jet fuel from alternative fuels by 2030*," such as woodland waste generated from forest management (Greene 2016). UPS has committed to fueling 40% of its ground operations by alternative fuels by 2025 (UPS 2021). The European Union's Renewable Energy Directive requires that 10% of road transportation fuels must come from renewable resources by 2020 (TransportPolicy.net, n.d.). The U.S. federal government's Renewable Fuel Standard requires 36 billion gallons of

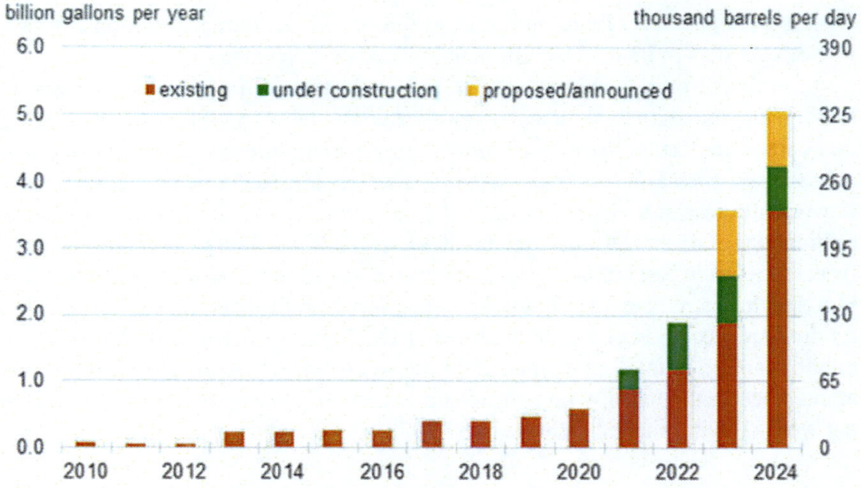

Fig. 8.2 U.S. renewable diesel production capacity. *Source* U.S. Energy Information Administration (2021)

renewable fuels, including biodiesel, to be blended into transportation fuels by 2022, a goal that has not been reached ("Alternative Fuels Data Center: Renewable Fuel Standard", n.d.). California's Low Carbon Fuel Standard calls for a 20% reduction in fuel carbon intensity by 2030 versus 2011 levels, mainly through biofuels. As a result, renewable diesel capacity in the US has doubled in just two years and stands at 1.2 billion gallons per year in 2022, and it's expected to rise to over 5 billion gallons per year by 2025 (U.S. Energy Information Administration 2021) (Fig. 8.2).

The oil and gas industry is not going to be left behind. ExxonMobil, for example, sees biofuels as the business segment with the highest global demand growth:

> ExxonMobil is focused on growing its lower-emission fuels business by leveraging current technology and infrastructure, in addition to continuing research in advanced biofuels that could provide improved longer-term solutions through upgrading lower-value bio-based feedstock. (ExxonMobil 2022)

Shell is even more specific:

> Achieving our target could mean that, by 2030, we are … increasing the amount of biofuels and hydrogen in the transport fuels we sell to more than 10%, from around 3% today. (Shell, n.d.)

BP has named "Bioenergy" as one of its three key areas for its low carbon investment focus (BP 2021).

The industry is already busily working towards its goals. Marathon Petroleum has converted a closed refinery in Martinez, CA to produce 730 million gallons of renewable diesel per year by 2023. A Phillips 66 facility in Rodeo, CA is being reconfigured to become the largest renewable diesel refinery in the world, producing 800 million gallons per year when it is completed. Most recently, Chevron has acquired biodiesel manufacturer Renewable Energy Group (REGI) for $3.15 billion as part of its strategic plan to invest $10 billion in low carbon businesses.

At the same time, governments and industries around the world are embracing hydrogen as the ultimate sustainable transportation fuel. As a highly combustible but non-carbon molecule, hydrogen could be used when high energy density and heat are required, such as heavy transportation, aviation, and steelmaking. Money is now flowing into research, development, and commercialization. The EU is targeting 10 million tons of renewable hydrogen by 2030, with 2.6% of transport fuels being made from "renewable fuels of non-biological origin" (European Commission 2021). The U.S. the Infrastructure Investment and Jobs Act has allocated $8 billion in direct funding and tax credits for four "hydrogen hubs." China has allocated $20 billion in public funding. In total, more than 350 large-scale projects worth $500 billion have already been announced, with hydrogen investments growing by roughly $1 billion per week (Environmental Defense Fund 2022).

8.2 Looking for the Right Oil

So how are we going to make all this new fuel?

Biofuels may sound futuristic, but they're not a new thing. Brazil has been promoting biofuels since 1975 with its Proálcool National Alcohol Program. By the mid-1980s, Brazil was producing 12 billion liters of ethanol, and 90% of new cars sold there were running on ethanol. In the U.S., biofuels took off with the Energy Policy Act of 2005 and the Energy Independence and Security Act of 2007, which created the federal Renewable Fuel Standards.

Decades of promoting biofuels has taught us some important lessons about the challenges for sustainable transportation fuels. First and foremost, these are new technologies which may not pan out as expected. For example, despite 15 years of research and billions of government research grants, cellulosic ethanol never took off. When the federal Renewable Fuel Standards were set in 2007, it called for 16 billion gallons of cellulosic ethanol by 2022. Today production is only 10 million gallons (DeCicco 2021). Similarly, jatropha was once touted as the biofuel feedstock of the future, but growing it was actually much more difficult and resource intensive than originally envisioned (Charles 2012; Verchot and Achten 2011).

Then there is the cost. Biofuels made from food oils are several times more expensive than petroleum-based fuels, for the simple reason that food oils are more expensive. There are 300 lbs in a barrel of oil. Even at $110 per barrel, petroleum is $0.37 per lb, which is about half the price of soybean oil at $0.74 per pound.

Finally, food oils are not easy to get. The thirst for biofuels is squeezing the food markets. Today, 40% of the corn grown in the U.S. is already used for ethanol (USDA Economic Research Service 2022) One third of the soybean oil in the U.S. is used for biodiesel (Charles 2021). All the additional renewable fuel commitments and mandates have sent prices for vegetable oils soaring. Soybean oil, for example, has risen from 26 cents per pound in May 2020 to 74 cents per pound today (Business

Insider, n.d.). Restaurants are feeling the heat. As John Healy, general manager for Columbus Vegetable Oils, Des Plaines, Ill., puts it: "*If there is a mandate out there, the mandate gets the oil first... You can't outspend a mandate*" (Gelski 2021). Longer term, though, such demand for food for fuel is creating a potentially serious problem: We only have so much land and more mouths to feed. By 2050, the world needs to produce 70% more food calories to feed a growing population that is also demanding higher quality foods, while it is losing 38,000 square miles of arable land each year (Trinastic 2015).

Meanwhile, whether biofuels are actually more sustainable is increasingly questioned. When the biofuels mandates were first introduced, governments were focused on energy security and economic development. It was generally assumed that since crops for biofuels could be grown and harvested repeatedly, they were naturally "renewable." Then, studies began to notice that biofuel production was in fact causing damaging land use changes. One study found that the palm oil biodiesel farms in Indonesia and Malaysia were converting rainforests and natural habitats to farms for biofuel (Verchot and Achten 2011). Indirect land use changes were also happening as farmland was converted to producing crops for biofuels, which then displaced people to clear rainforests, savannahs, and other natural habitats for growing food. In either case, these changes were releasing massive amounts of carbon from the soil into the atmosphere, negating the benefits of biofuels.

To avoid these negative consequences, the industry is switching to more advanced biofuels which are not made from food. For example, Scott Kirby, the CEO of United Airlines, explicitly told a panel at COP26 that they would not buy SAF made from food crops (Slavin 2022). Instead, "second generation" biofuels will be made from used cooking oils, animal fat, agricultural and sawmill residues, wood wastes, municipal waste sources, and dedicated energy crops such as miscan thus and bagasse.[1] Further on the horizon are "third generation" biofuels, such as algae, which are still in research stages and may be years away from reaching the market.

These second generation fuels, though, have supply chain challenges of their own. Used cooking oil, for example, must be collected from restaurants or industrial kitchens. Animal fat and other waste streams must similarly be collected through the different waste streams. The quantities may be small, locations are spread out, and collection is labor-intensive and slow. Even today, the amount of feedstock for biofuels is not enough for their planned use. The U.S. is already importing used cooking oil from Asia, and Europe is importing 70% of the advanced biofuels feedstock it's using. When Asian economies start their own biofuels mandates, they may start keeping this "precious" resource.

This is why there's so much interest in hydrogen. It's available wherever there is water. Just add energy, and you can turn water into hydrogen and oxygen. There are just a few things we need to figure out. First, while we have the technology to make hydrogen, the technology to use it safely is still futuristic. Cars and trucks have used bioethanol and biodiesel for decades, and even airplanes have flown successfully with sustainable aviation fuels. Cars powered by hydrogen fuel cells, however, are rare, and

[1] See https://lexparency.org/eu/32018L2001/ANX_IX/ for approved biofuel feedstocks in the EU.

hydrogen-fueled trucks are just starting to hit the road (Bomey 2020; FuelCellWorks 2021). Hydrogen aviation is still in R&D and at least a decade away (Ben-Achour 2021; O'Callaghan 2020).

If and when the technology is available, hydrogen will require massive investments in infrastructure to store and transport. Hydrogen is the lightest element in the periodic table, and its low density makes it much harder to store than fossil fuels. It can be transported through pipelines like liquid fuels, but the two could not be transported at the same time. To replace natural gas with hydrogen could require over $600 billion in investment by 2050, including storage facilities and pipelines (BloombergNEF 2020).

Furthermore, the costs to produce hydrogen will need to come down dramatically. Hydrogen is usually produced with electrolyzers, which uses energy to break down water into hydrogen and oxygen. The cost of electrolyzers came down 40% from 2014 to 2019 and is projected to continue to fall. At current prices, hydrogen costs $4 to $6 per kg to produce (Vickerset al. 2020) The goal is to get that down to $0.8 to $1.6 per kg by 2050, so that it would be competitive with natural gas at $6 to $12 per MMBtu in major countries (BloombergNEF 2020) (Fig. 8.3).

Finally, making hydrogen requires a lot of energy, so how hydrogen is made will determine its climate impact. Hydrogen is classified by its production processed, with four common labels: "black" if it's produced with coal, "gray" if it's produced with natural gas through steam methane reforming, "blue" if it's produced with natural gas and carbon capture, or "green" if it's produced with renewable energy. Today's dominant production process, steam methane reforming, produces a staggering 10 tons of CO_2 per ton of hydrogen (US Department of Energy 2020). Furthermore, since hydrogen by itself is a greenhouse gas with one hundred times the climate impact of CO_2, any leakage during transportation and storage could significantly

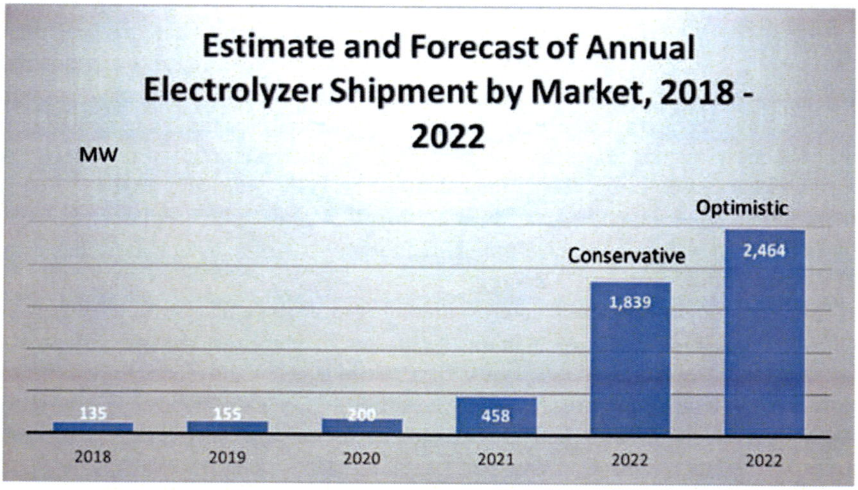

Fig. 8.3 Forecast of hydrogen electrolyzer market. *Source* BloombergNEF (2022)

increase its life cycle emissions (Ocko 2022). For these reasons, the market would probably demand strict certifications for hydrogen as it moves out of R&D and into production.

So why all this money and effort for hydrogen? Because the upside is simply mind-boggling: If produced sustainably, hydrogen has none of the food security or sourcing problems of biofuels, none of the GHG emissions of petroleum, and none of the storage limitations of batteries. It's an endless, clean, renewable source of energy that could be stored and transported much like petroleum or natural gas, perhaps even by repurposing some of the existing infrastructure. It might ultimately satisfy both environmentalists and the energy industry.

8.3 Energy 2.0

Transitioning to biofuels or hydrogen would fundamentally reconfigure the energy industry. First, it would mean that many of the oil and gas industry's assets, such as refineries, tankers, and pipelines, could be repurposed. Second, it would de-couple the assets of the industry, as the distribution of fuel would no longer be tied to petroleum reserves alone.

Finally, it would de-commoditize the industry. Whereas petroleum and natural gas produced anywhere is about the same to the end customers, biofuels and hydrogen are not. Although they could be mixed, transported, and eventually burnt together, the origin and production are critical in determining their climate impact. As climate change becomes increasingly important to the fuel users, these factors would in turn determine the marketability and pricing of fuels.

In other words, fuel would go from being a physical to a digital commodity.

This is already being happening with a new generation of biofuel standards. In 2016, the UN's International Civil Aviation Organization (ICAO)'s created a Carbon Offsetting and Reduction Scheme for International Aviation (CORSIA) program with the goal of reducing carbon emissions of airlines by 50% from 2005 levels by 2050. In 2018, CORSIA added SAF to its program (Prussi et al. 2021) It sees sustainable fuels being the most important tool for decarbonizing aviation, as airlines ramp up their use from 2035 on (Figs. 8.4 and 8.5).

To be eligible for CORSIA, though, any SAF must pass twelve criteria, including not just true GHG emissions reductions, but also protection for biodiversity, water, air quality, soil health, food security, labor, human rights, and local communities (ICAO 2021b) The fuels must be analyzed according to CORSIA's Life Cycle Analysis methodology, a 155-page document that compares the GHG emissions based on different models covering both direct emissions from fuel production and indirect emissions from land use changes (McCarl 2021). From these methodologies, CORSIA published default lifecycle emissions for fuels based on their region, feedstock, and conversion pathway (ICAO 2021a).

Meanwhile, California's Air Resources Board has been promoting alternative fuels through the Low Carbon Fuel Standard (LCFS) program since 2011. The

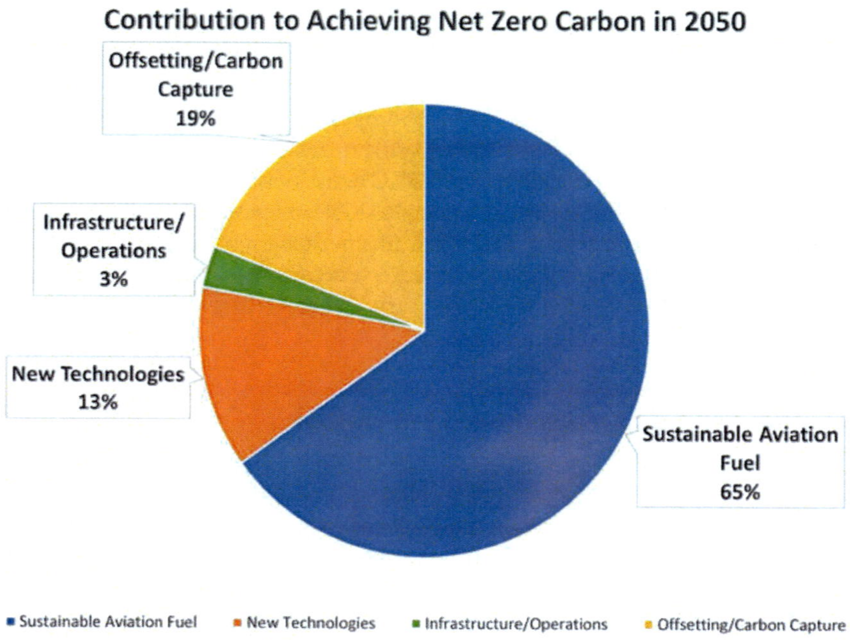

Fig. 8.4 CORSIA's plan for airlines achieving net zero. *Source* IATA (2021)

LCFS is "*designed to encourage the use of cleaner low-carbon transportation fuels in California, encourage the production of those fuels, and therefore, reduce GHG emissions and decrease petroleum dependence in the transportation sector*" (California Air Resources Board, n.d.). Its goal is to reduce the fuel carbon intensity of transportation fuels in California by 20% versus 2011 levels by 2030. The LCFS tries to do so through a cap and trade scheme where each fuel producer is required to submit a verified life cycle analysis for its product's carbon intensity. Producers with higher fuel carbon intensities must then purchase credits from those with lower fuel carbon intensities to reach the overall targets. The life cycle analysis performed under the CA GREET 3.0 model, which is a detailed analysis of every step of the supply chain, from feedstock to processing, including the source for petroleum and methane flaring during its production.

Both CORSIA SAF mandates and the California LCFS show an increasingly sophisticated market compared to the Renewable Fuel Standard days of 15 years ago, when all biodiesel was considered the same thing. As the alternative fuels markets grow and other countries and regions adopt mandates, we should expect ever more detailed requirements. For example, the CA GREET 3.0 model life cycle analysis required by the California LCFS is done on a very large spreadsheet. Furthermore, the CA GREET 3.0 model uses the 2006 IPCC GHG Inventory Guide (California Air Resources Board 2018). Since then several updated versions have been released, most importantly significantly increasing the emissions impact of methane which is

Milestones towards net zero

The below table illustrates a potential set of estimated milestones towards net-zero, including the mix of abatement measures ('pathways') and some noteworthy actions envisaged.

DATE	AMOUNT OF CO_2 ABATEMENT	PATHWAY	ACTION
2025	**381 megatonnes (Mt)** (2021-2025)	97% offsets, 2% SAF, 1% improvements above business as usual (BAU)	ICAO agree long-term goal for international aviation (2022); energy sector commits to at least 6 million tonnes SAF production; agreement of full implementation of Article of Paris Agreement
2030	**979 Mt** (2026-2030)	93% offsets; 5% SAF, 2% Improvements above BAU	Use of 100% SAF on aircraft. ANSPs fully implement ICAO Aviation System Block upgrades to deliver fuel efficiency improvements of 0.3% by 2030
2035	**1,703 Mt** (2031–2035)	77.5% offsets, 17.5% SAF, 3% improvements above BAU, 2% Carbon Capture Utilization and Storage (CCUS)	Evolutionary technology achieving 30% reduction in fuel burn, electric/hydrogen aircraft for regional markets (50-100 seats, 30-90 min flights) become available
2040	**3,824 Mt** (2036-2040)	44.5% offsets, 40% SAF, 7.5% non drop-in fuel (new propulsion technologies), 5% CCUS, 3% improvements above BAU	Feasibility of new aircraft such as blended-wing bodies demonstrated with full-scale working prototypes, electric/hydrogen for short-haul markets (100-150 seats, 45-120 min flights) become available.
2045	**6,153 Mt** (2041-2045)	55% SAF, 24% offsets, 10% non drop-in fuel, 8% CCUS, 3% improvements above BAU	Necessary infrastructure for new energy requirements (low carbon electricity/hydrogen) becomes available
2050	**8,164 Mt** (2046-2050)	65% SAF, 13% non drop-in fuel, 11% CCUS, 8% offsets, 3% improvements above BAU	Commercially viable annual SAF production of 449 billion litres available

Fig. 8.5 Schedule of targeted sustainable aviation fuel use for airlines. *Source* IATA (2021)

emitted during the production of some biofuels. While the California LCFS publishes the resulting carbon intensities of different fuels, the analysis backing up those results are not open for public review. Email requests for this information were not answered, and some of the phone numbers listed on its website were not in service. As emissions intensities of fuels becomes the basis for their pricing and their ability to meet government compliance requirements, all parties, including suppliers, customers, and general public, would probably demand greater transparency in the emissions intensities are calculated.

8.4 The Role for the Blockchain

So far, the sustainable fuels market has been a gold rush for production. There's been neither the need nor the time for anything else. As the market matures, though, both producers and users would get more sophisticated. Buyers would demand more rigorous validation of the attributes they are paying a premium for. Similarly, sellers would want to monetize more of the benefits for their products, from improvements in the original feedstock to production process efficiencies to using more renewable energy. Finally, both sellers and buyers as well as the general public may demand more transparent and up-to-date data about the fuels than the government agencies could provide. Renewable fuels and hydrogen must prove their emissions benefits to succeed, or fall to the wayside like bioethanol of yesteryear.

These requirements would lead to mechanisms similar to those in the carbon markets, including detailed product data, third party standards, independent verification and certification, digital certificates for trading, and procedures to prevent double counting of emissions benefits. Furthermore, these mechanisms would need to support a large range of producers of all different sizes in different locations using many different types of feedstock.

These are the challenges which are well suited for blockchain technology. We envision the following significant roles for blockchain in the sustainable aviation and transportation fuels space:

- Actual supply chain emissions calculations instead of LCA's
- Certification of origin
- Tradable certificates
- Markets for forward purchases
- Emissions reduction accounting

Since each alternative fuel has different raw materials and production processes, their embedded emissions footprint will differ widely. Not only will there be significant differences between the fuel types, such as soy versus used cooking oil versus hydrogen, but even different producers or lots of the same type of biofuels may have significantly different emissions. For example, because hydrogen is an important component for many biofuels, biofuels produced with hydrogen from electricity during peak hours of renewable generation would have lower emissions footprints than those produced during other times of the day or from on-site natural gas combustion. As the market grows and prices fall, producers would likely seek to differentiate their products based on the actual raw materials and production processes. Similarly, buyers may demand higher levels of assurance that their purchases truly met their promised quality.

To calculate the emissions footprint at this level of detail, we would need data from actual invoices for sales of feedstock materials and meters at the production facilities. Today's certification processes are not designed for this, since they rely on models and general aggregates. Furthermore, supply chains which include many participants of different sizes across different industries and countries are difficult for any single software solution to capture the full scope of the market. In such a case, the blockchain is a better solution that could span many participants of disparate sizes, industries, and jurisdictions. It is a lightweight integration channel where the minimum data common to all participants could be shared and maintained, without requiring custom integrations between particular participants. It could also be maintained collectively without the need for a single central authority that all parties must trust (Chen 2022).

At the same time, many companies consider such data confidential trade information and would only provide it to authorized parties, such as partners or auditors. The blockchain offers several potential solutions. First, the data could be encrypted using the public keys of trusted parties, so that only those parties could decrypt and access the data. Second, the data could be placed in permissioned blockchain networks, such as Hyperledger Fabric, or even inside private data channels of those networks, so that only trusted parties would have access to them. Finally, sophisticated algorithms such as "zero knowledge proofs" allow parties to prove claims without sharing the data behind them.

The right blockchain solution depends on the volume of data and the level of collaboration between the parties. Low volume data such as monthly utility bills could be stored and shared using encrypted files. High volume data such as live meter readings, however, would require a permissioned blockchain network, which provides a database to each participant. Low collaboration scenarios where only periodic total emissions footprints are needed could be on public or permissioned blockchains using tokens to represent the emissions values. High collaboration scenarios which require sharing granular data would need to be on permissioned networks where partners could be granted access to detailed data.

Such networks could include a broader range of partners than traditional software. For example, auditors and certification agencies could be granted access to data to certify the output for various regulators, customers, and markets. This could be important given the number of different certifications required in different countries or by different customers. These agencies could be given access to detailed data in permissioned data ledgers and then issue certifications in the form of Non-Fungible Tokens (NFT's.) Such an NFT could be attached to particular lots of output, for example all the fuel produced during a certain period of time, from a particular feedstock or raw material, or at a particular plant.

The NFT would be a token which could be transferred to customers to use in their emissions accounting or to satisfy regulatory requirements. NFT's could also be used to transfer the benefits of better quality products between parties without physical transfers, thus eliminating the need for expensive infrastructure and transportation costs. For example, a seller could exchange their NFT from a lot of fuel with a low emissions footprint with a buyer's NFT from a higher emissions lot. This is similar to the transfer of Energy Attribute Certificates or Renewable Energy Certificates, where a buyer could purchase claims to electricity generated from renewable sources without being physically connected to the wind turbines or solar panels. In such a transfer, the seller could no longer claim that their electricity is from renewable sources, but the buyer could now make that claim. The price paid for the transfer is thus the premium of renewable electricity over average grid electricity and varies depending on the supply of one versus the other. In the case of transferring NFT's of high and low emissions footprint fuels, the sellers and buyers are actually exchanging claims of two particular fuel lots' emissions footprints rather than one fuel lot versus an average. The value of the transfer should be based on the amount of the difference in emissions footprint and the value of the emissions reduction, for example in a particular regulated market such as the California LCFS or CORSIA, or by the premium customers are willing to pay for a product.

Several groups are already trying to implement such "book and claim" processes. For example, a "Sustainable Aviation Fuel Certificates" or SAFc, pioneered by the World Economic Forum or RMI, promises to scale the verified sustainable fuels:

> For the SAFc system to function smoothly and at scale, it requires a robust tracking and verification process as well as a registry to ensure that climate-related claims are legitimate and only claimed by a single party. (Hillyer 2021a)

Another project called Avelia, a pilot of Shell, Accenture, and American Express, is using the Energy Web Foundation's Ethereum-based blockchain to implement a book and claim process. Its initial launch in June 2022 offered 1 million gallons of SAF with the three companies' business travel departments as the initial customers (Shell 2022). According to Sabine Brink, the blockchain lead at Shell, this is just the beginning:

> With Avelia, we hope to demonstrate that the tracking of SAF data at scale can be delivered in a credible manner, thereby proving to decision-makers that a mechanism for corporations and airlines to book and claim SAF is an acceptable form of emission reduction. In turn, this creates increased demand signals to structurally scale the SAF production required to reduce emissions in aviation. (Danise and Marr 2022)

Such transfers, in turn, open up another important potential application for blockchains: Financing. Since NFT's of emissions footprint could be traded, a fuel producer could offer such NFT's from future production for sale to finance investment and production. Buyers could purchase this future production to meet their climate goals or regulatory requirements. While buyers today are already entering into long-term offtake agreements, such agreements are often negotiated directly between buyers and sellers. They also often involve custom negotiated standards of quality between buyers and sellers. This makes the process slow, expensive, and difficult to scale. As the supply of alternative transportation fuels increases, buyers may not feel the same urgency to lock in supply, making it increasingly important for sellers to streamline their transactions process.

It may be tempting to turn to commodity futures exchanges to create more streamlined long-term markets, but futures are not the right solution. Futures markets work best when there is large volume around a single commodity with standard characteristics. Whenever there are many different grades or variants, they try to use a standard commodity as the basis for trading and then set up pre-determined rules to calculate the price difference between the actual product and the standard commodity. This inevitably leads to two problems. Some products must not be accepted by the exchange as interchangeable or deliverable against the standard commodity. Other products may have prices which differ so significantly from the standard commodity that the traded futures do not really have much correlation with them. The need for pre-determined pricing rules between different grades and the standard commodity makes it virtually impossible to support all the different varieties of biofuels or hydrogen from different production processes. Finally, products with small volumes may simply not be economical for the futures exchange to include.

In a market with a wide variety of innovative new products like sustainable fuels, the blockchain is a better alternative to exchange traded futures markets because of its low cost and flexibility. The blockchain could be used to set up a peer to peer Over the Counter (OTC) market. Every type of biofuel and hydrogen could be offered by sellers as NFT's with their attributes and emissions profiles. The buyers could stake tokens of values, such as crypto stablecoins, for future purchases of NFT's. The values of NFT's would be individually determined between buyers and sellers based on their attributes and the parties' respective long-term supply and demand. Auditors with access to permissioned data of the producers would certify production based on agreed upon criteria and issue the NFT's. Performance guarantees for trades and even settlement could be managed by programs running on the blockchain, or smart contracts.

Therefore, as alternative transportation fuels grow, we expect that the market would evolve from long-term custom offtake agreements to OTC markets driven by blockchain. Blockchain OTC markets would be a major step forward for these fuels, as it would open up the base of potential buyers significantly and streamline the speed of transactions while lowering their costs.

Finally, the nascent alternative transportation fuels industry must address the double counting issue that has always been present with carbon credits. Already, observers have pointed out the flaws in the current alternative fuels markets that could lead to double counting. For example, when renewable energy is used to produce the fuels, could the renewable energy certificates also be counted as grid emissions reductions? (Jones 2020) Could emissions reductions be counted once under the California LCFS, and then a second time under another program? (Fu and Kruzman 2022) Could they be claimed as emissions reductions by the company which produces the low carbon fuels, and then again by the company which uses it later? Could the host governments of both companies then claim the low carbon fuel as part of their Paris Agreement contributions?

These possibilities make it important to create meta-registries, so that their emissions reductions claims could be tracked to prevent double counting. In the carbon markets, there are already efforts such as the Climate Warehouse to do just that.[2] Such efforts are impeded by the legacy of existing registries' data silos, even if they themselves use the blockchain. In the new, emerging of alternative fuels and hydrogen, a blockchain infrastructure could make creating such meta-registries much easier by avoiding the costly and time consuming integrations with legacy systems (Table 8.1).

---

[2] See our chapter on Carbon Markets for more details.

Table 8.1 Summary of blockchain technologies for sustainable fuels

Technology	Application	Benefits
Data Layer	Permissioned or public blockchain of data from	Trace supply chain inputs to fuels Calculate true emissions
DAOs and Oracles	Decentralized voting based on reputation	Certify emissions footprint Certify compliance with environmental, social goals
Non-Fungible Tokens (NFT's)	Tokens on public blockchains linked to specific lots of fuel	Immutable records on blockchain Store emissions footprint and sourcing compliance
Tokens	Tokens of carbon credits on public blockchains	Registry for ownership Trade benefits of fuels without physical transfer Pay for green premium
Smart Contracts	Executable programs on blockchain	Structure commitments to purchase fuels Funding mechanisms
Meta Registries	Public blockchain of from multiple registries	Transparency for market, regulators, general public Prevent double counting

8.5 Summary and Conclusion

Yes, there is an oil and gas industry beyond petroleum and natural gas.

It would be a very different industry, with digital certificates of provenance and emissions moving through blockchains while physical commodities are transported by pipelines and tankers. It would require not only a huge commitment by governments and industry, but also recognition of those efforts by customers and the general public.

While all these transformations are happening in earnest, we must not take their success for granted. Past attempts to create "renewable fuels" that had little or no climate benefits are already being passed over for the next generation of low carbon fuels. If the industry doesn't get it right this time, though, it may not get another chance. Electric buses and trucks are hitting the market, and even electric airplanes are being developed.

Therefore, even in the current gold rush, the industry and its customers must be rigorous about proving the true environmental benefits of the new fuels. Part of this would inevitably involve embracing a new way of thinking around digital technologies such as the blockchain.

References

Alternative Fuels Data Center: Renewable Fuel Standard (n.d.) energy.gov. https://afdc.energy.gov/laws/RFS.html. Accessed 1 Sept 2022

American Airlines (2021) American airlines makes additional commitment to sustainable aviation fuel. American Airlines Newsroom. https://news.aa.com/news/news-details/2021/American-Airlines-Makes-Additional-Commitment-to-Sustainable-Aviation-Fuel-CORP-OTH-12/default.aspx

Ben-Achour S (2021) How close are we to hydrogen planes, really? Marketplace.org. https://www.marketplace.org/2021/10/28/how-close-are-hydrogen-planes-really/

BloombergNEF (2020) Hydrogen economy outlook. Bloomberg Professional Services. https://data.bloomberglp.com/professional/sites/24/BNEF-Hydrogen-Economy-Outlook-Key-Messages-30-Mar-2020.pdf

BloombergNEF (2022) Hydrogen—10 predictions for 2022. BloombergNEF. https://about.bnef.com/blog/hydrogen-10-predictions-for-2022/

Bomey N (2020) Nikola, Toyota, Hyundai seek zero-emission hydrogen trucks. USA Today. https://www.usatoday.com/story/money/2020/10/26/hydrogen-trucks-nikola-gm-toyota-hyundai-zero-emissions/5981340002/

BP (2021) Sustainability report 2020. BP. https://www.bp.com/content/dam/bp/business-sites/en/global/corporate/pdfs/sustainability/group-reports/bp-sustainability-report-2020.pdf

Business Insider (n.d.) Soybean oil price today | soybean oil spot price chart | live price of soybean oil per ounce | markets insider. Markets Insider. https://markets.businessinsider.com/commodities/soybean-oil-price. Accessed 1 Sept 2022

California Air Resources Board (2018) CA-GREET3.0 supplemental document and tables of changes. CA.GOV. https://ww2.arb.ca.gov/sites/default/files/classic/fuels/lcfs/ca-greet/cagreet_supp_doc_clean.pdf

California Air Resources Board (n.d.) Low carbon fuel standard. CA.GOV. https://ww2.arb.ca.gov/our-work/programs/low-carbon-fuel-standard/about. Accessed 2 Sept 2022

Charles D (2012) How a biofuel dream called Jatropha came crashing down. NPR. https://www.npr.org/sections/thesalt/2012/08/22/159391553/how-a-biofuel-dream-called-jatropha-came-crashing-down

Charles D (2021) Biodiesel is booming, thanks to environmental incentives. NPR. https://www.npr.org/2021/10/28/1043413986/biodiesel-is-booming-it-may-help-the-climate-but-theres-a-big-environmental-risk

Chen S (2022) Blockchain mechanism for tracking ghg emissions through supply chain. SSRN. https://ssrn.com/abstract=4082449 or https://doi.org/10.2139/ssrn.4082449

Danise A, Marr B (2022) How shell is using web3 and blockchain for sustainability and energy transition. Forbes. https://www.forbes.com/sites/bernardmarr/2022/07/15/how-shell-is-using-web3-and-blockchain-for-sustainability-and-energy-transition/amp/

DeCicco J (2021) The US biofuel mandate helps farmers, but does little for energy security and harms the environment. The Conversation. https://theconversation.com/the-us-biofuel-mandate-helps-farmers-but-does-little-for-energy-security-and-harms-the-environment-168459

Environmental Defense Fund (2022) Hydrogen is booming: 3 things investors need to know to reduce their risk. Forbes, March 16, 2022. https://www.forbes.com/sites/edfenergyexchange/2022/03/16/hydrogen-is-booming-3-things-investors-need-to-know-to-reduce-their-risk/?sh=7249a3127398

European Commission (2021) Hydrogen fact sheet. European Commission. https://ec.europa.eu/commission/presscorner/detail/en/FS_21_3676

ExxonMobil (2022) Advancing climate solutions 2022 progress report. Exxon Mobil. https://corporate.exxonmobil.com/-/media/Global/Files/Advancing-Climate-Solutions-Progress-Report/2022-July-update/ExxonMobil-Advancing-Climate-Solutions-2022-Progress-Report.pdf

Fitch Ratings (2022) Transportation—long-term climate vulnerability scores. Fitch Ratings. https://www.fitchratings.com/research/corporate-finance/transportation-long-term-climate-vulnerability-scores-25-04-2022

Fu J, Kruzman D (2022) Is California giving its methane digesters too much credit? Grist. https://grist.org/accountability/california-methane-digesters-low-carbon-fuel-standard/

FuelCellWorks (2021) The first hydrogen trucks are rolling in Europe. FuelCellsWorks. https://fuelcellsworks.com/news/the-first-hydrogen-trucks-are-rolling-in-europe/

Gelski J (2021) Oil prices doubling, or even tripling, makes fry life a priority. Food Business News. https://www.foodbusinessnews.net/articles/18707-oil-prices-doubling-or-even-tripling-makes-fry-life-a-priority

Greene J (2016) FedEx commits to biofuels derived from healthy western forests. Forest2Market. https://www.forest2market.com/blog/fedex-commits-to-biofuels-derived-from-healthy-western-forests

Hebert J (2022) Sustainable aviation taking off thanks to inflation reduction act—third way. Third Way. https://www.thirdway.org/blog/sustainable-aviation-taking-off-thanks-to-inflation-reduction-act

Hillyer M (2021a) New certificates offer flyers a sustainable fuel option to cut CO_2. The World Economic Forum. https://www.weforum.org/press/2021a/06/new-certificates-offer-flyers-a-sustainable-fuel-option-to-cut-co2

Hillyer M (2021b) Clean skies for tomorrow leaders commit to 10% sustainable aviation fuel by 2030. The World Economic Forum. https://www.weforum.org/press/2021b/09/clean-skies-for-tomorrow-leaders-commit-to-10-sustainable-aviation-fuel-by-2030/

IATA (2021) Net zero carbon 2050 resolution. IATA. https://www.iata.org/contentassets/b3783d24c5834634af59148c718472bb/factsheet_netzeroresolution.pdf

IATA (2022) Net zero 2050: sustainable aviation fuels. IATA. https://www.iata.org/en/iata-repository/pressroom/fact-sheets/fact-sheet---alternative-fuels/

IATA (n.d.) Fact sheet: EU and US policy approaches to advance SAF production. IATA. https://www.iata.org/contentassets/d13875e9ed784f75bac90f000760e998/fact-sheet---us-and-eu-saf-policies.pdf. Accessed 1 Sept 2022

ICAO (2021a) CORSIA default life cycle emissions values for CORSIA eligible fuels. ICAO. https://www.icao.int/environmental-protection/CORSIA/Documents/ICAO%20document%2006%20-%20Default%20Life%20Cycle%20Emissions%20-%20November%202021a.pdf

ICAO (2021b) CORSIA sustainability criteria for CORSIA eligible fuels. ICAO. https://www.icao.int/environmental-protection/CORSIA/Documents/ICAO%20document%2005%20-%20Sustainability%20Criteria%20-%20November%202021b.pdf

Jones T (2020) Re: Comments of Center for Resource Solutions (CRS) on October 14–15, 2020 Low Carbon Fuel Standard (LCFS) Public workshop to discuss potential regulation revisions. Center for Resource Solutions. https://resource-solutions.org/wp-content/uploads/2021/03/CRS-comment-on-LCFS-Public-Workshop-on-Proposed-Revisions-11-4-2020.pdf

McCarl B (2021) CORSIA supporting document "CORSIA Eligible Fuels_LCA_Methodology". ICAO. https://www.icao.int/environmental-protection/CORSIA/Documents/CORSIA_Supporting_Document_CORSIA%20Eligible%20Fuels_LCA_Methodology_V3.pdf

McGarrity J (2021) US airlines pledge huge increase in use of sustainable aviation fuel by 2030. Fastmarkets. https://www.fastmarkets.com/insights/us-airlines-pledge-huge-increase-in-use-of-sustainable-aviation-fuel-by-2030

O'Callaghan J (2020) Quiet and green: why hydrogen planes could be the future of aviation. European Commission. https://ec.europa.eu/research-and-innovation/en/horizon-magazine/quiet-and-green-why-hydrogen-planes-could-be-future-aviation

Ocko I (2022) Climate consequences of hydrogen emissions. ACP. https://acp.copernicus.org/preprints/acp-2022-91/

Prussi M et al (2021) CORSIA: the first internationally adopted approach to calculate life-cycle GHG emissions for aviation fuels. Science Direct. https://www.sciencedirect.com/science/article/pii/S1364032121006833

Shell (2022) Shell, Accenture and Amex GBT launch one of the world's first blockchain powered digital book-and-claim solutions for scaling sustainable aviation fuel (SAF). Shell Global. https://www.shell.com/business-customers/aviation/news-and-media-releases/news-and-media-2022/shell-accenture-and-amex-gbt-launch-one-of-the-worlds-first-blockchain-powered-digital-book-and-claim-solutions-for-scaling-sustainable-aviation-fuel-saf.html
Shell (n.d.) Achieving net-zero emissions. Shell Global. https://www.shell.com/powering-progress/achieving-net-zero-emissions.html. Accessed 1 Sept 2022
Slavin T (2022) Airlines seek clearance for liftoff on sustainable aviation fuels. Reuters. https://www.reuters.com/business/sustainable-business/airlines-seek-clearance-liftoff-sustainable-aviation-fuels-2022-04-18/
TransportPolicy.net (n.d.) EU: fuels: biofuel policy | transport policy. TransportPolicy.net. https://www.transportpolicy.net/standard/eu-fuels-biofuel-policy/. Accessed 1 Sept 2022
Trinastic J (2015) The biofuel controversy. Scitable. https://www.nature.com/scitable/blog/eyes-on-environment/the_biofuel_controversy/
United Airlines (2022) Emissions reduction, sustainable fuel & innovation. United Airlines Corporate Responsibility Report. https://crreport.united.com/environmental-sustainability/emissions-reduction-sustainable-fuel-and-innovation
UPS (2021) 2020 UPS corporate sustainability report. About UPS. https://about.ups.com/us/en/social-impact/environment/climate/2020-ups-corporate-sustainability-report.html
USDA Economic Research Service (2022) Feedgrains sector at a glance. USDA ERS. https://www.ers.usda.gov/topics/crops/corn-and-other-feedgrains/feedgrains-sector-at-a-glance/
US Department of Energy (2020) Sustainable aviation fuel: review of technical pathways report. Department of Energy. https://www.energy.gov/sites/prod/files/2020/09/f78/beto-sust-aviation-fuel-sep-2020.pdf
U.S. Energy Information Administration (2021) U.S. renewable diesel capacity expected to increase significantly through 2024. EIA. https://www.eia.gov/petroleum/weekly/archive/2021b/210721/includes/analysis_print.php
Verchot L, Achten W (2011) Implications of biodiesel-induced land-use changes for CO_2 emissions: case studies in Tropical America, Africa, and Southeast Asia. Ecol Soc 16(4):38
Vickers J, Peterson D, Randolph K (2020) DOE hydrogen and fuel cells program record 20004: cost of electrolytic hydrogen production with existing technology. Hydrogen.Energy.gov. https://www.hydrogen.energy.gov/pdfs/20004-cost-electrolytic-hydrogen-production.pdf
Wolff C, Riefer D (2020) Clean skies for tomorrow sustainable aviation fuels as a pathway to net-zero aviation. weforum.org. https://www3.weforum.org/docs/WEF_Clean_Skies_Tomorrow_SAF_Analytics_2020.pdf

Mr. Si Chen is the president of Open Source Strategies, Inc. in Los Angeles, CA, which specializes in open-source software for climate finance and investing. He leads the development of open-source blockchain carbon accounting software at Hyperledger Labs. Previously, he has managed investment portfolios for institutional pension funds, central banks, and hedge funds and has been published in *The Journal of Portfolio Management*. He is also the co-founder and CTO of GraciousStyle.com, an online retailer.

Chapter 9
Sustainable Plastics

Si Chen, Katerina Serada, and Joseph Wyer

9.1 Introduction

Could oil and gas companies join the ranks of clean tech companies, riding a wave of growth while helping the world fix climate change?

The industry in fact does have a product that could reduce emissions while improving the lives of billions of people.

Plastics.

Cheap to make, easy to shape, and infinitely versatile, plastics are a true miracle of modern technology. They have come a long way since the 1960, where the material was popularly compared to a "cheap, ugly and sterile way of life" (Seabrook 2010). Today, we cannot imagine a world without plastic bottles for our medicine, plastic wrapping for groceries and food, plastic cases for electronics, plastic toys for children, and plastic packaging for just about everything. Plastics are also deeply rooted in the global economy. In 2019, the value of international trade in plastics reached over $1 trillion, or 5% of total international trade (UNCTAD 2021).

They are also playing important roles in fixing climate change. Plastics have life-cycle Greenhouse Gas (GHG) emissions that are between 10 and 90% of alternatives such as steel, aluminum, glass, fiberglass, paper, and cotton (Fig. 9.1).

They make buildings more energy efficient with better insulation, reduce food spoilage, reduce car and aircraft weight, make offshore wind turbines more durable, and reduce transportation weight of packaging. Carbon-fiber reinforced polymers are ten times lighter than the aluminum, steel, and titanium used in aircraft and

S. Chen (✉) · J. Wyer
Open Source Strategies, Inc., Los Angeles, CA, USA
e-mail: sichen@opensourcestrategies.com

J. Wyer
e-mail: wyer@usc.edu

K. Serada
SDG Hub, Milan, Italy

S. Saraji and S. Chen, *Sustainable Oil and Gas Using Blockchain*,
Lecture Notes in Energy 98, https://doi.org/10.1007/978-3-031-30697-6_9

% Difference in Total Greenhouse Gas Contributions, United States 2020 [1]

Comparison	Sector	Application	Plastic vs	Next-best alternative	
Plastics vs alternative materials	Packaging	Grocery bag	HDPE[3]	Paper	80
		Wet pet food packaging	PET/PP[4]	Aluminum or steel	70
		Soft drink container	PET	Aluminum	50
		Fresh-meat packaging	EPS/PVC[5]	Paper	35
		Industrial drum	HDPE	Steel	-30
		Soap container	HDPE	Glass	15
	Building and Construction	Municipal sewer pipe	PVC	Concrete or ductile iron	35-45
		Residential water pipe	PEX[6]	Copper	25
		Insulation	PU[7]	Fiberglass	80
	Consumer goods	Furniture	PP	Wood	50
	Automotive	Hybrid fuel tank	HDPE	Steel	90
		BEV[2] battery top enclosure	PP/glass fiber	Steel	10
	Textiles	Carpet	PET/nylon	Wool	80
		T-shirt	PET	cotton	15
Plastics vs plastics-enabled mixed materials	Packaging	Milk container	HDPE	Paper	-5
	Consumer goods	Water cup	EPS	Paper	0

Fig. 9.1 Plastics have a lower greenhouse gas impact than most alternatives analyzed, accounting for both direct and value-chain emissions. *Note* [1]Includes indirect emissions. [2]Battery Electric Vehicle. [3]High-density polyethylene. [4]Polyethylene terephthalate/Polypropylene [5]Expanded Polystyrene/Polyvinyl Chloride [6]Cross-linked Polyethylene [7]Polyurethane. *Source* McKinsey and Company (2022)

could reduce their weight by as much as 20%. Biobased plastics, such as biobased polyamides and biobased polyesters, are increasingly used for automotive parts such as seat cushions, bumpers, and fuel tanks to reduce vehicle weight and improve the performance of electric and hybrid vehicles (IMARC Group 2022).

For the oil and gas industry, plastics is a crucial growth area as demand for fossil fuels falls. In the growing economies of China, India, and Africa, per capita plastic use is less than half to one tenth that of the US or Europe (International Energy Agency 2018). ExxonMobil expects the chemicals market to grow by 160% from 2017 to 2030, exceeded only by biofuels, while other business segments are flat or declining (Exxon Mobil 2020). By 2050, the IEA expects that per capita consumption of plastics will exceed that of oil demand for passenger cars in the US, EU, China, and India (Fig. 9.2).

Unfortunately, much of the world does not want this to happen.

The key benefits of plastics–low cost and durability–have also become their own worst enemies. What do we do with our plastic bags, bottles, and toys when we're done with them? Nobody seems to know. So plastic waste just piles up. We see it on our daily walks and our vacations. Then we turn on the news, and we see mountains of it in city dumps and islands of it in rivers and oceans. While climate change seems far off, plastic waste is an immediate and visible problem. As a result, people are demanding, and governments are enacting bans on various types of plastics.

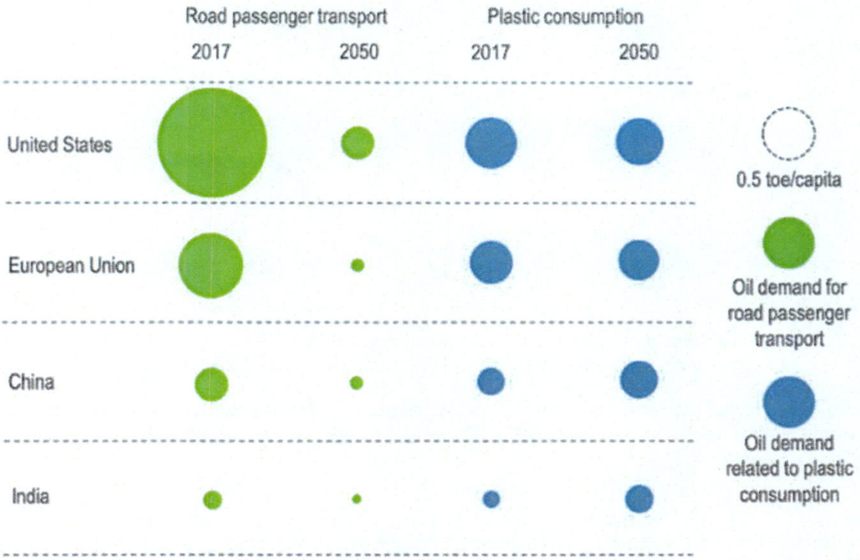

Fig. 9.2 Per capita demand for oil under "clean technology scenario." *Note* Toe/capita=tonne of oil-equivalent per capita. Oil demand for plastic consumption is calculated based on global average HVC energy intensities, including both feedstock and process energy, to reflect that plastic are widely traded. *Source* International Energy Agency (2018). IEA 2022; The Future of Petrochemicals; https://www.iea.org/reports/the-future-of-petrochemicals; CC BY 4.0

Can we make plastics sustainable, growing businesses instead of a hated public blight? In this chapter, we'll look at how that could happen and how blockchain technologies could help.

9.2 How Bad Is the Plastics Problem?

According to the recent OECD Global Plastic Outlook, only 9% of plastic waste was ultimately recycled (OECD 2022b). Meanwhile, 19% of the waste was incinerated and almost 50% went to landfills (Fig. 9.3). The remaining 22% was disposed of in uncontrolled dumpsites, burned in open pits or leaked into the environment. In contrast, 68% of paper is recycled (EPA 2021).

As a result, 109 million metric tons of plastics have accumulated in rivers, and 30 million metric tons in the ocean. Even if the dumping of plastic waste to rivers stops immediately, the existing build-up of plastics in rivers implies that plastic waste will continue to flow into the oceans for decades. The OECD further predicts the global growth of plastic waste streams under the business-as-usual scenario will cause a near tripling of plastic waste from 353 million tons in 2019 to 1.014 billion tons in 2060 (OECD 2022a). Short-lived applications in packaging, consumer products,

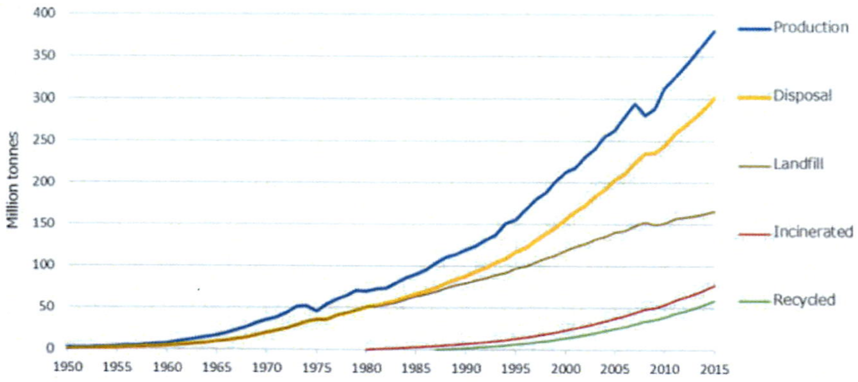

Fig. 9.3 Global production and fate of plastic waste, showing the estimated amounts disposed of via landfill, recycling and incineration. 'Disposal' represents the sum of those three processes. *Source* Geyer et al. (2017)

and textiles are projected to continue to dominate plastic waste, at 63% in 2019 and 59% in 2060. Plastic waste from construction and transport applications also remain important contributors, especially given the rapid economic development in many developing and emerging economies.

As if the environmental concerns weren't bad enough, plastics are starting to become a health issue as well. Microplastics, polymers with a diameter smaller than 5 mm, account for 12% of plastic waste (OECD 2022a). In March 2022, microplastics were detected in the human bloodstream for the first time (Turns 2022). This discovery is all the more concerning because some microplastics are small enough to cross the blood–brain barrier. Microplastics often carry toxic chemicals absorbed from the environment. In a study of accumulation in fish, microplastics alone were shown to cause neurotoxicity, oxidative damage, and energy-related changes (Barboza et al. 2018).

For years, much of the plastic waste problem was "out of sight, out of mind" thanks to trade. China alone imported nearly 60% of the world's plastic waste, with 16 million tons of plastics, paper, and other waste coming from the US alone in 2016. In December 2017, though, China banned the importing of most plastic waste as part of its National Sword Policy. The government cited the massive volume of "dirty" or "hazardous" waste which were not recyclable as a threat to its own environment. The plastics industry was thrown into upheaval. The US began sending plastic waste to Thailand, Vietnam, and Malaysia until those countries banned their import as well. Although we continue sending waste abroad to countries like Cambodia, Bangladesh, Ghana, Laos, Ethiopia, Kenya and Senegal, the bulk of plastic waste now has nowhere to go (Cho 2020).

As a result, plastic scrap prices collapsed. Municipalities ended up paying to remove plastic waste instead of getting paid for them just a year earlier. Franklin (NH), for example, went from getting paid $6 for its recyclables to either paying $125 per ton to recycle or $68 per ton to incinerate them. Stamford (CT) went from making \$95,000 from selling recyclables in 2017 to paying \$700,000 to remove them in 2018. Not surprisingly, over seventy US municipalities scrapped their recycling programs because they could no longer afford them (Cho 2020).

So the plastic waste piled up. The public got upset. Something had to be done.

9.3 Solving the Plastics Problem

The knee-jerk reaction is to simply ban plastics.

Today, 75% of people around the globe want single-use plastics (SUP) to be banned as soon as possible (Geddie 2022). Governments are starting to do just that. Canada is planning to ban a range of SUP products by 2023, including checkout bags, stir sticks, and straws. Maine, Oregon, and Colorado have recently passed 'extended recycler responsibility' laws, and California and New York are pushing new legislation around recycling and SUP (Schuerman 2022). The EU has already banned a number of SUP products, and it is taking steps to reduce the use of plastic products for which there are no alternatives and increase the amount of recycled material within plastic bottles. India has banned SUP's and mandated aggressive recycling targets for the rest in its "Extended Producer Responsibility" laws (Abraham 2022). Even in Africa, where plastic use is still low and demand is expected to grow significantly, 30 of 54 countries have enacted plastic bans (Kobo 2021).

Meanwhile, major plastics users are turning away from them as well. Companies such as Apple, Amazon, IKEA, Coca-Cola, and Unilever are all announcing initiatives to reduce plastic use or switch to recycled plastics. No trip to the local Whole Foods is complete without seeing a selection of bottled water in paper and metal containers marketed as "good for the environment." On a larger scale, NGOs such as the Global Commitment and the Plastic Pact Network of the Ellen MacArthur Foundation, ReSource Plastic of the World Wildlife Fund, and the Global Plastic Action Partnership of the World Economic Forum have formed to build coalitions of major corporate partners to eliminate unnecessary plastic use and switch to recycled plastics.

Together, these measures are starting to have an impact. From 2018 to 2020, plastics production from polymerization decreased by 10.3% worldwide. Quantities sent to landfill decreased 4.3%, and energy recovered from plastics remained the same for the first time since 2006. As a consequence, the supply of postconsumer recycled plastics increased by 11% compared to 2018, and their use into new products rose from about 4 million tonnes to 4.6 million tonnes—an increase of 15%. This demonstrates an initial shift towards a higher share of recycled plastics in the manufacturing of new products, from 7.2% in 2018 to 8.5% in 2020 (Tiseo 2021).

This may sound like good news, but it's just a drop in a very large plastic bucket. Even in Europe, which has been focusing on plastics since adopting its circular economy plan in 2015, only 14% of plastics were estimated to be recycled in 2021. According to the independent report ReShaping Plastics, "*The European plastics system is already adapting to address the challenges of climate change mitigation*

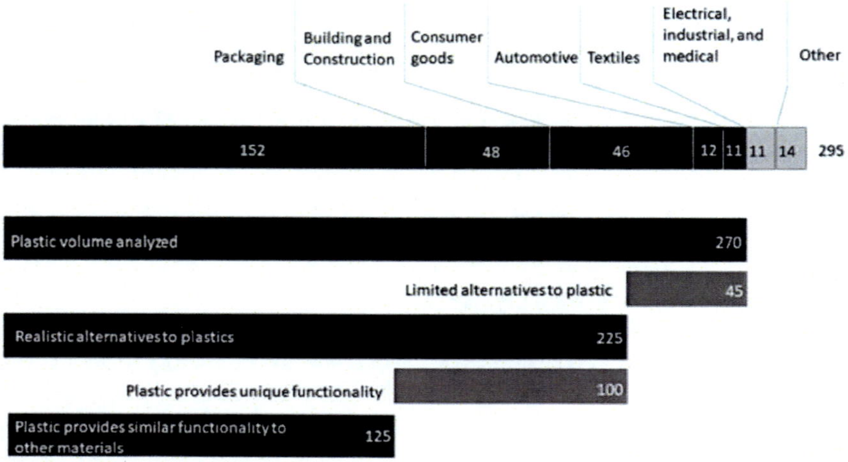

Fig. 9.4 2020 global plastic demand, million metric tons, with availability of alternatives considered for the five largest categories in terms of demand. *Source* McKinsey and Company (2022)

and circularity, but not yet fast enough to align with the goals of the Circular Plastics Alliance, European Green Deal, or the Paris and Glasgow climate agreements" (SYSTEMIQ 2022). Demands for greater sanitary protection during COVID-19 pandemic also increased plastic use and delayed some plastics bans.

The reality is that solving the plastic problem would require much more than just banning them. Plastics simply have too many uses, and replacements are often not available (Fig. 9.4).

More serious efforts have focused on designing a truly sustainable plastic economy. Given their role in global trade, including the now moribound plastic waste trade, it should be no surprise that the World Trade Organization (WTO) has taken a lead. From 2008 to 2019, WTO members proposed 128 measures related to the plastics trade, including everything from licenses to taxes, in addition to outright bans (Wolff 2020). The WTO has organized "Informal Dialogues" to promote reuse, recycling, and responsible use of plastics as well as better transparency and data (WTO 2022).

All the attention around plastics culminated in the UN Environmental Assembly passing a resolution in March 2022 to negotiate a binding international treaty on plastics by 2024. This will be the first international treaty covering all aspects of plastics. Like the Paris climate agreement, it will support a variety of national commitments towards its goals of not just reducing but ultimately ending plastic pollution. The treaty will focus on the full lifecycle of plastics, balancing its useful applications with pollution through a "circular economy" that emphasizes reuse and recycling (Gibson, Dunn & Crutcher LLP 2022).

Such a circular economy would probably be similar to the Ellen MacArthur Foundation's Plastics Vision (Ellen MacArthur Foundation 2021):

1. Elimination of problematic or unnecessary plastic packaging through redesign, innovation, and new delivery models is a priority.
2. Reuse models are applied where relevant, reducing the need for single-use packaging.
3. All plastic packaging is 100% reusable, recyclable, or compostable.
4. All plastic packaging is reused, recycled, or composted in practice.
5. The use of plastic is fully decoupled from the consumption of finite resources.
6. All plastic packaging is free of hazardous chemicals, and the health, safety, and rights of all people involved are respected.

So, what exactly is a circular economy? Instead of extracting new materials, it focuses on bringing products at the end of their life cycles back into the supply chain. If the original raw material could be recovered from waste sources and used again, it becomes part of a circular economy of the material. Alternatively, if it could be produced with feedstocks of biological origin instead of fossil fuel extraction, it could become part of a circular economy of carbon, where carbon is captured to grow the feedstock, released when it is used to create materials, and then captured again as new feedstock is grown. If neither could be done, a circular economy requires that materials have a clear end-of-life recovery pathway that aligns with existing recycling or organics processing infrastructure, such as commercial composting or anaerobic digestion.

Transitioning plastics to such a circular economy will not be easy. Plastics, like most things today, are part of a linear supply chain. The raw material is extracted somewhere, turned into finished products by manufacturers, transported to retailers, and sold to consumers. There, the product life cycle ends. The retailer, manufacturer, and original extractor do not know what happens to the product once it gets into the hands of the consumer. Even if they did, there's not much they could do about it. Their siloed business processes and software systems could not flow backwards, from the consumer back to the retailer or the manufacturer, to track the recycling of products and distribute payments backwards in the supply chain. Finally, even if consumers tried to recycle their plastics, many municipalities in developed countries do not have recycling infrastructure for them. In developing countries, many local governments have no waste collection at all, and all waste including plastics simply pile up in rivers or by the roadside.

Transitioning to a circular economy would therefore require deep changes throughout supply chains and include multiple industries plus governments. Some could adjust their product design and business models. Companies which use plastic in their product design could innovate to advance sustainability of the materials, reduce their carbon and environmental footprint, and at the same time to maintain or even improve the performance of their products. For these companies, sustainable plastics are increasingly emerging as central to their product development and brand identity. Sony, for example, spent a decade developing SORPLAS, a recycled plastic which consists of up to 99% recycled content and has 80% lower CO_2 emissions, while offering better durability and paint finish for its consumer electronics (Sony 2022).

For many companies, though, plastics are just used for utilitarian purposes such as packaging. Few could re-engineer their plastics, nor could they imagine any brand advantage from doing so. They would continue to depend on the existing plastics supply chain. For them, the key challenges will be the costs and availability of sustainable plastics, where plastics made from new fossil fuel feedstock are still the simplest and most economical choice.

Furthermore, the low cost of plastics derived from fossil-fuel based feedstock and the ease of plastic waste exports have also disincentivized investments in recycling, developing alternative feedstock, and other circular plastic technologies. As a result, the supply of recycled plastic is low. A 2021 report from AMERIPEN, which studied the corporate commitments of 35 consumer goods companies, stated that available recycled plastic resins (except HDPE) will fall short of demand. Supply is low "*because recycling markets do not follow typical supply–demand trends as consumers directly control the supply of recycled material ... Meanwhile, municipalities or local governments control how that supply is collected and how it enters the value chain*" (Schneider 2021). In addition, what supply is available often has material impurities. According to the report "Plastics recycling: challenges and opportunities", "*most plastic types are not compatible with each other because of inherent immiscibility at the molecular level, and differences in processing requirements at the macro-scale*" (Hopewell et al. 2009). These differences, when combined with current recycling practices where multiple types of plastic are mixed together, mean that using recycled plastics typically decreases quality, such as color or strength. All this makes it even harder to design around using recycled material, leading to a vicious cycle of low recycling, low availability of recycled materials, and low demand for them.

Finally, do we even know how much plastic is actually recycled? In some jurisdictions, for example, large companies must turn in invoices for recycling of their plastics. This has led to recyclers issuing multiple invoices to different companies for the same batch of recycled plastics. In other cases, the mass balance approach to accounting for recycled plastics could lead to confusing claims about which product actually used recycled versus new materials and how much (Tabrizi et al. 2021).

9.4 Solving the Plastics Problem with Blockchain

A 2006 OECD report identified several areas where new technologies could improve the circular economy and recycling markets (OECD 2006). First, new technologies are needed to address the lack of information exchange between participants, which inhibits buyers from evaluating the quality of recycled plastics. Second, transaction costs in the nascent recycled plastics markets are high, and this hampers their adoption. Finally, consumers have misperceptions regarding the relative quality of some goods produced from recycled materials, which might prevent their use even in applications where they are well suited.

Several startups are already using the blockchain to tackle these issues. One startup, Circularise, is trying to solve the problem of just knowing what's in a plastic and thus whether or how it could be recycled. According to the company, it provides a way for plastics producers to store an encrypted digital twin for their products anonymously on the public Ethereum blockchain. This digital twin stores all relevant data regarding a product or material, such as composition, use and abuse, and maintenance. It is updated throughout the value chain using a combination of QR codes, chemical tracers, and IOT devices (Peshkam 2019). Later, a user of the recycled plastic could query whether the product has certain chemicals through the blockchain, without being able to see the full original bill of materials or ingredients list. Circularise claims partnerships with Porsche, DOMO, Covestro and Neste to provide visibility for customers on the sustainability of their products and materials (Neste, Circularise 2022). Circularise uses the public Ethereum blockchain, which for many years used a slow, expensive, and high energy and GHG emissions proof-of-work algorithm. Even after the "Merge" of late 2022, which significantly reduced its energy use and GHG emissions, it is still more expensive than alternative blockchains.

Other startups are trying to address the plastics recycling supply chain more directly. The Finnish startup Empower claims to fund plastics cleanup by connecting sponsors, such as major brands like Dow, with local collectors in 15 countries. Its application helps existing recycling centers register and certify recycled plastic as they are brought in by collectors. This creates a digital record of the material collection on the blockchain. After collection and registration, the plastic can then be sold on a marketplace as certified recycled plastics, compliant with the EU's traceability regulations. Sponsors can also purchase recycled plastics credits at a fixed price to "offset" their plastics footprint. The funds are paid to collectors as plastic waste is collected, and the sponsors can view the results. For example, the urban furniture brand Vestre uses Empower to let customers view a product passport and see the material's journey from collection to recycling and finally integration into a new product (Vestre, n.d.). Empower plans to grow collection capacity to 1 million tons by the end of 2022 (Myrer 2021). Empower currently uses the Stellar blockchain, but plans to transition to a Cosmos-based blockchain (EmpowerChain, n.d.).

Finally, Plastic Bank uses a permissioned Hyperledger Fabric blockchain to support an end-to-end economy for recycling plastics while bringing better livelihoods to the collectors. According to the company, its platform verifies the credentials of the collectors, such as their age, and pays them above market rates for collecting plastics which would otherwise end up in the ocean. Collectors can then exchange the plastic for money, fresh food, clean water, cellular service, or other benefits (Plastic Bank, n.d.). The blockchain creates a full audit trail of the collectors, the plastics, and GPS locations of the collections. Payouts are sent to the recycling centers and collectors through the blockchain, supporting areas where payment services are not available. Plastic Bank offers a plastic credit, which is a claim on the amount recycled that could be used to offset the buyers' plastic use, to jumpstart recycling efforts in different locations. Once those locations scale up, it offers a fully audited recycled plastic that is directly incorporated into customers' products. Companies such as SC Johnson have partnered with Plastic Bank's Social Plastic™ feedstock program to integrate recycled plastic feedstock into their Mr. Muscle and Windex brands (Plastic Bank, n.d.). Plastic Bank reports that it has so far stopped 2 billion plastic bottles from entering the ocean and provided more than 22,000 people with additional income (Plastics Today 2022). The blockchain provides an integral, audit proof verification that the plastics were recycled under fair working conditions (Frankson 2022).

These efforts are the first steps in creating a supply chain for recycled plastics, where the plastics could be verified as actually recycled and then sold in efficient, digital markets to buyers. By creating an audit trail, they enable brands to make more credible claims about their use of recycled materials. Over time, they could support more uses for recycled plastics, such as construction material, aggregates for road building, or even waste to energy power plants.

These use cases show the usefulness of the blockchain where there are no existing central governing authorities. Today's recycled plastics market is reminiscent of the early days of carbon offsets, where Empower and Plastic Bank act as the project developer, standards organizations, auditor, and broker/retailer of credits. The blockchain is used to improve the credibility of their claims by certifying that the material was truly recycled. Over time, we could expect this market to evolve to a structure similar to carbon offsets. For example, the standards for certifying whether material is recycled would be separated from the project developers who oversee their collection. The sale of recycled materials or credits would happen with specialized brokers or transaction networks. The blockchain could adapt to such changes over time, with other organizations joining the networks in new roles.

9.5 A Sustainable Plastic Economy with Blockchain

A fully sustainable plastics economy must address both the disposal of plastics and their GHG emissions footprint, which includes both producing virgin plastics and reusing them through their entire lifecycle. Those emissions are a combination of both energy-related and process emissions. Energy-related emissions happen when energy is used to produce the chemicals, such as ethane cracking of natural gas. Process emissions happen because the carbon content of the outputs is less than those of the inputs, implying that some amount of carbon was lost in the process.

According to the IEA, the chemicals industry was responsible for 2.2 gigatons of CO_2 equivalent GHG emissions in 2017, greater than the iron, steel, and cement industries and nearly twice that of the world's airlines. If these trends continue, plastics are expected to account for 5–10% of global emissions by 2050 (Laville 2021). Under the more strict 1.5 degree Celsius scenario, their lifecycle emissions would make up approximately 13% of global carbon budgets (Center for International Environmental Law 2019).

Reducing these emissions will require the usual improvements in energy efficiency, conversion of coal to energy sources with lower emissions, or carbon capture utilization and storage (CCUS), as well as recycling and using alternative, non-petroleum feedstocks, as shown in Fig. 9.5. While this combination will be by no means easy, it is similar to issues discussed in other chapters of this book.

Attempts to improve plastic waste collection could also lead to higher emissions if the plastics are reprocessed or incinerated. When waste cannot be prevented or recycled, recovering its energy content is usually better than landfilling it. In line with the EU's waste hierarchy, the EU member states are increasingly shifting away from landfilling waste to incineration with energy recovery (European Commission, n.d.). The Commission has developed a waste-to-energy initiative to improve the recovery of energy in the incineration of waste. Such initiatives, however, increase system-level GHG emissions because plastics are burnt. This does not mean we should not do so when necessary. Rather, it simply means that both emissions and waste should be accounted for properly.

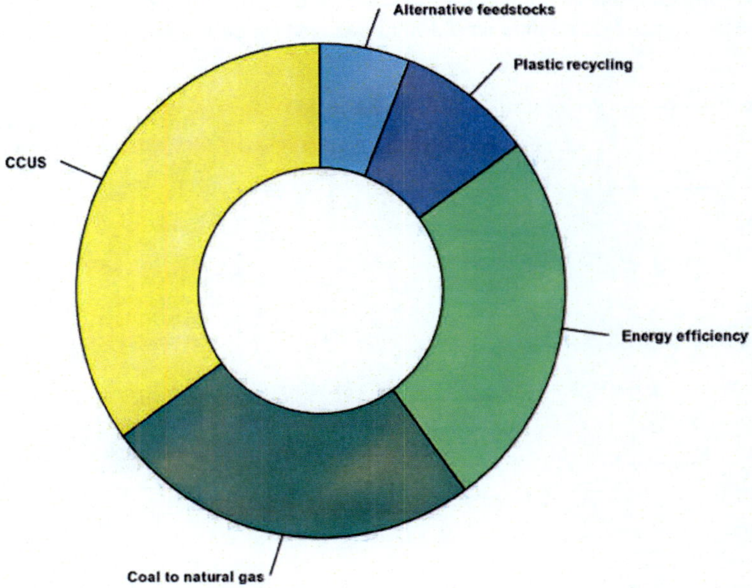

Fig. 9.5 Reducing plastics emissions. *Source* International Energy Agency (2018). IEA 2022; The Future of Petrochemicals; https://www.iea.org/reports/the-future-of-petrochemicals; CC BY 4.0

Blockchains could support a complete accounting for sustainable plastics by tracking both energy and materials use. A workflow of the process is given in Fig. 9.6. Similar to other use cases, the foundation would be a data ledger that tracks all the steps of the plastics lifecycle, from energy use to feedstock to recycling to final use or disposal. When plastics are first produced, the manufacturers would record the quantity of feedstock used if new (virgin) materials are used but not if the recycled feedstock is used. Both types of feedstock would come with their associated emissions footprint. This creates records for Materials (M) and Emissions (E) in Fig. 9.6. The energy used for processing the feedstock is then entered, along with any renewable energy or CCS used to reduce emissions, as additions to the Emissions (E) account. Next, when plastics are recycled, they would be entered from sources such as Circularise, Empower, or Plastic Bank that could be trusted to provide proof of recycling. When recycled material is used instead of virgin feedstock, it would reduce the amount of original virgin feedstock used. The emissions of recycling and reprocessing the plastic would be added to the data ledger as well. Finally, for plastics which are not recycled, their final disposal emissions should be included in the ledger, netted against any emissions captured from CCS.

From this data ledger, Non-Fungible Tokens or NFT's could be issued which record the total emissions footprint and recycled materials content of plastics. These NFT's would be immutable records that could be transferred in supply chains as plastics are transported, used, and reused. They could be linked to particular lots of plastics produced, and the sizes of the lots could vary depending on the needs. The NFT's should be on public blockchains so that they are universally accessible, but only information which is important to their emissions and circularity attributes should be stored. Private data would not be shared on public blockchains, and even

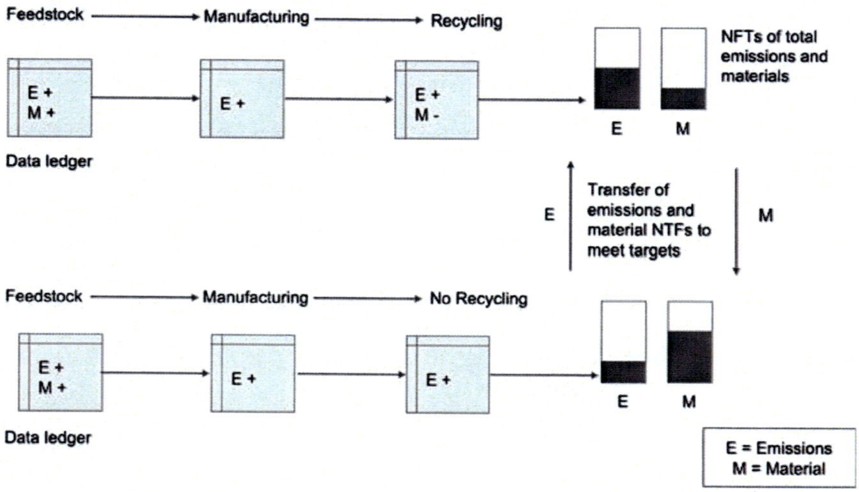

Fig. 9.6 Plastics emissions and materials tracking and exchange with blockchain under a cap and trade program

the identities of the parties transacting in them could be made pseudo-anonymous with cryptographic accounts.

The goal is to achieve a circular plastic economy by getting both emissions and virgin feedstock use to zero over time. This could be done through a "cap and trade" scheme used by governments to reduce carbon emissions. A customer, a group of customers, or even a government regulator could set a declining target for the amount of virgin feedstock allowed. Members of the network would commit to meeting this target or paying into a fund if too much new material is used. Such a scheme would be implemented with a smart contract with limits for the amount of new materials and energy used. The smart contract could compare the NFT's of all the manufacturers versus the preset limits. If they are below the limits, the company would be issued tokens, which could be sold to companies that exceed the limits. Smart contracts could certify each company's compliance with its circular plastic commitments.

By setting the amount paid higher than the cost of using recycled materials, it would even out the cost advantage of new materials. Payments would then be used to fund additional cleanup, through services similar to what Empower or Plastic Bank offer, and disposing of the recovered material. Over time, such a scheme could help solve the fundamental economic problem of the circular economy, where it's cheaper to use new materials than recycle and reuse existing products. By making firm commitments to reach zero new materials used, we could "close the circle." At the same time, it would support the most economic way to reach a sustainable economy by specifying the most important parameters–zero emissions and zero new materials–without locking us into a particular pathway.

Bringing both material use and emissions together allows us to finally compare plastics on equal footing with other materials. Is it better to recycle plastics to make more packaging and bottles or to use them for construction or road building materials? Is it better to replace construction materials such as steel and cement with recycled plastics? Plastics may finally have their day in the sun as a way to reduce emissions versus other carbon intensive materials, if they are used in responsible lifecycles.

Finally, tying together material use and emissions makes it possible to link materials markets with carbon markets. For example, if pathways which use less materials also reduce carbon emissions, then they could be marketed as carbon credits as well as recycled materials. Gold Standard, for example, has a voluntary carbon offset project for PET recycling in Romania.[1] They could also be counted as part of Scope 3 emissions reductions for oil and gas companies. Finally, plastics would probably fall under compliance carbon programs at some point due to the sheer amount of their GHG emissions. When that happens, reducing emissions from recycling would create valuable carbon credits.

To prevent double invoicing or double counting of recycled plastics, a meta registry of the plastics supply chain would be needed. This meta registry would be similar to the Climate Warehouse currently being implemented in the carbon markets.[2] It would

---

[1] See https://marketplace.goldstandard.org/collections/projects/products/plastic-recycling-romania-europe.

[2] For more information, see our chapter on carbon markets.

integrate the plastic supply chain data from the different public and permissioned blockchains, such as those used by Empower and Plastic Bank.

Any trading scheme for recycled plastics would need to be designed carefully. Already, environmental groups are protesting that the EU's mass balance approach to accounting for plastic recycling could lead to "*a loose method for determining recycled content is applied, as a result of industry pressure to circumvent the rules of the game and allow for creative accounting*" (Zero Waste Europe 2021). Their concerns would sound familiar to anyone who has worked with carbon credit markets: What good are claims that recycled materials were used if they're not actually in the physical products labeled as "recycled," but instead come from virtual certificates of recycling in other regions, industries, or product types?

These concerns must not be ignored. Just as concerns about carbon credits have undermined carbon markets in the past, concerns about the truth of recycling claims could undermine efforts to create markets for them. Rather, all participants, including the oil and gas industry, should consider these questions:

- Are transferable certificates or tokens of recycled materials really leading to decreasing virgin materials use?
- Given the low levels of recycling around the world, could we consider these recycling activities "additional," in that they would not have taken place without sale of transferable certificates or tokens?
- Given the lack of infrastructure for using recycled materials, could transferable certificates or tokens get more materials to be recycled without investing in expensive infrastructure first, or causing emissions by transporting heavy, bulky, low value recycled materials?
- Is it better to prioritize collections in high cost, advanced economies with high plastic use or low cost, developing economies with high growth potential–and lower costs?

A reasonable approach, based on examples from the renewable energy and carbon markets, may be to allow certificates or tokens for meeting recycled content standards, as a way to jump start the recycling of plastics. Then, as the amount of plastics recycling reaches a higher level and recycling infrastructure is scaling up successfully, to switch from transferable claims to actual recycled content for meeting recycling goals The blockchain could support both use cases through fungible tokens of recycled plastics for trading and non-fungible tokens that represent digital certificates of recycled content. We provide a summary of the blockchain technologies that could be used for sustainable plastics in Table 9.1.

9.6 Summary and Conclusion

Plastics bring out the issues for the oil and gas industry in the starkest terms.

The products are enormously useful and could well be the future of the industry, yet their negatives are also immediate and visible. People see plastic trash on their daily

Table 9.1 Summary of blockchain technologies for sustainable plastics

Technology	Application	Benefits
Data Layer	Permissioned or public blockchain of supply chains	Trace plastics through their lifecycle Record use of materials and energy use
DAOs and Oracles	Decentralized voting based on reputation	Certify plastics were properly recycled Certify meeting other environmental and social goals
Non-Fungible Tokens (NFT's)	Tokens on public blockchains linked to specific lots of plastics	Immutable records on blockchain Store certified emissions, materials content, recycled content, compliance with standards Book and claim transfers to fund green premium
Materials Tokens	Tokens to track new materials used on public blockchains	Registry of use of new materials Regulate materials footprint of plastics lifecycle
Emissions Tokens	Tokens to track emissions footprint on public blockchains	Registry of emissions from production Regulate emissions footprint of plastics lifecycle
Smart Contracts	Executable programs on blockchain	Purchase commitments for sustainable plastics Funding mechanisms
Meta Registries	Public blockchain of data from supply chains	Transparency for market, regulators, general public Prevent double counting

walks and on the news and social media. The specter of health hazards hangs in the air. Unless the industry could make its products acceptable, society as a whole could reject them. Bans are already happening. Lawsuits could be next. Major customers like food and pharmaceutical companies are looking at alternatives, even if they are more expensive and less suitable, just to look better to their customers.

A sustainable plastic economy may be unfamiliar, yet many innovative companies in other industries are embracing the circular economy wholeheartedly as a way to increase profits and customer loyalty. Car manufacturers have long accepted trade-ins as a way to sell more new cars for decades, and used cars, services, and financing are more profitable than new cars. Stockbrokers are moving from the old "churn 'em and burn 'em" style of selling stocks towards "wealth management" models of long-term relationships and fees. Apple aggressively tries to buy back its products and reuse the components while providing gift cards to consumers to buy more. It has transitioned from selling machines to subscriptions and repeat purchases, and that business model has propelled it to be one of the most valuable companies in the world.

The oil and gas industry should rethink plastics in similar terms. Is the business model of selling its output at the lowest price trapping the industry in low-margin, commodity businesses? Could the industry profit from the recycling of plastics, instead of giving that up to potential competitors? Could the industry innovate like Sony to produce higher-value plastics that also produce higher profits? Could a circular economy which encourages higher prices, repeat purchases, and reprocessing fees deliver more predictable returns and, therefore higher valuations?

The oil and gas industry cannot fix the plastics problem on its own. Yet, given its enormous resources and the long-term importance of plastics to the industry, how could it not take the lead in fixing it? When there is no recycling option, local governments' only way to fix the plastic waste problem is to ban its use. Even in Africa, where plastics use is one-tenth that of the U.S. or Europe, governments are enacting plastics bans because they know of no other way. It's clear that society could only accept plastics which are part of a circular economy. If the oil and gas industry has invested in petrochemicals in the past, it could invest in the recycling facilities that will return plastics to the circular economy, or at least provide a viable pathway for their disposal. Thus, ExxonMobil's recent plans to build large scale plastics recycling plants are probably just the beginning of a trend in this direction (ExxonMobil 2022).

Long-term, the industry will need to work with many layers of manufacturers, retail consumers, recyclers, and governments in countries around the world to create a sustainable plastics economy. The blockchain is designed for collaboration at scale, across countries and industries, without relying on one central authority. It could help organize a circular plastic economy in ways that traditional systems could not.

References

Abraham, B (2022) EPR for plastic packaging waste is a step in the right direction but has set impossible targets. Indiatimes.com. https://www.indiatimes.com/news/india/epr-guidelines-for-plastic-packaging-recycling-562665.html

Barboza LG, et al (2018) Microplastics cause neurotoxicity, oxidative damage and energy-related changes and interact with the bioaccumulation of mercury in the European seabass, Dicentrarchus labrax (Linnaeus, 1758). PubMed. https://pubmed.ncbi.nlm.nih.gov/29287173/

Center for International Environmental Law (2019) Plastic & climate: the hidden costs of a plastic planet. https://www.ciel.org/project-update/plastic-climate-the-hidden-costs-of-a-plastic-planet/

Cho R (2020) Recycling in the U.S. is broken. How do we fix it? State of the Planet. https://news.climate.columbia.edu/2020/03/13/fix-recycling-america/

Ellen MacArthur Foundation (2021) Our vision for a circular economy for plastics. Ellen MacArthur Foundation. https://ellenmacarthurfoundation.org/plastics-vision

EmpowerChain (n.d.) A new impact driven Cosmos Blockchain. Coming soon. https://empowerchain.io/. Accessed 6 Sept 2022

EPA (2021) National overview: facts and figures on materials, wastes and recycling | US EPA. Environmental Protection Agency. https://www.epa.gov/facts-and-figures-about-materials-waste-and-recycling/national-overview-facts-and-figures-materials#Trends1960-Today

European Commission (n.d.) Waste prevention and management. European Commission. https://ec.europa.eu/environment/green-growth/waste-prevention-and-management/index_en.htm. Accessed 6 Sept 2022

Exxon Mobil (2020) Advancing climate solutions. ExxonMobil. https://corporate.exxonmobil.com/Climate-solutions/Advancing-climate-solutions

Exxon Mobil (2022) ExxonMobil to build its first large-scale plastic waste advanced recycling facility. ExxonMobil. https://corporate.exxonmobil.com/news/newsroom/news-releases/2021/1011_exxonmobil-to-build-its-first-large-scale-plastic-waste-advanced-recycling-facility

Frankson S (2022) Co-Founder of Plastic Bank, July 8, 2022. https://www.linkedin.com/in/shaunfrankson/

Geddie J (2022) 75% of people want single-use plastics banned, global survey finds. Reuters, February 21, 2022. https://www.reuters.com/business/environment/75-people-want-single-use-plastics-banned-global-survey-finds-2022-02-22/

Geyer R, et al (2017) Production, use, and fate of all plastics ever made. Sci Adv 3, e1700782. https://doi.org/10.1126/sciadv.1700782 See https://www.science.org/doi/10.1126/sciadv.1700782

Gibson, Dunn & Crutcher LLP (2022) Update on UN roadmap for a new global plastics treaty. Gibson Dunn. https://www.gibsondunn.com/update-on-un-roadmap-for-a-new-global-plastics-treaty/

Hopewell J, Dvorak R, Kosior E (2009) Plastics recycling: challenges and opportunities. NCBI. https://www.ncbi.nlm.nih.gov/pmc/articles/PMC2873020/

IMARC Group (2022) Automotive plastics market: global industry trends, share, size, growth, opportunity and forecast 2022–2027. Research and Markets. https://www.researchandmarkets.com/reports/5577994/automotive-plastics-market-global-industry

International Energy Agency (2018) The future of petrochemicals—analysis. IEA. https://www.iea.org/reports/the-future-of-petrochemicals

Kobo K (2021) Why Western-style plastic bans aren't working in Africa—Quartz Africa. Quartz. https://qz.com/africa/2007658/why-western-style-plastic-bans-arent-working-in-africa/

Laville S (2021) Twenty firms produce 55% of world's plastic waste, report reveals. The Guardian, May 18, 2021. https://www.theguardian.com/environment/2021/may/18/twenty-firms-produce-55-of-worlds-plastic-waste-report-reveals

McKinsey & Company (2022) McKinsey, July. https://www.mckinsey.com/~/media/mckinsey/industries/chemicals/our%20insights/climate%20impact%20of%20plastics/climate-impact-of-plastics-v2.pdf

Myrer W (2021) 2021: >3000% growth. Medium. https://medium.com/empowerplastic/2021-3000-growth-d1623e5c4851

Neste Circularise (2022) Press. Circularise—press. https://www.circularise.com/press-releases/transparent-and-traceable-neste-teams-up-with-circularise

OECD (2006) Improving recycling markets. OECD iLibrary. https://www.oecd-ilibrary.org/environment/improving-recycling-markets_9789264029583-en

OECD (2022a) Global plastics outlook: policy scenarios to 2060. OECD iLibrary. https://read.oecd-ilibrary.org/view/?ref=1143_1143481-88j1bxuktr&title=Global-Plastics-Outlook-Policy-Scenarios-to-2060-Policy-Highlights&utm_source=Adestra&utm_medium=email&utm_content=GPO

OECD (2022b) Global plastics outlook: economic drivers, environmental impacts and policy options. OECD. https://www.oecd-ilibrary.org/sites/de747aef-en/index.html?itemId=/content/publication/de747aef-en

Peshkam M (2019) Transforming plastic pollution using blockchain. Blockchain Research Institute. https://www.blockchainresearchinstitute.org/project/transforming-plastic-pollution-using-blockchain/

Plastic Bank (n.d.) Frequently asked questions. https://plasticbank.com/faq/. Accessed 6 Sept 2022

Plastic Bank (n.d.) Social plastic program. https://plasticbank.com/social-plastic-program/. Accessed 6 Sept 2022

Plastics Today (2022) Plastic bank diverts 2 billion plastic bottles from the ocean. https://www.plasticstoday.com/sustainability/plastic-bank-diverts-2-billion-plastic-bottles-ocean

Schneider S (2021) US plastics recycling struggles to meet demand, highlighting investment gap. S&P Global. https://www.spglobal.com/commodityinsights/en/market-insights/blogs/petrochemicals/022621-us-plastic-recycling-pet-bottles-packaging-waste-investment

Schuerman M (2022) New York state is looking for a new solution to plastic waste. NPR. https://www.npr.org/2022/05/27/1101522591/we-never-got-good-at-recycling-plastic-some-states-are-trying-a-new-approach

Seabrook J (2010) Plastics. The New Yorker, September 13, 2010. https://www.newyorker.com/news/news-desk/plastics

Sony (2022) SORPLAS | Post-consumer recycled plastic Sony products. Sony. https://www.sony.com/en-ma/electronics/sorplas-recycled-plastic

SYSTEMIQ (2022) ReShaping plastics: pathways to a circular, climate neutral plastics system in Europe. Actu Environnement. https://www.actu-environnement.com/media/pdf/news-39392-ReShaping-Plastics-Executive-summary.pdf

Tabrizi S, Crêpy M, Rateua F (2021) Recycled content in plastics: the mass balance approach. Zero Waste Europe. https://zerowasteeurope.eu/wp-content/uploads/2021/05/rpa_2021_mass_balance_booklet-2.pdf

Tiseo I (2021) Plastic cumulative production globally 2050. Statista. https://www.statista.com/statistics/1019758/plastics-production-volume-worldwide/

Turns A (2022) The chemicals that linger for decades in your blood. BBC, May 13, 2022. https://www.bbc.com/future/article/20220512-the-chemicals-that-linger-for-decades-in-your-blood

UNCTAD (2021) Global trade in plastics: insights from the first life-cycle trade database. UNCTAD. https://unctad.org/webflyer/global-trade-plastics-insights-first-life-cycle-trade-database

Vestre (n.d.) Product passport VSTR45. https://vestre.product-passports.com/vstr45

Wolff A (2020) Speech—DDG Alan Wolff—WTO informal dialogue on plastic pollution and environmentally sustainable plastics trade. World Trade Organization. https://www.wto.org/english/news_e/news20_e/ddgaw_17nov20a_e.htm

WTO (2022) Plastics dialogue launches three workstreams to advance discussions. World Trade Organization. https://www.wto.org/english/news_e/news22_e/ppesp_21mar22_e.htm

Zero Waste Europe (2021) Avoid the 'mass balance approach' to block ambitions for increased recycled content in plastics. https://zerowasteeurope.eu/wp-content/uploads/2021/02/Feb_2021_Joint_Letter_Mass_Balance_Approach.pdf

Si Chen is the president of Open Source Strategies, Inc. in Los Angeles, CA, which specializes in open-source software for climate finance and investing. He leads the development of open-source blockchain carbon accounting software at Hyperledger Labs. Previously, he has managed investment portfolios for institutional pension funds, central banks, and hedge funds and has been published in *The Journal of Portfolio Management*. He is also the co-founder and CTO of GraciousStyle.com, an online retailer.

Joseph Wyer is a research analyst with a background in mechanical engineering and data analytics. He has worked on decarbonization from a variety of angles, including blockchain and circular economy, and currently conducts policy research to support the clean energy transition. He is based in San Diego, California and enjoys surfing and hiking.

Chapter 10
Carbon Credit Markets

Si Chen

10.1 Introduction

Will forests allow us to keep on burning oil and gas?

Shell thinks so. It's been promoting carbon offsets, or credits for reducing or removing Greenhouse Gas (GHG) emissions from the atmosphere, as a way to meet one of the most aggressive climate ambitions in the oil and gas industry—net zero emissions:

> Becoming a net-zero emissions energy business means that we are reducing emissions from our operations, and from the fuels and other energy products we sell to our customers. It also means capturing and storing any remaining emissions using technology or balancing them with offsets. (Shell USA, Inc., n.d.)

Meanwhile BP is setting up for the other side of the trade. It purchased a majority stake in Finite Carbon, "*the largest developer of forest carbon offsets in the US*", in December 2020, stating that:

> Global demand for offsets is likely to grow as more companies use them to achieve their climate-related goals, and we intend to offer offsets to our customers to help them meet their goals. (BP 2021)

It doesn't stop with forestry either. The industry is looking seriously at carbon credits for funding everything from carbon capture to offshore emissions storage to permanent coal mine sequestration.

The appeal of carbon credits is obvious. It allows companies to raise funding for climate action by trading its benefits without the time and expense of building out physical assets. It also allows the oil and gas industry a way to support those customers who could not reduce emissions directly by purchasing emissions reductions or even removals elsewhere.

S. Chen (✉)
Open Source Strategies, Los Angeles, CA, USA
e-mail: sichen@opensourcestrategies.com

S. Saraji and S. Chen, *Sustainable Oil and Gas Using Blockchain*,
Lecture Notes in Energy 98, https://doi.org/10.1007/978-3-031-30697-6_10

When Shell tried to do exactly that with "carbon–neutral LNG," where shipments of LNG are paired with voluntary carbon offsets, though, it brought a hailstorm of negative publicity. Bloomberg called it "*The Fictitious World of Carbon Neutral Fossil Fuel*" (Stapczynski et al. 2021). Carbon Market Watch, an NGO which specializes in research on carbon markets, called it "*Net-zero pipe dreams*" (Carbon Market Watch 2021). Even industry publication Upstream called it a "*controversial solution*" (Klinge 2021).

Environmentalists seem dead set against the whole idea. Project Drawdown calls offsets "*problematic*," "*dubious*," a "*shell game*", and "*imaginary*" (Foley 2021) Greenpeace is more blunt: Carbon Offsets are a Scam, they say, "*a bookkeeping trick*" and "*the next big thing in greenwashing*" that "*feigns compassion*," "*preys on fear*," "*takes advantage of uncertainty*," and is driven by "*greed*" and their ultimate *bete noire*, "*Big Oil and corporate polluters want to keep putting profits over people and the planet*" (Greenberg and Colombo 2021). Even John Oliver devoted a full show criticizing them (HBO 2022).

Given the skepticism, could carbon markets help the oil and gas industry reach its Net Zero goals? Could they help finance novel technologies such as carbon capture, methane reduction, and sustainable transportation fuels? What role, if any, could blockchain technology play in solving the problems of carbon markets?

10.2 How Carbon Markets Work

The basic idea of a carbon market is pretty simple. One party could do something to reduce or remove GHG emissions, and then it could sell those reductions or removals to another party. The buyer could then count them as if it had reduced or removed the emissions themselves. To standardize the market, different GHG emissions such as methane and refrigerants are converted to CO_2 equivalents, so that all transactions happen in tons of CO_2 equivalent GHG emissions. Hence, "carbon" markets.

These transactions trace back to the Clean Air Act of 1990, which created a successful market-based mechanism for reducing sulfur emissions. Thus inspired, the Kyoto Protocol introduced the world's first carbon market, the Clean Development Mechanism (CDM), in 1997. Today there are government-operated compliance carbon markets in over 50 jurisdictions around the world with a traded value of over $850 billion, plus a voluntary carbon offsets market and industry carbon mechanisms (EDF 2018) (Chestney 2022).

Compliance markets include the European Union Emissions Trading Scheme (EU ETS), California's Air Resource Board Cap-and-Trade Program, the Regional Greenhouse Gas Initiative (RGGI) of northeastern US states, China's new emissions trading program, and proposed cap and trade programs everywhere from India to the state of Washington. All these programs share some common characteristics. They are run by government agencies, who target industries and companies that generate the highest emissions. Companies that fall under the program are required to submit regular emissions reports through accredited auditors (European Commission, n.d.).

The programs then allocate each company an allowed amount of emissions. Companies who do better than the target would earn credits, while those that do worse than the target would need to buy credits to meet their targets, either from companies which earned credits or from the government at a set rate. The amount of allowed emissions is set to fall each year so that the program, as a whole, reduces the total emissions of the covered companies.

The voluntary carbon offset market is much smaller, with about $1 billion in total annual volume. Voluntary carbon offsets are not issued by any government agency, nor is any buyer required to purchase them. They are also not based on the total emissions of any organization, but rather are claims that a particular project reduced or removed emissions. Voluntary offsets are usually issued by third-party standards organizations, such as VERRA, The Gold Standard, Plan Vivo, Climate Action Reserve, and the Global Carbon Council. To get offsets issued, a project developer must submit a detailed description of the project to the standards organization, which checks if it complies with one of its methodologies. If so, the standards organization would certify the project and issue carbon offsets in its registry system based on audits performed by third-party verification bureaus. Once issued, the offsets are usually sold by brokers and eventually bought by both companies and individuals, who could claim that they offset their own emissions. At that point, the voluntary offset is considered "retired" in the standards organization's registry, which means it's permanently removed from further transfers or use.

Whereas compliance carbon markets are based on calculations of total emissions versus mandated targets, voluntary offsets are based on comparing the results of a particular project, such as a wind farm or a forest, to counterfactuals, or what would have happened without the carbon offsets. Furthermore, while compliance carbon markets typically target large companies in high emissions industries, such as power generation, steel, and transportation, the voluntary carbon offsets market supports a much greater range of activities. For example, in 2020, there were voluntary carbon offsets issued for everything from cookstoves to landfill to biodigesters to waste heat recovery (Fig. 10.1).

A close relative of these carbon markets are Energy Attribute Certificates (EAC), or Renewable Energy Certificates (REC.) Even though they're not denominated in tons of CO_2 equivalents, they also allow the transfer of climate-related energy claims between parties. For example, a wind farm could issue a REC for one megawatt hour of electricity generated from wind power and sell it to a utility. The utility could then use the REC to meet its regulatory renewable power standards, even though it was not connected to the wind farm directly. Alternatively, corporate buyers could purchase REC's to claim that the energy they used came from renewable sources.

The boundaries between these carbon markets are not set in stone. Voluntary carbon credits could be turned in for compliance credits in some jurisdictions. In other cases, REC's could be used to meet compliance requirements, essentially turning them into compliance credits. Given the large difference between carbon pricing in different countries, this creates a rich opportunities of trading and arbitrage. Some oil and gas companies have already been buying offsets to meet compliance requirements or in anticipation of greater carbon regulation in the future.

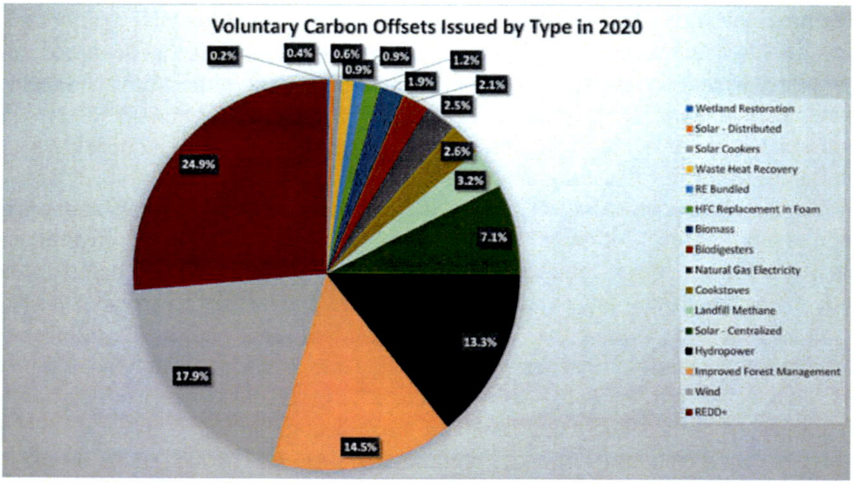

Fig. 10.1 Types of projects in the voluntary carbon market. *Source* Chen et al. (2021)

Finally, several new carbon market mechanisms may ultimately prove very important to the oil and gas industry. One is California's Low Carbon Fuel Standard (LCFS.) Administered by the California Air Resource Board, this standard covers the transportation fuels sold in California with a goal to reduce the carbon intensity of transportation fuels by 20% in 2030 versus 2010 levels. Credits are issued for fuels based on their certified carbon intensity versus a target, which declines over time (California Air Resources Board 2018).

For the industry, LCFS is important because it covers transportation fuels, and because it is based on the attributes of fuel, rather than the overall emissions of a company or the complicated criteria of voluntary carbon offsets. In this way it's similar to REC's. For more on how this would be relevant to the industry, see our chapter on Sustainable Transportation and Aviation Fuels.

A second important carbon market mechanism is the Carbon Offsetting and Reduction Scheme for International Aviation (CORSIA), a consortium organized by the UN's International Civil Aviation Organization (ICAO). This industry consortium was organized in 2016 for international airlines to purchase carbon offsets against their emissions. CORSIA requires airlines to submit their emissions reports and publishes an official list of approved offset standards for airlines to purchase from. Originally, CORSIA called for voluntary purchases starting in 2021 but mandatory purchases after 2027 of offsets above the average emissions from 2019 and 2020. The reduction in air travel and emissions due to the COVID-19 pandemic, however, has caused CORSIA to adopt 2019 as its baseline, and the implementation may be delayed as well (Gordon 2020).

CORSIA is significant to the oil and gas industry because the airline industry is an important user of fuels, and because it is also the first major industry consortium set up at least in response to government climate regulation. When the EU ETS cap and trade program began to regulate aviation emissions in 2012, including airlines from outside the EU created a firestorm of international protests. China threatened to pull orders from Airbus. The US exhibited a rare moment of bipartisan climate policy when President Obama signed the EU Emissions Trading Scheme Prohibition Act to ban US airlines from participating in the EU ETS. As more countries threatened retaliations, the EU backed down. Today, only intra-EU flights are subject to the EU ETS (Lewis and Volcovici, n.d.) (Lowe 2013). Meanwhile, to stay ahead of the regulations, the airlines industry banded together to form CORSIA, which would rely primarily on voluntary carbon offsets until 2040 (IATA 2021). Did it work? While airlines are rallying behind the sustainable aviation fuels, they are not required to purchase any carbon offsets until 2027. Even when they do, they could buy CORSIA eligible voluntary carbon offsets for $1 to $5, while EU ETS allowances are trading near $90.

This battle over international aviation will probably be just a first attempt to resolve supply chain emissions. Today's supply chains extend across national and industry boundaries. Government regulators must either include supply chain emissions into their emissions mechanisms or risk companies shifting production and jobs outside their jurisdictions. Similarly, major corporations could only achieve their climate goals by reducing emissions from their supply chains and business partners. They may well turn to internal or industry emissions trading schemes using carbon credits. Both are already happening. In June 2022, the EU approved a Cross Border Adjustment Mechanism (CBAM), which will require emissions credits for embedded emissions for a number of imported products, including iron and steel, cement, fertilizer, aluminum, and electricity (Simões 2022). Meanwhile, large corporations including Wal-Mart and Apple are already working on their supply chain emissions reductions with everything from training programs to renewable energy investments for their suppliers.

In time, we may see the oil and gas industry creating carbon credits for compliance and supply chain emissions reductions targets based on their investments in carbon capture, sustainable fuels, and hydrogen. At the same time, it could also be a major user of both voluntary and compliance carbon credits for both itself and its customers.

10.3 Key Problems of Carbon Markets

Or we might never get there.

Today's carbon markets are plagued by well-known problems which have limited their usefulness to both users and suppliers:

- High administrative costs
- Long delays
- Lack of transparency
- Low liquidity
- Oversupply of low quality credits

- Low prices for credits
- Double counting

Compliance carbon markets, such as the EU ETS, often issue too many free allowance credits, so that companies end up buying very few credits to meet their emissions targets (European Court of Auditors 2020). The government agencies who run compliance carbon programs often lack adequate resources and are slow and unresponsive. The California Low Carbon Fuel Standard, for example, offers a large spreadsheet for calculating the emissions intensity of fuels.

Meanwhile, the standards organizations which issue credits in the voluntary carbon markets continue to rely on PDF's, spreadsheets, and site visits. As a result, the validation and verification of carbon offsets is so slow that the average offset is issued nearly three years after the activity occurred and so inefficient that 70% of the price of offsets is eaten up by cost (Chen et al. 2021) (Union Square Ventures 2020).

Once issued, most carbon credits are traded "over the counter" through brokers and other intermediaries. Only the largest and most standardized credits, such as the EU ETS allowances, are traded on futures exchanges. In the majority of carbon markets, there is little transparency about what credits are available or what prices they traded at. As a result, the only information available may be through informal surveys. Transactions are difficult for both suppliers and users of carbon credits.

Even with all the money spent on validation and verification, carbon credits suffer from serious concerns about their validity. Critics have identified many problems with both compliance and voluntary carbon credits:

- A study found that 85% of the credits issued under the CDM were in fact not additional and therefore contributed no real climate benefits (Cames et al., 2017).
- Offsets for renewable energy projects such as wind and solar power are the biggest sector of the voluntary carbon offsets market, even though most renewable energy projects are economically viable on their own.
- Credits were issued for forestry projects even when they did not change the existing land use (Elgin 2022), or where the landowners simply shifted their logging to different forests (Song and Temple 2021).
- Credits were also issued for commercial forests which had no biodiversity benefits and may have been viable without the credits (Carbon Market Watch 2012). This practice continues today in the voluntary carbon offsets market, with projects such as commercial teak plantations.[1]
- Credits issued for HFC-23 refrigerants gas destruction actually created an incentive for producing more HFC-22 refrigerants just to earn such credits (Carbon Market Watch 2021).
- Credits were even issued for massive coal plants (Fogarty 2011).

Finally, the specter of "double counting" hangs over the market. Some are concerned that the same carbon credits have been sold multiple times to unsuspecting buyers. The greater risk, though, is that all the different standards organizations, markets, and

[1] See for example https://marketplace.goldstandard.org/products/nicaforest-high-impact-reforestation-program.

compliance schemes could allow the same project to be counted multiple times. For example, a project developer could issue voluntary carbon offsets with more than one standards organization. Or it could issue voluntary carbon offsets and obtain compliance carbon credits. Finally, it could issue carbon credits and Renewable Energy Certificates or obtain credits in fuel carbon standards such as the California LCFS. Such double counting would be very hard to track today, given all the different carbon markets and their separate databases.

No wonder environmentalists have nothing good to say about carbon credits, even if they promise money for trees. In their eyes, they are simply greenwashing which delays real climate action. It does not help that some of the biggest offset users are oil and gas companies (Table 10.1).

It would be foolish to brush off such concerns. Questions about environmental integrity at least pushed, if not altogether drove, the EU to end the use of CDM

Table 10.1 Top 20 companies using voluntary carbon offsets in 2020

	Top buyers	Offsets retired (tons CO_2e)
1	Anadarko	7,274,107
2	Takeda Pharmaceutical Company	4,130,879
3	Telstra Corporation Limited	2,253,924
4	Delta Air Lines	1,716,145
5	Shell	1,666,484
6	Eni Upstream	1,490,000
7	The Boeing Company	1,350,000
8	Greenchoice	1,087,463
9	DPDgroup	1,046,184
10	Guccio Gucci Spa	1,000,000
11	CHEVRON PETROLEUM COMPANY	950,000
12	3Degrees Group, Inc	942,053
13	Natural Capital Partners Americas, LLC	786,205
14	Element Markets Emissions, LLC	701,566
15	Chanel	630,000
16	Butagaz	600,835
17	Lyft	600,000
18	ESWE Versorgungs AG	594,000
19	Volkswagen Brand Passenger Cars	550,000
20	Brisbane City Council	526,499

Source Chen et al. (2021)

carbon credits in its EU ETS compliance programs in 2012. That in turn led to a near total collapse of the CDM, as its number of new projects fell to virtually nothing. Environmentalist groups are not stopping there. Ahead of the Glasgow COP26 summit, Greenpeace executive director Jennifer Morgan called for nothing short of abolishing them altogether:

> There's no time for offsets. We are in a climate emergency and we need phasing out of fossil fuels. (Prentice 2021)

Similarly, such concerns may explain why the Science Based Targets Initiative (SBTI), which certifies climate targets of major companies, specifically excludes the use of carbon credits in its Corporate Net-Zero Standard:

> **Exclude the use of carbon credits**: Carbon credits do not count as reductions toward meeting your science-based targets. Companies should only account for reductions that occur within their operations and value chain. (Science Based Targets 2021)

A blog entitled "Science-Based Net-Zero Targets: 'Less Net, more Zero'" further explains their thinking while calling out the oil and gas industry for using offsets:

> Net-zero = offsets instead of reducing emissions.
>
> Net-zero delays the urgent action we need before 2030.
>
> Burn now, pay later' strategy of fossil fuel industry:
>
> The Net-Zero Standard will require ambitious near-term emission reduction targets, at least 90–95% absolute emissions reductions by 2050 at the latest, and reducing the dependency on carbon removals for remaining residual emissions in the long term. This is the exact opposite of many self-defined and offset dominated, net-zero targets for 2050 from fossil fuel intensive companies and sectors, such as oil and gas.
>
> An overreliance on carbon offsets in net-zero targets creates multiple problems around land use, equity, fairness and climate justice. ... [The SBTi Net-Zero Standard] will help reduce demand for low quality carbon offsets being used as a substitute for emissions reductions. (Dowdall 2021)

The combination of overissuance and low acceptance lead to one thing: Low prices. That, unfortunately, is exactly what we see in the market. For nearly a decade, the EU ETS emissions allowances traded for less than 10 Euros per ton CO_2 equivalent, before finally lifting off in 2020 to over 90 Euros per ton today (Trading Economics 2022). In most markets, carbon prices still hover under $20 per ton (Citizens' Climate Lobby 2021). Voluntary offset prices are even lower, with many in the $1 to $10 per ton range (Ecosystem Marketplace 2021) (Fig. 10.2).

For those in the industry looking to fund climate investments with carbon credits, the current prices for carbon credits are simply too low. Add on top of it the high costs to issue and monitor the projects and the difficulties of transacting in an illiquid market, and it's hard to see how the carbon markets could meet the oil and gas industry's needs without significant improvements.

For those in the industry looking to use carbon credits, today's markets pose a different set of problems. First, they lack the depth and liquidity to meet the industry's needs. The available credits are often from years ago and from projects which are now out of favor. Finding credits of reasonable quality is so difficult that many users

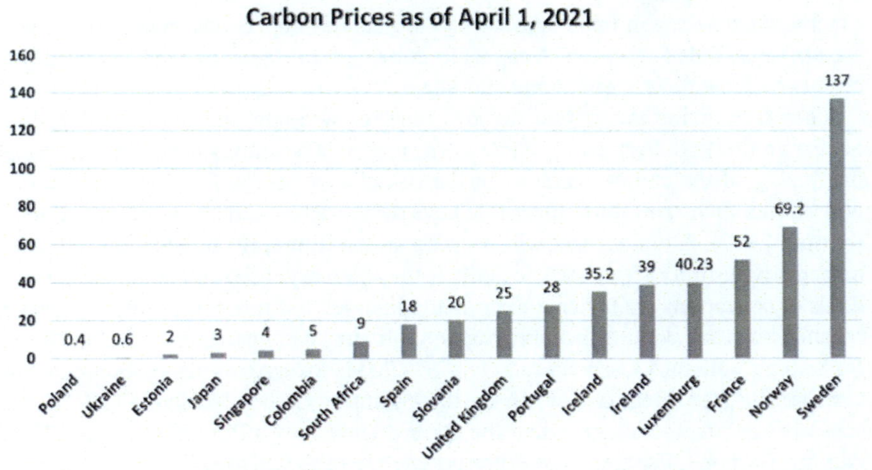

Fig. 10.2 Carbon prices around the world (USD). *Source* Citizens' Climate Lobby (2021)

are investing in carbon projects directly. To obtain the volumes that it would need, the oil and gas industry would practically have to be in the forestry business as well.

Second, the low prices alone discredit their use: How could buying carbon credits for $1 look like anything other than greenwashing, when carbon capture costs over $100 per ton?

Finally, the lack of credibility creates potential liability for the industry. Currently there are no consequences if carbon credits are later challenged. As the industry uses carbon credits like BP and Shell are planning, however, things could be very different. Could it be liable for their customers' loss of reputation if they supplied credits which caused negative publicity? Could it be liable to investors for net zero corporate climate targets using credits which are discredited?

10.4 Fixing the Carbon Markets

Clearly, a lot must be improved. To fund the industry's massive investments in carbon capture and low carbon fuels in the coming decades, carbon credit prices need to be much higher than they are today. Similarly, unless there is some general acceptance that the carbon credits are valid, buying them would at best do little good and at worst cause further reputation loss and invite potential liability.

The two challenges are in fact two sides of the same coin. Carbon credits of low credibility also have very low prices. This is not only because buyers avoid them, but because there is no price floor for sellers issuing them. If a project is economically viable without carbon credits, then anything from the sale of those credits would just be free money for the project developer. A value-maximizing developer would simply issue as many of them as possible as long as prices are above zero. This is exactly what we see in the market today: Project developers continue to issue new credits for as little $1 per ton. As a result, these carbon credits markets are awash in cheap credits with little environmental integrity.

Therefore, fixing the carbon markets requires weeding out the low value, low quality credits first. Fortunately, there are a number of initiatives aimed at improving the quality of the carbon markets. In the compliance sector, for example, the EU will be ending the free allowances that have depressed its carbon credits for so long (Simões 2022). Although that will be years in the future, its anticipation probably helped drive up prices from 2020 to 2022. In the voluntary carbon markets, major standards organizations are revising their methodologies and processes. Finally, several organizations are working on improving the voluntary carbon markets. The Taskforce for Scaling Voluntary Carbon Markets (TSVCM) is focused on market liquidity and transparency, the Integrity Council for the Voluntary Carbon Markets (ICVCM) on standards and methodologies, and the Oxford University's Principles for Net Zero Aligned Carbon Offsetting on the appropriate use of the offsets.

The TSVCM has proposed that carbon credits should follow these "core carbon principles" to ensure their validity:

- Real—They must have really occurred. This means the project must be verified by an independent auditor.
- Permanent—They must be permanent. For projects that promise carbon removals, such as forestry, this means a minimum duration of 50 to 100 years. Offset projects try to protect against losses, such as fires or droughts, by having additional credits kept in reserve.
- Additional—They must not have occurred without the sale of the offsets. In other words, if a project is economically viable on its own, then it cannot issue carbon offsets as well.
- Baseline—They must be measured against the correct baseline emissions amounts.
- Leakage—There must be accounting for the potential indirect increases in emissions which result from the project. For example, if a forestry project protects the trees in one area, it may cause farmers to cut down more trees outside the protected area (The Taskforce on Scaling Voluntary Carbon Markets 2021).

The ICVCM has proposed a similar set of principles, with more focus on the organizations issuing the offsets:

- Additionality—same as TSVCM.
- Mitigation Activity Information—Detailed information about the project should be made available.
- No double counting—Emissions reduction or removal should only be counted once towards targets or goals. No double issuance, double claiming, or double use.
- Permanence—combines the TSVCM's Permanence and Leakage requirements.
- Program governance—Requires the standards organization to have governance to "*ensure transparency, accountability, and the overall quality of carbon credits.*"

- Registry—Requires the standards organization to have proper systems to track its credits.
- Robust independent third-party validation and verification.
- Robust quantification of emissions reductions and removals.
- Sustainable development impact and safeguards—Requires that projects meet standards such as the United Nations Sustainable Development Goals, which extend beyond emissions reductions.
- Transition towards net-zero emissions—Requires the offset project not to lock in emissions which "*are incompatible with achieving net zero emissions by mid-century*" (Integrity Council for the Voluntary Carbon Markets 2022).

"The Oxford Principles for Net Zero Aligned Carbon Offsetting" focuses on the last point: How do offsets relate to the overall need to reduce emissions? It proposes the following:

- Prioritise reducing your own emissions—Minimise the need for offsets in the first place.
- Ensure environmental integrity—Use offsets that are verifiable and correctly accounted for and have a low risk of non-additionality, reversal, and creating negative unintended consequences for people and the environment.
- Maintain transparency—Disclose current emissions, accounting practices, targets to reach net zero, and the type of offsets you employ (University of Oxford 2020).

Only when that is done do they see a place for offsets:

> Users of offsets should increase the portion of their offsets that come from carbon removals, rather than from emission reductions, ultimately reaching 100% carbon removals by midcentury to ensure compatibility with the Paris Agreement goals.
>
> An immediate transition to 100% carbon removals is not necessary, nor is it currently feasible…
>
> In other cases, the emission reduction requires physically storing the carbon whose emission was averted, for example, installing Carbon Capture and Storage (CCS) on industrial point sources or gas power stations. (University of Oxford 2020)

For the oil and gas industry, participating in such initiatives could be important steps in improving the quality of voluntary carbon offsets and therefore their ultimate usefulness.

Meanwhile, the focus on reducing emissions at major corporations will open up a new and far more important carbon market: Supply chain emissions credits.

Note again that the SBTI's guidelines stated that "*Companies should only account for reductions that occur within their operations and value chain*" (Science Based Targets 2021). The "value chain" here may be interpreted to include both companies' upstream and downstream supply chain emissions. For the oil and gas industry, that would include airlines, transportation companies, and utilities, and eventually their end uses such as business travel and building energy. Within value chains, Energy Attribute Certificates or Renewable Energy Certificates have become a standard vehicle for meeting emissions targets such as the Renewable Portfolio Standards.

The use of such certificates is also accepted under the Greenhouse Gas Protocol for companies' own emissions accounts, as if the company had installed renewable energy directly at its facilities (Sotos 2012). These certificates also form an integral part of the renewable energy targets for Google, Apple, IKEA, and other major companies. Thus, similar forms of "attribute certificate" for the emissions reductions of carbon capture and sequestration, advanced biofuels, and hydrogen should be transferable.

The advantages of carbon credits based on energy attributes are that they rely on hard data rather than subjective analysis, tap into much larger markets, and, most importantly, command much higher prices. While voluntary carbon offsets hover between \$1 to \$10 per ton and most compliance markets are below \$30, the price of CO_2 under the California LCFS is as high as \$200 per ton. This may well be the preferred route for the oil and gas industry.

10.5 Role for Blockchain in the Carbon Markets

At its heart, a supply chain carbon credit is simply a "book and claim" transfer of captured emissions or lower carbon intensity of fuel from a site where it is produced to a user somewhere else. Yet such a transfer must follow the same rigorous standards of voluntary carbon credits to be legitimate to customers, regulators, and the general public. This means all the standards set by the ICVCM, requiring detailed records, robust quantification and verification, no double counting, and a proper registry of the credits and their transfers. Furthermore, once issued, such credits must be transferable with greater ease and liquidity than the current over-the-counter markets.

For both existing carbon markets and new supply chain carbon credits, blockchain technologies could significantly reduce administrative costs, cut the time required to issue credits, improve liquidity and transparency, and police against double counting. They could also operate at smaller scales, thus allowing individual companies or groups of companies to transact. Yet the blockchain is not without its own share of problems, and we must make sure the cure is not worse than the disease.

The first and most obvious use for blockchain is digital Measurement, Reporting, and Verification (MRV), where automated processes replace the Word documents and Excel spreadsheets that are the standard fare for validating offsets today. Proponents envision a future where data from sensors, drones, and satellites feed directly into the cloud, which could compute the amount of forest cover or energy savings and issue offset credits in real time. Since many offset projects involve multiple parties, including the project developer, the standards organization issuing the offset, and one or more verification bureaus, a blockchain where all the parties could share data is a natural solution. Both VERRA and Gold Standard have recently formed groups to explore digital MRV and blockchain technologies (The Gold Standard, n.d.) (Verra 2022a, b). The enterprise-focused Hedera blockchain has already developed a Guardian workflow which supports standards organization, project developers, and validation and verification bureaus working under a defined policy to issue carbon offsets (Science Based Targets 2021). Other groups, such as the InterWork Alliance, have put forward similar frameworks for building more efficient carbon credit market mechanisms using digital MRV and blockchain.

A further possibility is to replace the centralized organizations in the offset process with decentralized structures, again using the lockchain. For example, could a decentralized network of multiple parties collectively validate the data instead of one bureau? Chainlink, a network that specializes in oracles which connect off-chain data to blockchains, has tried to do exactly this in an experiment to make payments to farmers for regenerative agricultural practices based on satellite data (Zhou 2021). Similarly, a Decentralized Autonomous Organization (DAO) could rely on multiple parties to vote on whether carbon credits should then be issued, replacing the centralized organizations such as Verra and the Gold Standard. This could change the market from one where the standards organizations are responsible for both maintaining the standards and issuing credits based on them to one where standards are independently maintained from organizations that issue credits based on the standards.

Could we then go a step further and replace the registries of offsets with blockchains? Today, compliance market carbon credits are registered with government agencies, while voluntary carbon credits are registered with the standards organization that issued them. Each of these organizations maintains records of the credits issued, their owners, and whether the credits have been used or retired. Standards of the systems used for the registries vary. While larger agencies and organizations use software from APX or IHS Markit, smaller government agencies still keep the information in Excel spreadsheets or worse.

The appeal of blockchains is obvious, at least to the proponents. They believe that carbon credits should be tokenized on blockchains instead of held in registries' databases. Tokenized carbon credits on blockchains will be more secure against possible data loss or tampering, but also easier to trade and more liquid, which in turn should draw greater capital into the offsets market. Arianna Simpson, partner of the venture capital firm a16z that invested in a blockchain offset startup called Flow Carbon, founded by no less than Adam Neumann of WeWork fame, stated on twitter:

> If there's a financial incentive (ie tokenized credits can be sold more easily via a transparent market, and for a high price) that's likely to increase the supply. People will be more incentivized to build offset projects. (Simpson 2022)

Several startups promising to tokenize carbon credits on the blockchain, including Flow Carbon, Moss, and Nori, have been funded. Some use the blockchain to develop new carbon projects, while others seek to "bridge" existing offsets into blockchain tokens. There are two ways to do this. One is for an intermediary to hold the offset while creating mirror tokens for them on a blockchain. This is the approach of the Universal Carbon (UPCO2) token (Uphold, n.d.).. The issue with such an approach is that it relies entirely on the intermediary to hold the offset in trust for buyers of the token. This creates a centralized, single point of failure that makes the tokens DINO, or Decentralized In Name Only, in the eyes of blockchain developers.

A more decentralized approach is the Carbon Bridge of the Toucan Protocol, which asks users to retire their offsets on the original registry and register them with Toucan. Toucan will then verify the original offsets and issue tokens against them to the user:

> The Toucan Carbon Bridge is a one-way bridge: carbon credits can be brought on-chain but can't go the other way. The reason for that is simple: we want to prevent double-counting and value the deterministic nature of public blockchains. This is why users of the Carbon Bridge need to retire the credits on the source registry before bringing them on-chain. This way, we can guarantee that a carbon token is unique and that burning a carbon token on-chain is equivalent to retiring a carbon credit and doesn't require a centralized entity to go retire the "real-world asset" in a carbon registry. After all, a public blockchain is probably a safer and more transparent way to keep track of carbon assets than any centralized ledger system. (Toucan, n.d.)

This methodology, however, is problematic precisely because it is one-way. Toucan has no communication back to the original offset registry. Instead, it asks users to "retire" the offset on the original registry so it could no longer be used on that registry. Retiring an offset, though, means that the user has used it in their own carbon accounts. Thus, this creates the very problem of double-counting that Toucan claims to solve. For example, a user could retire its offset on the registry and then submit the retired offset to be tokenized on Toucan and other blockchain networks as well. It should be no surprise then that Verra publicly denounced this practice:

> Verra will, effective immediately, prohibit the practice of creating instruments or tokens based on retired credits, on the basis that the act of retirement is widely understood to refer to the consumption of the credit's environmental benefit. (Verra 2022a, b)

The correct solution should require putting the tokenized credit into a locked or transferred state, so that they can no longer be sold on the original registry. This would involve two-way coordination between the registries and any blockchain that holds tokens of their offsets. In other words, blockchains need to work in sync with existing offsets standards organizations and registries, not try to replace them. Both Verra and Gold Standard have now started working groups on this topic.

Whatever the methodology, tokenizing offsets raises another question: What is the real benefit for doing this? Voluntary carbon offsets are currently traded on over the counter through brokers and notoriously illiquid, with the lack of transparency being one of the top concerns, especially amongst regulators (Chen et al. 2021). The Taskforce for Scaling Voluntary Carbon Markets has proposed a futures market to improve transparency and visibility of the market (The Taskforce on Scaling Voluntary Carbon Markets 2021). Proponents of tokenized offsets see blockchain trading as the better alternative.

Greater liquidity would uncork the genie of financial innovation. It could reduce transactions and make them faster, but it could also bring speculation and volatility. Buyers view carbon credits as a "tax" and resist speculators driving up prices. The rise of EU ETS carbon credits from about 20 Euros at the beginning of 2019 to 84 Euros at the end of 2021 drew out protests against speculators from the governments of Spain, Poland, and Czechia, with the latter two even calling for a suspension of the

EU ETS altogether (Simon 2021) (Krzysztoszek and Zachová 2021). We could only wonder what they would say about the volatility of tokenized offsets: The Toucan Base Carbon Token, or BCT, rose as high as $8 and sank to $1.56 per ton. KLIMA, which used financial engineering on top of the BCT to offer investors yields of over 1000%, debuted in October 2021 at $2,000 per ton, rose as high as $3,600, and eventually fell back to $4.98 per ton as of May 25, 2022 (CoinMarketCap, n.d.).

Finally, immutable records on blockchains are supposed to prevent double spending. This has always been touted as a key benefit of cryptocurrencies. Many advocates for tokenizing carbon credits similarly believe that such immutable records would prevent the double counting problem described earlier.

In reality, the blockchain could help solve this problem, but not easily. Contrary to what the advocates say, tokenizing the credits from one registry is not much better than keeping the credits on the registry's own database. Since solving the double-counting problem would require a central repository of all the registries' credits, including both government agencies and voluntary standards, the blockchain could only be helpful if it created such a meta-registry of registries.

The Climate Warehouse project aims to do just that with a blockchain that connects multiple carbon registries. Originally developed by the World Bank, it is planned to go live with the International Emissions Trading Association (IETA) in October 2022. It will initially include ten government compliance registries and five voluntary registries, adding more registries over the next two years. The ultimate goal of the Climate Warehouse is to provide a public data source on all the voluntary carbon registries. As a result, it will not judge the quality of different offsets standards or their credits but instead try to incorporate as many as possible. Nor does it plan to support the Paris Agreement's National Defined Contributions (NDC's), the tokenization of offsets on blockchains, or the transfers of offsets between different registries. So while it is an important step forward for policing the integrity of the carbon markets, it is really just the beginning (Table 10.2).

10.6 Summary and Conclusion

In the years to come, the oil and gas industry could become major players on both sides of the carbon markets. Some sectors, such as power generating utilities and energy intensive industries, would become increasingly subject to government compliance carbon markets. Others, like airlines, may turn to voluntary offsets to stay ahead of regulators. In both cases, carbon credits could be the instrument for the oil and gas industry to support their customers' businesses. The key question is whether the general public will accept this as legitimate climate action.

On the other side, the oil and gas industry will need financing to support its own massive decarbonization efforts, such as reducing methane emissions, carbon capture and sequestration, and sustainable fuels and hydrogen. Carbon credits could be the mechanism for the industry to realize higher prices for its low carbon products, but only if there is confidence in carbon trading mechanisms in general.

Table 10.2 Summary of blockchain technologies for carbon markets

Technology	Application	Benefits
Data Layer	Permissioned or public blockchain of data from carbon projects	Digital Measurement, Reporting, Verification (MRV) Improve time to issue carbon credits Lower cost of issuing credits
Tokens	Tokens of carbon credits on public blockchains	Registry for ownership and use (retirement) Trading Funding
DAOs and Oracles	Decentralized voting based on reputation	Validate data from carbon projects Issue carbon credits based on standards
Meta Registries	Public blockchain of data from multiple carbon registries	Transparency for market, regulators, general public Prevent double counting

The outcry against carbon neutral LNG is a cautionary note for the use of carbon offsets. How could anyone believe an industry that on one hand claims that $1 to $5 per ton could negate the emissions of LNG, and on the other hand is looking to fund carbon capture at $100 or more per ton? If the oil and gas industry is to successfully incorporate carbon credits into its business model, it must follow the standards which are requiring transparency, proper quantification and verification, and proper use of the offsets. It must also support a market of carbon credits with higher prices and greater legitimacy. Blockchain is a technology that could help, but the industry must be ready to use it for the right purpose.

References

BP (2021) Sustainability report 2020. bp.com. https://www.bp.com/content/dam/bp/business-sites/en/global/corporate/pdfs/sustainability/group-reports/bp-sustainability-report-2020.pdf

California Air Resources Board (2018) Appendix H low carbon fuel standard (LCFS) program and examples. CA.GOV. https://ww2.arb.ca.gov/sites/default/files/barcu/regact/2018/ict2018/apph.pdf

Cames M, Harthan RO, Füssler J, Lazarus M, Lee CM, Erickson P, Spalding-Fecher R (2017) How additional is the clean development mechanism?" https://ec.europa.eu/clima/system/files/2017-04/clean_dev_mechanism_en.pdf

Carbon Market Watch (2010) Briefing paper: HFC-23 offsets in the context of the EU emissions trading scheme. Carbon Market Watch. https://carbonmarketwatch.org/2010/07/15/hfc-23-offsets-in-the-context-of-the-eu-emissions-trading-scheme/

Carbon Market Watch (2012) Forestry/land-use projects in the CDM. Carbon Market Watch. https://carbonmarketwatch.org/2012/05/30/forestry-projects/

Carbon Market Watch (2021) Net-zero pipe dreams: why fossil fuels cannot be carbon neutral. Carbon Market Watch. https://carbonmarketwatch.org/publications/net-zero-pipe-dreams-why-fossil-fuels-cannot-be-carbon-neutral/

Chen S, Marbouh D, Moore S, Stern K (2021) Voluntary carbon offsets: an empirical market study. https://ssrn.com/abstract=3981914 https://doi.org/10.2139/ssrn.3981914

Chestney N (2022) Global carbon markets value surged to record $851 bln last year-Refinitiv. Reuters, January 31, 2022. https://www.reuters.com/business/energy/global-carbon-markets-value-surged-record-851-bln-last-year-refinitiv-2022-01-31/

Citizens' Climate Lobby (2021) Carbon pricing around the world. Citizens' Climate Lobby. https://citizensclimatelobby.org/laser-talks/carbon-prices-around-world/

CoinMarketCap (n.d.) KlimaDAO price today, KLIMA to USD live, marketcap and chart. CoinMarketCap. https://coinmarketcap.com/currencies/klimadao/. Accessed 25 Aug 2022

Dowdall T (2021) Science-based net-zero targets: 'less net, more zero.' Science Based Targets. https://sciencebasedtargets.org/blog/science-based-net-zero-targets-less-net-more-zero

Ecosystem Marketplace (2021) Ecosystem marketplace's state of the voluntary carbon markets 2021. Ecosystem Marketplace. https://www.ecosystemmarketplace.com/publications/state-of-the-voluntary-carbon-markets-2021/

EDF (2018) The world's carbon markets: a case study guide to emissions trading. Environmental Defense Fund. https://www.edf.org/worlds-carbon-markets

Elgin B (2022) U.S. public forests are cashing in on dubious carbon offsets. Bloomberg Law. https://news.bloomberglaw.com/environment-and-energy/u-s-public-forests-are-cashing-in-on-dubious-carbon-offsets

European Commission (n.d.) Monitoring, reporting and verification of EU ETS emissions. https://ec.europa.eu/clima/eu-action/eu-emissions-trading-system-eu-ets/monitoring-reporting-and-verification-eu-ets-emissions_en. Accessed 24 Aug 2022

European Court of Auditors (2020) Special report 18/2020: the EU's emissions trading system: free allocation of allowances needed better targeting. European Court of Auditors. https://www.eca.europa.eu/en/Pages/DocItem.aspx?did=54392

Fogarty D (2011) Carbon credits for India coal power plant stoke criticism. Reuters. https://www.reuters.com/article/us-india-carbon-coal-idUSTRE76B1XI20110712

Foley J (2021) Opinion: the world needs better climate pledges. Project Drawdown. https://drawdown.org/news/insights/opinion-the-world-needs-better-climate-pledges

The Gold Standard (n.d.) Digitising MRV." The Gold Standard. https://www.goldstandard.org/our-story/digitising-mrv. Accessed 8 Sept 2022

The Gold Standard (n.d.) Optimising markets: digital monitoring reporting + verification (MRV). The Gold Standard. https://www.goldstandard.org/our-story/digitising-mrv. Accessed 25 Aug 2022

Gordon G (2020) Policy brief: CORSIA baseline adjustment in response to COVID-19: a blessing or a curse?|SDG Knowledge Hub|IISD. SDG Knowledge Hub. https://sdg.iisd.org/commentary/policy-briefs/corsia-baseline-adjustment-in-response-to-covid-19-a-blessing-or-a-curse/

Greenberg C, Colombo L (2021) Carbon offsets are a scam. Greenpeace. https://www.greenpeace.org/international/story/50689/carbon-offsets-net-zero-greenwashing-scam/

HBO (2022) Carbon offsets: last week tonight with John Oliver (HBO). YouTube. https://www.youtube.com/watch?v=6p8zAbFKpW0

IATA (2021) Media kit—fly net zero. IATA. https://www.iata.org/contentassets/a686ff624550453e8bf0c9b3f7f0ab26/flynetzero_mediakit.pdf

Integrity Council for the Voluntary Carbon Markets (2022) "Part 2: core carbon principles. Integrity Council for the Voluntary Carbon Market. https://icvcm.org/wp-content/uploads/2022/07/ICVCM-Public-Consultation-FINAL-Part-2.pdf

Klinge, Naomi. 2021. "Nature-based offsets bring controversial solution to carbon-emissions problem." Upstream Online. https://www.upstreamonline.com/energy-transition/nature-based-offsets-bring-controversial-solution-to-carbon-emissions-problem/2-1-1023726.

Krzysztoszek A, Zachová A (2021) Czech minister would support Poland's call for suspension of EU ETS. EURACTIV.com. https://www.euractiv.com/section/politics/short_news/czech-minister-would-support-polands-call-for-suspension-of-eu-ets/

Lewis B, Volcovici V (n.d.) Insight: U.S., China turned EU powers against airline pollution law. Reuters. https://www.reuters.com/article/us-eu-airlines-climate/insight-u-s-china-turned-eu-powers-against-airline-pollution-law-idUSBRE8B801H20121210. Accessed 24 Aug 2022

Lowe P (2013) U.S. officially prohibits ETS participatio: Business aviation news. Aviation International News. https://www.ainonline.com/aviation-news/aviation-international-news/2013-01-03/us-officially-prohibits-ets-participation.

Norkin D (2022) Guardian v2.0: the next generation of ESG marketplaces built.... Hedera. https://hedera.com/blog/guardian-v2-0-the-next-generation-of-esg-marketplaces-built-on-hedera

Prentice A (2021) Greenpeace calls for end to carbon offsets. Reuters. https://www.reuters.com/business/sustainable-business/reuters-impact-greenpeace-calls-end-carbon-offsets-2021-10-06/

Science Based Targets (2021) SBTI CORPORATE NET-ZERO STANDARD. Science Based Targets. https://sciencebasedtargets.org/resources/files/Net-Zero-Standard.pdf

Shell USA, Inc (n.d.) Our climate target|Shell United States. shell.com. https://www.shell.us/energy-and-innovation/our-climate-target.html#iframe=L3dlYmFwcHMvY2xpbWF0ZV9hbWJpdGlvbl9VU19OZXRfemVyb18yMDIyLw. Accessed 23 Aug 2022

Simões H (2022) Carriage details|legislative train schedule. https://www.europarl.europa.eu/legislative-train/theme-a-european-green-deal/file-carbon-border-adjustment-mechanism

Simon F (2021) Europe's energy price hike fuelled by speculators, Spain and Poland say. EURACTIV.com. https://www.euractiv.com/section/emissions-trading-scheme/news/europes-energy-price-hike-fuelled-by-speculators-spain-and-poland-say/

Simpson A (2022) If there's a financial incentive (ie tokenized credits can be sold more easily via a transparent market, and for a high price) that's likely to increase the supply. People will be more incentivized to build offset projects. Twitter. https://twitter.com/AriannaSimpson/status/1529205712710250499

Song L, Temple J (2021) The climate solution actually adding millions of tons of CO_2 into the atmosphere. ProPublica. https://www.propublica.org/article/the-climate-solution-actually-adding-millions-of-tons-of-co2-into-the-atmosphere

Sotos M (2012) GHG protocol scope 2 guidance. Greenhouse Gas Protocol. https://ghgprotocol.org/sites/default/files/standards/Scope%202%20Guidance_Final_Sept26.pdf

Stapczynski S, Rathi A, Marawanyika G (2021) The fictitious world of 'carbon neutral' fossil fuel. Bloomberg.com. https://www.bloomberg.com/news/features/2021-08-11/the-fictitious-world-of-carbon-neutral-fossil-fuel

The Taskforce on Scaling Voluntary Carbon Markets (2021) Task force on scaling voluntary carbon markets final report. Institute of International Finance. https://www.iif.com/Portals/1/Files/TSVCM_Report.pdf

Toucan (n.d.) Bridge. Toucan. https://docs.toucan.earth/protocol/bridge/carbon-bridge . Accessed 25 Aug 2022

Trading Economics (2022) EU carbon permits—2022 data—2005–2021 historical—2023 forecast—price—quote. Trading Economics—EU Carbon Permits 2022 Data—2005–2021 Historical. https://tradingeconomics.com/commodity/carbon

Union Square Ventures (2020) UInion square ventures energy & climate market map. Union Square Ventures. usv.com

University of Oxford (2020) The Oxford principles for net zero aligned carbon offsetting 2020. Smith School of Enterprise and the Environment. https://www.smithschool.ox.ac.uk/sites/default/files/2022-01/Oxford-Offsetting-Principles-2020.pdf

Uphold (n.d.) UPCO2—carbon credits. Uphold. Accessed August 25, 2022. https://uphold.com/en/assets/environmental/buy-upco2

Verra (2022a) New digital MRV working group: invitation to apply. Verra. https://verra.org/new-digital-mrv-working-group-invitation-to-apply/

Verra (2022b) Verra addresses crypto instruments and tokens. Verra. https://verra.org/verra-addresses-crypto-instruments-and-tokens/

Zhou A (2021) Blockchain can help us tackle climate change. Here's how. The World Economic Forum. https://www.weforum.org/agenda/2021/06/blockchain-can-help-us-beat-climate-change-heres-how/

Si Chen is the president of Open Source Strategies, Inc. in Los Angeles, CA, which specializes in open-source software for climate finance and investing. He leads the development of open-source blockchain carbon accounting software at Hyperledger Labs. Previously, he has managed investment portfolios for institutional pension funds, central banks, and hedge funds and has been published in *The Journal of Portfolio Management*. He is also the co-founder and CTO of GraciousStyle.com, an online retailer.

Interviewees

Interview 1: Aaron Lohmann, Karl Osterbuhr, Dan Cearnau (EarnDLT Team)

Interviewees:	Aaron Lohmann, Founder and CEO, EarnDLT Karl Osterbuhr, President, EarnDLT Dan Cearnau, Co-Founder & CTO, EarnDLT
Interviewer:	Dr. Soheil Saraji, Associate Professor of Energy and Petroleum Engineering, University of Wyoming
Interview Date:	22 June 2022

Soheil Saraji:	Let's start by a short introduction for the purpose of this interview.
Aaron Lohmann:	Sure, I can kick that off. My name is Aaron Loman. I'm the founder and CEO of Earn DLT. We are an enterprise application that provides permissioned-blockchain access to large and medium-sized businesses.
Karl Osterbuhr:	I'm Karl Osterbuhr, the acting president of Natural Resource and Energy Group. My background is in geology, and I've been in this space for most of my career.
Soheil Saraji:	Great. Thank you, gentlemen. Aaron and Carl, I understand you had completely different backgrounds before you joined forces, and it would be interesting to get your different perspectives. How do you think this energy transition that we're going through will impact the energy industry as a whole and the oil and gas industry particularly?
Aaron Lohmann:	In my opinion, the energy industry plays a very significant role in in the future of the energy transition, and in management of carbon and methane in that transition. I believe that they will play a major part and that system like the blockchain that will enable

S. Saraji and S. Chen, *Sustainable Oil and Gas Using Blockchain*, Lecture Notes in Energy 98, https://doi.org/10.1007/978-3-031-30697-6

these participants, these energy companies, to accurately qualify and quantify their activities will help them to improve on that effort.

Karl Osterbuhr: I think hydrocarbons are going to play a very significant role in our energy needs, especially worldwide. Instead of phasing out fo sil fuels in the near term, which seems impractical at this point, we need to come up with a solution to dilute our impact to the greenhouse gas situation. racking carbon and coming up with ways to allow producers to offset and minimize their impact is going to be crucial to the energy transition.

Soheil Saraji: What do you think would be the challenges that this energy transition would bring to the industry? And what would be the opportunities?

Aaron Lohmann: I think the challenges are numerous for these players in this transition. But some of the major challenges are the public's perception, in terms of the activities, and these companies representations behind their activities around carbon management and methane mitigation. I think that the opportunity though, to your point exists in being able to provide a very transparent source of truth that's validated not only by the energy company reporting what they're doing, but also by reputable third parties in the space that are also contributing to that dataset representing the carbon management and methane management practices.

Soheil Saraji: Thank you. I think Dan has just joined us. Hi, Dan. Can you introduce yourself if you don't mind?

Dan Cearnau: I'm Dan. I'm the CTO at my firm. I've been involved in software for around 12 years, initially starting with mobile applications. Then in 2014, I started working on blockchain applications, especially on one of the first smart contracts on the platform "Ethereum." During the 2016–2017 ICO bubble, I built products ranging from cryptocurrency exchanges, launch pads, and generic blockchain applications, like identity or supply chain management.

It was very different then. I had a software development company, so we were mostly doing outsourcing for different clients. and I would say the market was very different. It was reminiscent of the Wild West, especially in 2016 and 2017. IPOs, ICOs, IDOs, etc. Because I had a software development company, a lot of people came to us to develop the blockchain product. At some point, I believe, there was a couple of articles about blockchain, and people describing things that definitely were not possible to implement. They were coming to software development companies asking to do all kinds of weird things. It was pretty strange.

Soheil Saraji: So, back to Karl; we were discussing the energy transition. Let's talk about the challenges and opportunities that the energy transition can bring to the oil and gas industry. Karl, you've been in the industry for many years, and you're capitalizing on some opportunities within the industry by changing directions. What are your thoughts on the industry's problems and challenges, as well as the opportunities that this energy can bring to the industry?

Karl Osterbuhr: In my mind, some of the largest challenges are just the corporate mindset overall, to adopt the mindsetto take on a problem like this, reducing your emissions on a voluntary level. Some industry players still don't recognize that climate change is an issue and they're resistant to adapt. And I think that's a huge challenge. The technology is available and can rapidly accelerate the transition, it's just getting the mindset necessary to make the change.

Soheil Saraji: I agree with you there. Dan, there is the concept of energy transition, and that impacts the energy. I know your background is not related to the oil and gas industry, what do you believe the challenges will be for the oil and gas industry in the coming decades of the energy transition? And what could be the opportunity for the industry?

Dan Cearnau: I think it's mostly about transparency. And this is what blockchain is bringing to the table. The blockchain has had two big advantages. One is transparency and the fact that you can't hide things on a chain. Once something is issued, it is distributed over the network. You can't cook the books, you can't erase documents, so it increases the transparency. And I think that will matter a lot in this area and this market. The second is, of course, the traceability; the fact that when you're buying a token that represents RSG gas, you know who you're buying it from. And that it has that environmental impact, is validated, is accredited to do that. Because otherwise, they think there's going to be a lot of fake carbon credits, people faking in order to reduce their environmental impact,

Soheil Saraji: I agree with you, and I do have a related question later. Now, let's talk about the concept of carbon economy. It's not a reality yet. Certain implementations exist, such as the European Union's ETS system or California's cap and trade system, but they are not universally accepted or implemented. I guess my question for you, Aaron. What do you think of the carbon economy concept? And What do you think it's going to look like in the future?

Aaron Lohmann: I think it will be successful. I don't think it's going to be a smooth transition, though. I don't think we as a species have a choice but to incorporate carbon somehow into our economy. The economy and how we trade and transact with one another is such a deep component of our society. And energy is such a deep component of our society. We have to tie those together somehow. So, I do believe carbon is going to be in the future a major component of the economy, but the transition is not going to be smooth. I think there are a lot of bumps that need to be worked out along the way. And a system like blockchain with the transparency and authenticity offered by that technology is going to help in order to bring this to life.

Soheil Saraji: Thank you. So, Karl, the same question for you: what do you think about the carbon? And how do you imagine it will look, say, in a decade?

Karl Osterbuhr: Well, I think the carbon economy as a whole, is a huge challenge for us getting to net zero and maintaining the temperature one and a half degrees C [above the pre-industrial levels]. If this is humanity's goal, we need action sooner rather than later.. If we intend to manage greenhouse gas emissions, we need a way to effectively track, or account for them. That's where blockchain technology can help. Utilizing blockchain in carbon accounting can bring transparency, efficiency, and trust to the entire energy supply chain.

Soheil Saraji: From another point of view, some people criticize the carbon economy concept because they believe it is just another game to delay or stop emissions. One radical way to reduce emissions is to go cold turkey and just stop emitting. However, the carbon economy is a gamified version of that, allowing the economy to set the parameters that allow it to push us in that direction.What are your thoughts on that?

Dan Cearnau: First of all, I think the carbon economy is a global economy, because we live on the same planet. This has its own challenges, because you can't [develop] rules that apply on a global level. So, I think, the main advantage of blockchain here is to have accountability. On the global level, we all decide to have certain rules, and then how do we enforce these rules? How do we make sure that for example, an automaker in Germany doesn't fake the amount of carbon dioxide emissions from their cars? So, I think in the end, you need an accountability system. A global accountability system with the stakes for each party that is involved, that they have something to gain or lose. In the end, it can be gamified. Most of the people that want to reduce their carbon footprint, in the end, want to feel better about it, and they want to gain something. So, I think gamification of the global carbon economy would make a great impact, if you provide certain gimmicks. You can [receive] NFS if you reach a certain level, you could get certain tokens if you reach certain goals and things like that. All of this needs to be integrated, all the parties that are monitoring it, the parties that are reducing the carbon footprint. [There are] quite a lot of different parties that need to collaborate and [be] accountable.

Aaron Lohmann: Yeah, I think when you look at human nature, the ultimate question that everybody has is, "What's in it for me?" to get them to change anything about their activity or their habits. Therefore, you do need to incentivize. I believe, to change humans activity, you could require them to go cold turkey, as you were mentioning earlier, but I don't think that's a very likely scenario, or very successful scenario. So, by gamifying, this carbon economy, so to speak, it may not be the absolute best ultimate solution, but I think it probably is going to be the most successful in terms of changing human behavior quickly by incentivizing humans to actually change their behavior.

Soheil Saraji: Yeah, I agree with you guys on the gamification, and I guess the next step for us is to just ask what the oil and gas industry can do in this new carbon economy that is forming. There is responsibly sourced gas, for example. And there are oil and gas companies who announced that they want to go carbon neutral by 2030, 2050. What does that mean? Is it even possible? And, in general, what do you think the oil and gas industry can do in this carbon economy?

Aaron Lohmann: I think it really comes down to transparency, and how they're representing their activities to the market. The unfortunate reality is these energy companies don't have the best reputation when it comes to public perception in many circles. So, they are going to face a very real challenge when announcing to the market that they intend to go carbon neutral, or that they intend to certify their activities as green. And this is where the blockchain in combination with reputable science-based third parties come into play. I don't think the oil and gas industry can do it by themselves, I think they are going to need reputable science-based organizations looking over their shoulder and reporting in a reliable way to the public, what they actually are doing. And they're going to need the rails, technological rails, or systems like blockchain to allow for all of those parties to interface and effect that data in a reliable and with integrity that the market can believe in and get behind. Those will be the challenges but I think this is how oil and gas companies can participate, not just saying that they're going to do it, but really showing that they're going to do it by collaborating with these trusted groups to help make that possible.

Soheil Saraji: Karl, what is the role of the oil and gas industry in this regard? And is it currently fulfilling its role? What are the challenges in your mind?

Karl Osterbuhr: The oil and gas industry needs to assume the leading role in decarbonizing the planet, and not wait for regulation. Many companies are taking a proactive approach to the transition with aggressive plans to lower their carbon emissions... One of the challenges will be the extra capital expenditures required to reduce emissions. Another will be quantifying emissions in near real-time, although there are a lot of studies and initiatives in place to facilitate this goal. Our greatest challenge will be coordinating a massive shift in the way we do business in the most complex supply chain on the planet. Again, this is where blockchain technology can play a huge role.

Soheil Saraji: Yes, that's great. Also, there is, as you said, a market component to it. Does the market reward the companies that are making efforts in this regard or not?

Karl Osterbuhr: You can actually make a case, if you reduce your methane emissions to a certain level, by capturing all the gas that you lose, you can easily pay for your implementation efforts. We've run the numbers on some of those cases. It depends on how low you can get your methane emission numbers to but you can actually pay for it by preventing the loss of that gas.

Aaron Lohmann: It's the best practice. I mean, it is what these companies should be doing anyway.

Karl Osterbuhr: Implementing the kind of change it will take to meet the challenges of a low-carbon future isn't going to be easy. Many larger companies are very resistant to change, and a lot of the smaller companies aren't in the best position to finance the changes needed. Currently, the marketplace is exploring options for differentiated natural gas. We've heard of European, Asian markets, and some US based utilities that are interested in purchasing natural gas with a lower methane intensity.

Soheil Saraji: Yeah, that's a typical problem in the industry. When you introduce a new concept or technology —for example, blockchain—, you probably don't have the skill in the workforce. It will take a long time for the US energy industry accept blockchain because they don't even have in-house experts who knows the technology. That's a very slow process. In my opinion, oil and gas industry, is a little bit conservative industry. The change in oil and gas industry happens very slow. Let's go to Dan and see what he has to say about this.

Dan Cearnau: I am going to emphasize what I said before. In the last years, we've seen a lot of companies claiming that they will become carbon neutral by a certain date, and I think most of it is PR [(Public Relations)], in my opinion. Of course, it's very hard to track these things, and the way they claim that certain amount of energy is renewable. I would like to give you an example, I think last summer, when Tesla decided to accept Bitcoin as a payment method and then Elon Musk came and said, they will not accept

Bitcoin, because Bitcoin has a big environmental impact. And of course, there were a lot of people that were very revolt about that. They have started doing the math of how much the Bitcoin network is impacting the carbon footprint, and it was very difficult because I saw numbers ranging from 90 to 10% of the Bitcoin network is powered by renewable energy. So, basically nobody knew. I think they made a consortium with Elon Musk, and there was the CEO of Twitter and they tried to figure out how [much] was the environmental impact of the Bitcoin network. So, I think we need to create a traceable way of [tracking] renewable energy that needs to be registered. So, we know exactly what the energy is used for, and how the energy is produced. And there is also a need to provide certain privacy for the both the producer and the consumer. But I think blockchain can help here by implementing a way in which you can prove that the energy you're using is renewable, because you get the token. Nobody needs to know exactly who you're buying that energy from, at what price and so on. This could be a market advantage, but to be able to prove yourself. You're doing an investment. You have a Bitcoin miner or whatever energy business, and you want to prove that you are using renewable energy for 70% or so. So, this needs to be calculated by the industry, because it's very hard to know exactly the environmental impact of certain energy.

Soheil Saraji: Especially Bitcoin, which is illegal in some countries. So, people may hide their IP, which makes it difficult to track their energy mix.

Dan Cearnau: Yeah, definitely. I guess because of the energy crisis, like what's happening, especially in Europe. Governments will give subsidized [energy] to people, because you want the government to pay for the people to heat their homes, and not to leave them to freeze in their homes, but you don't want to pay subsidize for someone [who is] mining Bitcoin. So, that's the big problem. I think Kazakhstan was having a very cheap energy is very because the government basically subsidized most of it. So, all the miners went there in order to mine Bitcoin, because the energy is cheap. It doesn't matter where it came from. So, you need to consider both environmental impact, and also the economic impact.

Soheil Saraji: Yeah. Maybe it is a good time for us to focus more on blockchain. We've already discussed some applications that blockchain could and some issues in the energy industry. Tell us a little about your project and how it is helping the industry.

Aaron Lohmann: I can give a high-level overview of what our product is doing and our project for the industry. And then Dan and Karl can probably dive into more of the technical aspects about how we're interfacing with the industry. So essentially, what EarnDLT provides is the ability for a user and energy producer in this case, to document all of their green energy production in a data room, that has been authenticated by the blockchain, by the cryptography and the blockchain. Further to that we allow for that data room, or that cluster, or that group of data, to be unitized, and tokenized. So that the owner of that data can transfer the data and even sell that data to another user. And the way that energy companies will use it is basically by monitoring their energy production on wells that have been certified by a third-party science-based organization as low methane intensity or as responsibly sourced gas on that well, and what our platform will allow for is the producer to report directly to the blockchain on their production numbers, but then allow for that third-party scientific organization to report directly on that unit of energy regarding the methane intensity and the certification of that as green energy. And so, in combination with the blockchain, what you have is that immutable, permanent record around that unit of energy that MMBtu of gas, responsibly sourced gas, and the methane that was produced or leaked in the production of that unit of energy.

Dan Cearnau: First of all, we're running on a private blockchain. So, the data is not shared on a public ledger like Ethereum, or Bitcoin network. This offers [features like] privacy as well as the ability to have certain control over the network. Now, like I told you, blockchain has a few characteristics: transparency, decentralization, and the fact that nobody controls it. Now, in the case of enterprise blockchain solution, like our consortium of enterprises that use a blockchain, it is not about the total distribution or the uncontrollable part of the chain itself. As I said, when there is accreditation, it's basically someone at the company that created that. So, we're not a fully decentralized solution. And this also makes the product of our company more enterprise friendly. As you know, having public blockchain integrated into certain enterprises is difficult because of the way you cannot recover fees that are lost. So, if you are on the Ethereum blockchain, public blockchains and your application does something or someone loses the keys, and there's no way to get it back. This is a problem. So, that's why a lot of corporations are very afraid of using public blockchain solutions.

Soheil Saraji: If I may ask, what is the private blockchain? Is it your blockchain? Or are you using an existing available enterprise solution?

Dan Cearnau: It is a fork of Quorum, a blockchain powered by consensus. It was formerly built by JPMorgan; they started building the blockchain solution, then it was taken over by consensus. It's one of the biggest companies that provide infrastructure for blockchain. Behind the scenes, it uses Ethereum as a virtual machine for smart contract execution, but it doesn't have the energy-intensive proof of work. It has proof of authority, meaning that the chain works like a consultant; you have different companies that agree on certain rules, and the rules are set by the chain. So, each member of the consortium is running a node and thus validating the network, increasing its security, and providing validation to the network. And they don't need to run energy-intensive nodes like Bitcoin or Ethereum to have miners, if all they need is a basic node, and the algorithm automatically selects one of the members to mine the next block.

Soheil Saraji: Right. Who are inside the consortium as validators?

Dan Cearnau: Members of the could be the energy companies, auditors that want to verify it, or an authority that that does the certification. It's about accessing the data and securing the network. Now, of course, you can add new members if consortium decides, you can remove certain members right.

Soheil Saraji: That's perfect. Thank you. Karl, do you want to add anything to this? Or maybe if you have a perspective on how the oil and gas companies work in relationship with your company, please share with us.

Karl Osterbuhr: EarnDLT is offering a solution to mint your own digital asset, put it on the blockchain, create smart contracts to improve efficiency and accuracy. And then bundle that up, roll out ownership, if you will, to permission parties, and get that onto an exchange where you could see value for that asset you've created. I think that's the exciting thing about what EarnDIT is doing, and it has the potential to be a game changer in the voluntary credit market.

Soheil Saraji: I guess there are a few things that are not clear to me. You are going to put the data on blockchain that is relevant to the carbon emissions, for example, or the methane emissions of a particular company. And then whoever has the token will have access to that information. I believe that's clear. But how does the carbon credit fit into this scheme?

Aaron Lohmann: Yeah, correct me if I'm wrong because Karl has sort of perfected this mechanism on our end. What we're doing to generate that carbon offset equivalent value for the methane is taking the certified methane intensity that's assigned to that unit of energy by the third-party scientific organization and comparing that to the EPA national average of methane intensity for a well, and then measuring that delta between the benchmark and the actual, computing the weight of the carbon that was avoided between those two numbers. And that's how we arrive at that CO2 equivalent.

Soheil Saraji: I see, But as time passes, that might become a little bit harder, because if everybody starts switching to low-emission schemes, then your average is going down. As a result, it become increasingly difficult to have a baseline.

Aaron Lohmann: I think that's probably how we should be. I mean, we shouldn't just allow these companies to claim that they're avoiding carbon when the entire industry is already green, because at that point, hey're not avoiding carbon. They're doing what everybody else is doing. And that's where we want to drive them all, ultimately.

Soheil Saraji: This makes sense to me. Now that we are on the topic of blockchain, I want to get a little bit deeper into the. We talked about all of the benefits that blockchain can provide, particularly in the carbon economy, and how carbon accounting, green energy generation, carbon offsets, and all of that could work. Dan also talked a little bit about the sustainability of blockchain and its challenges. Are there any other applications that you are excited about?

Also, there are several challenges, which I assume are out there, that slows down the adaptation of blockchain in the energy industry or specifically the oil and gas industry. In your opinion, what are the challenges and potential solutions.

Dan Cearnau: I like to start on this. Yeah, of course in the oil and gas, and a lot of other industries, Blockchain is not used at the full potential. You can use blockchain as a financing metadata, a way of raising funds for different kinds of projects. And this is the same on oil and gas, for mining, for real estate. If you have a new project, and you want to raise funds from your investors, and you want to give them a security token. It is similar to bonds but it provides more flexibility than the traditional one because it also has market liquidity, allowing you to use the secondary market that you can implement. So definitely in the financing sector of the oil and gas industry, there's definitely a lot of potential.

Second [application] is in the supply chain management, because they have a lot of heavy machines, that are manufactured and distributed and sold from one manufacturer to distributor, and so on.

You can use it for inventory management, that's a very important thing. I don't know a lot about oil and gas but think, for example for an airplane, it has certain part. You have a machine, that machine is being used with certain parameters throughout its life. It has maintenance on it, and let's say after a few years you want to sell that machine for millions of dollars. Any buyer would like to get the maintenance log, how much that equipment was used, in what conditions and so on. Having all of these on a blockchain, you can think of an NFT representing the machine, provides a lot of transparency, flexibility for the secondary market equipment for oil and gas.

Soheil Saraji: So, one example that comes to mind is a drilling rig. Its value is dependent on the usage; the more you use it, the lower its value.

Dan Cearnau: Especially the maintenance, right? I think it's very important. If you did maintenance on time, and kept the log of what parts was changed, it would be easier to make transactions here.

Soheil Saraji: Karl probably knows better than me. But I guess that there are a lot of intermediaries in between that make a lot of money out of this, so they are the people that connect the rigs to the companies. I don't know the US market, but I was in the Middle East for some time and knew a little about it. You would essentially go to someone without any technical background but whose expertise is knowing people, and they would find the right drilling rig for you, and they would get a good percentage of the deal—even 1% or half a percent, which is a lot of money.

Dan Cearnau: So that's part of the operation of oil and gas companies. And, of course, the greatest part is that when you produce the oil, you could also have products, and I know they're already in the financial world: financial instruments to hedge against prices. So, let's say you start drilling for oil and the price of oil is $100 per barrel. I think you'd like to hedge if the price of oil in a barrel goes down. What do you do with your project? So, you could do that with blockchain [and develop] instruments that allow for the financing of the operation as well as hedging against market volatility, allowing companies to lock in the price of the product they are drilling based on the stage of the project, in order to make sure the project is still viable after discharge.

Soheil Saraji: Thank you. Karl, what do you think would be the other application of blockchain, or maybe even some futuristic applications that blockchain could have in our industry?

Karl Osterbuhr: I'll have to think about that one. I did work on a project around six years ago that we were trying to raise capital for. There was a lot of interest in the project however it was very difficult for investors to wrap their head around the technology piece (blockchain and tokenization).. A lot of oil and gas executives, and senior-level people are distrustful of the crypto space in general. So, there has been some resistance to adoption.. I think the big potential here is the paradigm of web 3.0 with the ability to own and monetize your own data. Taking control of your data, getting it on to the blockchain and into a marketplace where you can monetize it will create a lot of value for oil and gas companies in the future. High quality data integrated throughout the supply chain will bring much needed transparency. Reserves forecasts, title opinions, Operating Agreements, and other sophisticated contracts could be handled with smart contracts much more efficiently than they are today..

Soheil Saraji: That's one of the challenges, which is distrust, and the other thing is that the industry has a tradition of hiding information. For example, if you buy a piece of land somewhere that has good potential for oil development, you don't want everybody else to know about it until you buy enough land. So, there are things that the industry doesn't necessarily want to share early on. But, as you mentioned, there are many things they don't have such an impact on the company's economy, but they can monetize it for whatever future applications.

Aaron Lohmann: Or it can help to contribute directly to enhancing the value of the core product that they're representing. In Dan's scenario where you're tracking the maintenance record of a rig, you could deduce that the token with the maintenance records enhances the value of that rig significantly if it's included in the transaction.

Soheil Saraji: Yeah, It's like a car, and if you know all the information about the car, then you can decide which one is right for you or your application. And do you have any other applications in mind, Aaron?

Aaron Lohmann: Yeah, we've been thinking about a few other applications using Blockchain to help streamline the fundraising mechanism for these oil and gas industries and tying that fundraise to the 45Q, Carbon offset. One of the challenges with the 45 Q is that it's a nonfungible tax credit. And it presents a lot of challenges for the producer of that tax credit to actually monetize those tax credits and make a viable business out of that. What the blockchain would allow for with some creative legal structures is the ability for the oil and gas industry to essentially syndicate an offering, a financial offering, with the expected return to those investors being these 45Qs that would be in a much more fungible form, being tokenized. And with the legal structure that we're intending to implement, those investors would be to transfer those 45Q credits digitally within the pool.

Soheil Saraji: That's exciting. We are working on a similar project. I think carbon sequestration could benefit a lot from the whole carbon economy. It is technically proven but economically challenging to implement. And I think blockchain, or blockchain in combination with carbon economy, can help finance that sector. And that's exactly what oil and gas companies can invest in very easily, because carbon sequestration is a very similar process to oil production. I do think there's a huge potential there. But there are also a lot of technical issues with applying blockchain solutions.

Dan Cearnau: As you said about the more science fiction application, I mean, if you have a plot of land, and you can prove from a judge report that you have a certain amount of oil underneath it, you could be incentivized not to develop for that oil, and get the token that represents all that you already have on the ground. You're not doing it in order to, of course, to have lower the environmental impact.

I remember there was a project, I think, it was a goldmine. They had a certain amount of gold in the ground. They said, okay, in order to get the gold out of the ground, let's say, we have a million ounces of gold, we need whatever amount of money, say $2 billion. It has a big environmental impact, and this gold eventually ends up in a vault. So, what's the point of it? We can just tokenize the gold that's already in the ground. And not do develop a huge mine around it. And the value is still there in the end.

Soheil Saraji: Yeah, that's interesting.

Aaron Lohmann: As a result, it is even better for an oil producer to leave the oil in the ground rather than extract it.

Soheil Saraji: Ok. As the owner of the land you might decide, "I'm not going to lease it out to the oil and gas industry if I can have a financial benefit from doing it." That's interesting.

I'm going to highlight one of the problems that I see with the blockchain as it applies to oil and gas and I want to get your perspective on it. We call it the "oracle problem." Blockchain is a great digital technology and works perfectly with any natively digital asset. As soon as we bring blockchain to the real world, we create a weak link or an attach point. The bridge between the real-world and blockchain, where the data enters the blockchain could be manipulated for a variety of different reasons. How can we ensure that only correct data ends up on the blockchain? I want to give you a concrete example: a third-party company that sets standards for say methane emissions works with a lot of oil and gas companies to certify their wells as low emission wells. But who pays this third party to certify the wells? Typically, the company who is responsible for emissions. We immediately have misalignment of incentives. Now, if this third-party is also responsible to put the methane emission data on the blockchain to certify Responsively Sourced Gas, then we create a trust problem. Would a customer who buys RSG certificates trust this scheme? What do you think about this problem? Do you have any solutions?

Aaron Lohmann: I believe it's a very real problem, what you've described, and I think the blockchain can actually offer a really clean solution for that problem that would anonymize those certification entities to the producer. I think the way that it's structured right now, the producers have already engaged the certification entities prior to the blockchain solution being in place, so we're sort of coming in after the fact. But in the future, what I'm envisioning is when a producer wants to have a well certified, they would actually issue a smart contract on this private permissioned blockchain into a marketplace setting where certification entities that fit the parameters of what we would require to be represented on the blockchain would then anonymously bid for that contract to certify that well, and the producer would effectively approve that anonymous bid allowing for that certification entity to provide their data to the blockchain directly on the production numbers of that well as it's being used.

Soheil Saraji: But do we have a pool of trustworthy certifiers that can participate? I mean, we kind of have to give away a little bit of decentralization to make this happen. What do you think?

Dan Cearnau: I think it needs a bit more decentralization, definitely. Because, in the end, like you said, someone could have an incentive to lie about the data. And this is a problem of corruption and how you can fix corruption is by [taking advantage of] blockchain theory and game theory. To make the cost of cheating so big that does not worth cheating. So, by distributing the voting power, and basically in order to lie about the data, you will need to convince a lot of people, to bribe a squad of people.

So, to make that as difficult as possible, you can decentralize each part of the signatures. I think that was part of the idea because also the third party will have auditors. It's not just them doing their own thing, they have auditors on a higher level. So, I think adding as many entities to the chain as possible, would be great: auditors, banks, etc. So that we have a clear record of all the signatures and the processes, which make it as difficult as possible. You can't really make it zero, there's always a chance of the data being manipulated.

Soheil Saraji: That makes sense. I've been thinking about it a lot and I do have the same line of thoughts. Karl, what do you think about this Oracle problem?

Karl Osterbuhr: I've been thinking about it a lot, trying to relate it to your reserve report. I mean, those are audited by one of the large four independent reserve firms. And they're still not perfect, but it's the best we have. I think one of the things I agree with what's been said is you've got auditors along the way. Maybe you get ISO certified yourself. I'm not sure if you're familiar with ISO standards.

I mean this a little off topic, but basically ISO 9000 was a principle introduced by Motorola back in the day, and it's about manufacturing quality in as opposed to inspecting it in. So, ISO 9000 is all about Six Sigma manufacturing. And what six sigma is 3.4 defects per million opportunities. Back in the pre-Six Sigma manufacturing, people would make parts, and they would inspect every part or every 10 parts and throw two or three away and keep the rest of them. ISO 9000 came in and now you have ISO

9000 manufacturing for the space industry, the aircraft industry, phones, any type of precision manufacturing. And the mindset is you build in precision, you don't inspect the precision after the parts made; you build it in before the parts made. So, you control all your processes that build the quality into the part. Ultimately, you end up with 3.4 defects per million opportunities.. Building quality into the process will help and allowing third-party certification to qualify the emissions data with digital signatures on the blockchain should improve data integrity and quality.

Soheil Saraji: That's a great point. Redundancy in data is very important. You must trust multiple sources of data because one source of data may be less accurate than another, or someone may be able to manipulate one but not all.

Soheil Saraji: Great. I am involved in academia, and I have the luxury of considering a very idealistic solution to the problem. Having multiple sources of data verification is a key. There is a type of drone that can be flown over a field to determine the mission. Although the data solution is different when using a sensor in the field, it is possible to correlate and determine how accurate you are. In addition, there are satellites specifically tasked with emission monitoring. You can get a very broad picture of carbon emissions for a very large field, but it's unlikely that you'll be able to narrow it down to a precise level per well. One can combine all of these into different data layers that correspond to different companies. That is one way to do it. Also, I believe an organization, a DAO, or an organization comprised primarily of nonprofits must be a part of the verification and validation process. I'm hopeful that universities and academia could play a role there because they understand the data, they could analyze the data; this could be part of a Ph.D. or master's project that allows students to analyze data from different sources and determine how they match. Universities or other scientific-based institutions, such as research centers or nonprofits, could obtain the data, analyze it, and then sign the final report; For instance, if Project Canary submitted this data, a group of auditors or scientific entities could examine the data to confirm or refute the findings. As Dan mentioned, they are providing multiple layers of protection to reduce the risk that a single entity could manipulate the entire dataset. However, it is to be seen if we move in that direction. Some oil and gas companies are either developing their own blockchain solutions or acquiring startups that are developing blockchain solutions. I simply disagree with this approach. The entire concept behind blockchain is that no one owns it; if you start owning it, you're just creating a trust problem.

Aaron Lohmann: We are also working with PwC on a new audit process that's been developed by the American Institute of Certified Professional Accountants (AICPA), and they're using the process right now to audit the numbers behind stable coins. So, a large company like Circle hires PwC to audit the fact that what's being represented as a stablecoin is true. But what PwC is working on with us is a process to audit these data sources. And to ensure that after we double-check the data source, like Karl was saying, what was reported to the state versus what was reported to the blockchain, we're building a reconciliation mechanism into the system that will help keep the integrity of these inventory numbers as true as possible. And that should be fairly automated so that PwC can then come in and audit our system instead of going out to each user. And, by auditing our system, we should be able to represent this as authentic data.

Soheil Saraji: That's a good solution, too. Some sort of standard is another thing we lack; we don't have a good start, for example, when it comes to what constitutes a carbon offset or a carbon credit. We don't necessarily have a very clear idea about those things, especially if you look at a lot of startups that are simply attempting to create carbon offsets based on trees and vegetation. How do you even prove that the temperature changes and the amount of carbon that gets out of the atmosphere is different for different vegetation? I believe companies like yours could be initiating this type of standards in the oil and gas sector.

Is there anything that I have missed about the topics that we discussed today? please feel free to add to the discussion.

Aaron Lohmann: I just want to go back to one question that you had asked earlier that I don't think was directly answered, or maybe it was, but I wanted to address it. I think one of the reasons that blockchain has been adopted relatively slowly over the last several years that it's been available in the industry is because the tools don't really exist, I believe, to properly use the blockchain within the industry yet. And that's what we're trying to bring to market now. It almost reminds me a bit, it may be a bad example, when the internet first came to life. Previously, you could access the internet using user-friendly browsers and tools. You know, not many people were considering it as part of their business. Not many organizations were considering it as part of their business. But when email became easier to exchange because of the tools that were developed on the Internet, it also became easier to browse and search because of the tools that were developed. That's when these organizations started to incorporate it. And now you can't even imagine doing business without the internet firmly involved within its operation. So, in my opinion, this is why it has been so slow to be adopted.

Soheil Saraji Very good point.

Dan Cearnau I recall some projects from four or five years ago that attempted to tokenize energy and kilowatts. And I see a lot of people, who are instaling photovoltaic panels to the point where they are no longer only consumers, they produce energy. You don't want to invest in storing the energy yourself by buying a Tesla battery pack for your home. When you're producing energy, you put it into the network, and you should get a token back for each kilowatt. And then you can store those in your digital wallet; you can trade them if you want, or you could use it later to buy electricity. Therefore, you can create a full economy where people generate electricity, others consume electricity, and other people just store the energy. I saw an article a few years ago that said that Bitcoin mining is the biggest battery in the world because it moves energy around. If you have cheap energy in one place, you can mine Bitcoin there and stop mining in a different place. If you think about it, you transfer energy from Iceland to New Zealand and back and forth at no cost or you can store energy. You would only configure your Bitcoin mining farm to use it when energy is cheap or someone who produces energy by solar panel can only mine during the day. So, you can get a full economy around energy if you tokenize the kilowatt at the end and have smart contracts on the network. The distribution gets a share; you could have all kinds of games around it.

Soheil Saraji We have companies in Wyoming that have portable mining stations. They go to oil and gas fields where they can get cheap or free gas to power their mining stations. To me, the best solution is to simply stop flaring and try to collect and sell the gas. But if that's not possible, mining Bitcoin would be at least the next best solution, which adds value.

Karl Osterbuhr: I have a friend who's doing that and she's got a test project in Colorado. We looked at it for some stranded gas wells, my brothers and I did in Kansas, where there's a lot of stranded gas wells around. And the challenge is that the there are two knobs on the ROI, potentially: it was bitcoin price and hash rate. For example, the hash rate when all the miners got pulled from China went down and your ROI went up. But when recently bitcoin price fell below $30,000, the return on investment went way down and the hash rate was still climbing. It's going to work well, in the Bakken Shale. We have 90% crude oil, and you're barely getting any capital return on selling your gas, essentially, a lot of the producers out there give away their gas just to get rid of it. But they don't care because most of the revenue comes from oil because the stream is about 90% crude oil. So, the gas is kind of a waste product. In that situation or scenario, the Bitcoin mining would work really well, but stranded gas alone not quite as yet. Well, you're really sensitive to the bitcoin price and hash rate.

Interview 2: Bryan Hassler

Interviewee: Bryan Hassler, Chairman, PureWest Energy
Interviewer: Dr. Soheil Saraji, Associate Professor of Energy and Petroleum Engineering, University of Wyoming
Interview Date: 27 June 2022

Soheil Saraji: Thank you for making time for this interview today. Please start by giving us a little bit of background about yourself.

Bryan Hassler: Certainly. I got my bachelor's and master's degree in engineering from the University of Wyoming. I spent the first two years of my career in Rock Springs Wyoming, in production and drilling engineering, and transferred to Salt Lake City though I don't know what they are called now. At that point in time it was Mount Fuel Supply, and now QStar in reservoir engineering. I did a little work in an asset in divestiture type work for the oil and gas company, after I did some pipeline ratemaking and then late in my career at QStar, as the interstate transmission system was going to open access, I started marketing trading. Next, I transferred to Denver with a small company, at that point in time, was Barrett Resources Corporation, and worked on midstream infrastructure development and building what was at that time the largest physical marketing natural gas team in the Rockies region. After that, I left Barrett and went to work for a small methane producer, Pinnacle Energy. We ultimately sold that to Marathon. It was Powder River Basin coal bed methane development. I then took some time off and did midstream work for Alberta Energy Company, which is now Ovintiv. They sold that midstream business to TEPCO partners, which was an early MLP. And then, I rejoined Paul Ready whom I work with in Barrett Resources and Pinnacle Energy, and we started Antero Resources. We sold Barnett Shale development in 2005, to XTO Energy Inc., and since that point in time, I've been opportunistic. Folks recruit me into small opportunities, we build the opportunities and sell them. Out of banking, physical commodities trading business, through a series of companies like RBS Sempra commodities, and Freepoint Commodities. I did a short period of time with BP and enterprise products; BP on the trading side of the business in what they call long-term strategic origination, enterprise products, assisting them with routing gas to some very large processing plants that they built here in the Rockies, one in Maker, and one just outside of Willpower, Wyoming or Pioneer Plant. Now, I'm chairman of PureWest Energy, which was Ultra Petroleum. We are a leader in methane emissions monitoring and the largest natural gas producer in Wyoming. We focus westward to customers in California, Oregon, Washington, and also British Columbia. All those states are embracing renewables, low-carbon intensity hydrocarbons, as well as no carbon intensity energy. The energy transition is on its way. It is the future, and we know that there are some issues there.

Soheil Saraji: Well, that's a great summary.

Bryan Hassler: I'm sure I skipped some bits and pieces in there, but we can clarify as we go on.

Soheil Saraji: Absolutely. I was looking at your LinkedIn page and going through all your experiences. You're one of the few people that have done so much in the energy sector, from upstream petroleum engineering work to midstream, energy finance, and the business aspect. That's great, and it means that you have a lot of perspectives from different parts of the energy sector to share.

Bryan Hassler: So, this is very broad but very shallow.

Soheil Saraji: This is actually ideal for upper management. I think at that level, you need to know all aspects of the energy sector from technical to economics and business so that you can make the right decisions. I think that's the advantage of having those experiences.

Bryan Hassler: I guess one interesting aspect is that I did this for a little over a year and a half with Excel Energy as vice president of asset development. So, I looked at everything. I looked at Coal, wind, solar, combined cycle turbines, and energy infrastructure that they owned. My wife likes to say that I set the path toward Excel's investment in renewables. They were purchasing power from renewables, but they had very low development expertise.

Soheil Saraji: Great. So that makes it perfect for this interview because we are going to really look at the entire energy sector, mostly from an oil and gas perspective. So, we both agree that there is an unevitable energy transition happening, and there's a lot of disagreement on how it would turn out in a few decades. We agree that it would impact the energy industry as a whole, and the petroleum industry as a part of the energy industry would probably be one of the sectors most impacted by this energy transition. What do you think this impact will look like? Or what are the main areas that we can see development?

Bryan Hassler: Well, certainly jobs in the future. But, what is the future? Is the future the next five to 10 years? Or is it the next 20 to 30 years? 30 years from now we're looking at many companies saying they want a net zero carbon footprint. And I ask a lot of these companies, what does a net zero carbon footprint look like? And essentially, I think they still will anticipate using hydrocarbon fuels in what resources they've got remaining. Most of those will be natural gas-fired generation units, there is a transition toward hydrogen. I think, from an oil and gas perspective, you're already seeing some of the very large majors, and some of the intermediate-sized companies try to evaluate what does that energy transition look like it? It will still incorporate natural gas and crude oil as reigning feedstocks in the electric generation stack of fuels. I think you'll still see coal, but the environmentalists have pretty much put the nail in the coffin of the future of coal. Also, carbon capture and sequestration may factor into some of that, but as you know, it's very expensive, heavily subsidized with tax incentives, and very difficult to establish. But there is a developing carbon capture sequestration and utilization industry developing, and I think that'll factor into the extension of utilization of crude oil, natural gas, certainly in the feedstock for the future. And you'd get natural gas and crude oil factor into many of the everyday things we utilize in life; or we depend upon food, feedstock fertilizers, plastics, clothing, you name it. So, hydrocarbon molecule sits in our everyday lives, and we'll continue to do so.

Soheil Saraji: Yes, I very much agree with what you said. I think any changes that happen in any industry or any sector typically bring challenges and opportunities. What opportunities the oil and gas industry would have in this transition?

Bryan Hassler: Well, certainly. In a low-cost environment that we've had, say, for the last half-decade, sub $60 crude oil, sub $3 natural gas, the costs/benefits associated with cleaning up your emissions have been skinny to nonexistent. But I think those companies that are embracing the future have taken that step over the course of the last three to four years to reduce methane emissions, therefore methane intensity. I think you're going to see those companies that pay attention to detail and reduce their emissions to as close to zero as possible [will benefit]. I think you're going to see a re-ranking in terms of how markets look at hydrocarbon production, where they purchase their supplies from, and how they dispose of those assets that they're acquiring. I think you're seeing a lot of it. Europe is driving a lot of the transition. Certainly, Asia has driven a lot of the transition. On the West Coast of the United States, certain states start embracing low-carbon intensity [fuel], I think often, our politicians and our regulators aren't looking at the entire solution. They're looking at what their constituents are telling them like we want to be net zero versus what's most cost-effective to the consumer to continue to reduce that carbon intensity in the future.

Soheil Saraji: Yeah, the net zero seems to be a hyped concept but then when you go a little bit deeper, it's very hard to define. Do we have enough data to even know if a company is net zero or not? There are technologies and standards that we need to develop in the coming years. For example, methane emission monitoring and control will be key in the future of oil and gas. So, what can we do as an oil and gas industry to become the leader in the energy transition? How can we adapt to become a part of the energy transition rather than the resistance to it?

Bryan Hassler: It has to do with the education of regulators and politicians, that there is a different way to get to the same point more cost-effectively. We've hired some folks in the state of California, which is probably the most progressive in terms of no hydrocarbons in my backyard. But it has certainly left them in a precarious position in terms of [the] reliability of electric supply and fuel supplies to customers in the state of California. I think they'd been most progressive in terms of fuel standards, they're embracing biogas. We're trying to keep the level playing field. [For example,] biogas is nothing more than methane, that's been emitted into the environment from municipalities, waste facilities, you name it. We've got to embrace that cleaner energy from the oil and gas sector can be much more cost-effective. And I think what we're all looking for is just standardization of measurement of our carbon intensity or methane intensity, and standardization of the monitoring techniques, and [development of] firm's that are going to independently verify and validate emissions. [Also,] I think we want to get a level playing field; a set of rules and regulations that everybody is familiar with and abides with. I think, at least as a methane producer and natural gas producer, we want to make the case that in the future of energy in whatever the energy transition is to produce responsibly and to assist in lowering carbon and methane emissions across North America and across the world for that matter.

Soheil Saraji: That's true. This transition will not happen overnight. We cannot shut down all fossil fuels and switch to renewable today. There's going to be a transition period. I agree with you that oil and gas will play a major role in the future. I heard the news that Germany is going to open their coal-fired power plant again. Germany was one of the most progressive countries to move towards solar and wind. But now, with all uncertainties in terms of energy in Europe, they decided to step back and then take it slow.

Bryan Hassler: Yeah, if you look at European nation's energy prices, $35 to $40 per MMBtu, here in the United States for the month of July, [it is] $6.5 per MMBtu. Countries are reliant upon our low-cost supplies to give them alternatives to issues like the Russia-Ukrainian War, where natural gas supplies to Europe have, in essence, gone to zero or close to zero. Russian crude oil kind of shifted, at least out of the European markets in the North American markets. It has put a constraint on the system. What happens when you put a constraint on the system? Even if you've got every great intention of lowering your carbon intensity and carbon footprint as rapidly as possible, it puts a burden on the populace, so to speak, in terms of inflation in energy, fertilizer, food, and down the chain.

Soheil Saraji: Definitely, yeah. We want to make sure that our energy sector is antifragile. I think the oil and gas industry has shown that it can support and provide reliable baseload for the electric grid. This brings us to the economy of energy. There are a lot of companies moving towards net carbon zero, and some companies have started investing more in carbon sequestration to reduce their net emissions. The others, like your company, are looking into developing certified low-carbon fuels. My question is: from an economical point of view, does this make sense in today's world to pursue such a venture? Is it viable for a company to invest in such activities? What are the incentives for the companies to move in that direction?

Bryan Hassler: One, I think it's the right thing to do. Oil and gas companies, whether they're small mom-and-pop operators or very large, super majors, they need to evaluate what they're doing in this world and find a way to do it better. And I think in the 40-plus years I've been in this industry, there has been a continuous improvement. When I first started, tank bottom [fuels]s were taken out of the bottom of the tank and pumped into a truck, which pumped the tank bottoms on a dirt road to keep the dust down. It's not happening today, but it was very effective to keep the dust down on your oil field roads. So, we've moved to the better economy [and] better efficiencies. You start with it's the right thing to do, but then you take a look at how you must economically do it. If you're drilling from [multi-well] pads, you can monitor and evaluate many more wells, than you can do on single-well pads, and it's more efficient. So, you're bringing costs downs effectively. There's a learning curve through every basin in terms of how you do that. When you look at PureWest, or Jonah Energy on the other side of the fault divide [Jonah Field/Pinedale Anticline, Wyoming], we operate mostly on federal lands, close to national parks, ozone nonattainment areas. So, our license to operate was built around the lowest cost emissions so that we could continue to operate. It's tough in a $2 [per MMBTU] environment; it's much easier to take a bit of your profitability and invest it in continuous monitoring, independent third-party evaluation and find a way to tokenize say the responsibly sourced gas that you ultimately produce. I would hope that there's a breakeven point associated with what we're doing. We're also seeing markets request more and more; what's your methane intensity? What's the benefit of your supply or other supply? And we're starting to get that message out there. We're not seeing folks paying for that, but that's because their public utility commissions are incenting them to reduce their emissions in that manner.

Soheil Saraji: This gives us a good segue to the concept of carbon economy. In your opinion, is something like a global carbon economy possible? And what does it look like if it's implemented?

Bryan Hassler: Well, I think you could look at in North America and Canada that is shifting more towards [that]. There's no tax on carbon; there's no government forcing you here in the United States to dispose of carbon or pay for carbon; there's no carbon trading exchange, really, it's all voluntary. And it's an opaque market, you're starting to see some exchanges stand up, that are validating carbon offsets, and you're starting to see an economy where carbon offsets can be utilized as one way to transition into the future, and lower your methane intensity, your carbon intensity as a producer to effectuate that investment in the future.

We've bought carbon offsets. We've issued those in conjunction with our independent third-party verified low-methane intensity and sold them into the Pacific Northwest. But you've got to have somebody on the buying side that really wants to show their net zero, in essence.

Soheil Saraji: So, it seems like there is some interest, but maybe the interest is not high enough yet to make the entire industry move in that direction. But there is an interest.

Bryan Hassler: There's an expanse, certainly, towards independent validation. And there's a cost to listing your responsibly sourced gas, your renewable gas, and your carbon offsets, onto an exchange, and sometimes the cost exceeds the profit that you're going to receive at this point in time.

Soheil Saraji: So, it seems like the reason a lot of companies are moving in that direction is that they are anticipating the future to be different. And there would be some incentives in the future, right?

Bryan Hassler: Yeah. And the Freeport is down, so we've lost two BCF a day of export capacity through LNG. But a lot of the International LNG players being European and Asian of origin, are starting to press their LNG totaling facilitators to find the lowest methane intensity gas out there, because they're pressed by European standards and developing Asian standards.

Soheil Saraji: Moving beyond responsibly sourced gas, I think that the oil and gas industry could play a role in the carbon economy also via CCUS. We have been talking about it for maybe a couple of decades now or even longer and technically, like at least carbon storage is viable, But it has not become a reality yet. Do you think the carbon sequestration or the entire CCUS would work?

Bryan Hassler: Yeah, I've been part of some of these carbon symposiums. You look at ethanol facilities and cement facilities that have high-purity

CO_2 that can be captured. I think you're seeing Tallgrass do that with their Trailblazer pipeline system. [There is] a lot of ethanol facilities along their pipeline quarter. There's pipe duplication in that ditch now with the Rockies Express pipeline, not moving gas West–East. They can convert the natural gas pipeline built in the late 70 s into something that can move carbon from Nebraska, Iowa, and other points in between and capture it directly in the state of Wyoming. The state of Wyoming is blessed with some tremendous carbon-capable deep saline aquifers that you can be assured that things will rest. If you could convince regulators to keep coal running, it helps you do two things, one, huge emitters of carbon dioxide, a lot of those coal facilities in the state of Wyoming, are co-located with viable deep saline aquifers to where carbon capture could work. Two, you keep the electric reliability that coal provides that renewables don't. So, it's too bad there was not more support in terms of extending the life of coal, historically, because I think we'd be in a much better situation to capture carbon from those coal units, sequester it and keep those units running economically well into the future. I've had discussions with Holly [Krutka, Director of the School of Energy Resources, University of Wyoming], I said, we'd be a great carbon economy, but we're shutting all our emitters down because Pacific Northwest, California, and Colorado, are requiring coal to be eliminated from the production stack on the electric side.

Soheil Saraji: It seems that is the direction we are going right now. There is also another exciting development in the energy sector around hydrogen. As you know, carbon sequestration is a key component of blue hydrogen generation. This could be another way to push for a more economic carbon sequestration process. What's your opinion about hydrogen, do you think there is a possibility for hydrogen to play a role in the future?

Bryan Hassler: Yeah, there's a transport issue. I think hydrogen will be located at hubs that can consume it directly. So, [for example] the Magnum project in Utah, I'm sure you're familiar with. They've got salt that can store hydrogen. They've got plenty of renewables and water to generate hydrogen, [and] fill those hydrogen caverns and, and co-run 800 megawatts of power, initially on 30% Hydrogen-70% natural gas, but the technology is there for the turbine to be retrofitted to burn 100% hydrogen in the future. From that perspective, I think you could do that. [In] Southwest Wyoming, they could build a hydrogen hub that co-locates with the facility in Utah, but also feeds the Pacific Northwest [via] Ruby pipeline, [which is a] fairly new vintage and internally coated [pipeline]. I talked to El Paso, probably a year, year and a half ago, and they're testing the internal capability of all their pipes to handle hydrogen. I think that Ruby's [pipeleine] got 42 inches in diameter, and it's hitting those states that are embracing hydrogen or non-hydrocarbon bearing fossil fuels to meet their energy transition. So, Ruby hits Northern California, Oregon, [it also] can hit Washington. [This] could be an interesting thesis for [University of] Wyoming.

And then you get into blue hydrogen or green hydrogen. I've had some discussions with Bloomberg's hydrogen folks. Bloomberg does a lot of work on behalf of their client base. But the question I had [was about] a shown huge growth in hydrogen, and it's all green hydrogen, because that's what everybody thinks they want. But at the end of the day, I said, how much incremental wind and solar do you need to install to make sure that you can generate the green hydrogen that doesn't meet 100% of anybody's needs, but your projections? The renewables folks said it's physically impossible to put that much renewable in place to generate hydrogen. So, we have to look at blue hydrogen. I think, certainly, what the school of energy resources, your department, and the Wyoming Energy Authority can do is to work with the industry to get a thesis out there [investigating if] blue hydrogen can be economic. It's [probably] more economic than green hydrogen, it's [also] more reliable. [Regarding] the carbon associated with the generation, the carbon dioxide molecule, can be sequestered effectively.

Soheil Saraji: Yes, I agree with that. If we have hydrogen haub, we need to figure out storage. And the storage might not be as simple of the solutions as they have in Utah, because we might not have as many salt caverns in Wyoming. The other place to look is the aquifers that we are considering for carbon storage, they maybe could be retrofitted for hydrogen storage. They have their own challanges, but the problem is that there is robust study.

Bryan Hassler: But if you look at depleted gas reservoirs as potential, you could store it, we could do that.

Soheil Saraji: Definitely. There's the mixing problem with the currently existing gas storage, aquifers seem to be a cleaner solution. But both of them need to be studied. One of the issues, for example, that have been raised by current pilot projects in Europe is that if a trace amount of CO_2 exists, it reacts with hydrogen and creates methane *in-situ*. This is exacerbated by living organisms, some microbes actually digest hydrogen very quickly. Those are types of issues that need to be studied. So, if we want to generate a huge amount of hydrogen and we cannot transfer most of it, we need to think about storage.

So, we discussed the oil and gas role in the energy transition, we talked a little bit about the carbon economy and hydrogen. I want to switch the topic to other technologies. There are a few new technologies that oil and gas industry is experimenting with, like AI and machine learning. There is also blockchain technology. What are the new technologies that you find interesting?

Bryan Hassler: I think there is a role for blockchain in terms of attaching attributes to physical production that's continuously monitored, to give folks assurance that what we say we're doing and what the independent third-party validator says they're doing is completely documented. A lot of, I guess, cheaper storage will assist in managing, the huge amount of data that is collected. I just had a discussion with the PureWest team this morning. We're doing a number of things, we're evaluating different sensors out on location now, and trying to see which modules more effectively monitor emissions. And early data says, they all do a pretty good job of what they do, they use pretty much the same basic technology. Software development and visualization of the data seems to be the differentiator with respect to the six or seven manufacturers that they're working with. And they think that will be a bigger differentiator, to take the data that's relevant, condense it pictorially into something that goes beyond something you or I might visualize, and use [it for] market. [For example] Portland General Electric, or Pacific Gas and Electric, they can take a look at that picture and say, yeah, what we're buying is what we're saying, and what we're getting. So, simple action and data assimilation into something that's usable, not only to professionals but to counterparties, that may not have the basic understanding, but they can at least visualize the data and say that what they're buying is what they're getting.

I think that the software and technology and storage and ability to track all that data in a manner that is quick, easy, and usable is good. I'm starting to see end-use markets, worried about SEC requirements, emissions requirements, and how they report that. So, assimilation of that data into a very usable format, standardization of data, or standardization of protocol and emissions reporting, I think is going to help everybody. But there's no singular body that is providing that broad North American standardization, so I think everybody's excited to see that. Certainly [there are] ways in which we can reduce our footprint, [like] new technologies, and ability to use a gas package on a rig versus diesel-burning engines. Those [are] things that we're looking at. [Also,] Reclamation technologies in the oil and gas sector. PureWest and Jonah are involved in a project up on the Palmdale Anticline and it was just reported this weekend, the reclamation to reintroduce bugs, honey bees, and that type of environment that can help propagate even further reclamation of land that may have been disturbed. They're seeing some pretty good results with respect to changing the seed mixtures and the reclamation mixtures. Things as simple as that are things that I never think of, but when you see it reported, it's pretty interesting.

Soheil Saraji: That's really interesting. The example of blockchain that you brought up is not a widely accepted technology in our industry yet. There are companies that are experimenting with it, but there are obviously certain obstacles that just slow down the adaptation of this technology in the oil and gas sector. What do you think are the problems or challenges that blockchain faces in our industry?

Bryan Hassler: What do people want to see blockchain capture in the future? I think you can build a blockchain that'll capture your methane intensity, your emissions, and your production monitoring consistently. I think monetization is key. I'm going to implement blockchain to track all my data. Once again, it's a huge amount of data. How do I keep my mind around it? And how do I introduce those there Looking for the capabilities that blockchain offers to embrace the technology? And then your platform and standardization, how do I get in front of somebody with capabilities of monitoring blockchain? My validators? It's listed on an expansive platform, how to get the word out?

And it sounds like [there are] plenty of producing companies out there that are independent, using third-party, monitoring and evaluating what our production profile looks like. But the market isn't there because the market been embedded with biogas. That's the solution to the future. I've got a good friend who puts it quite succinctly, "there's just not enough shit to go around out there to meet everybody's RNG goals and objectives". You're probably familiar with cowboy clean fuels introducing a different nutrient in the coal-bed methane seems up there that the bugs can digest easier than the coal. So, they'll digest these nutrients produce methane that can be geo-chemically monitored, and produce renewable natural gas in-situ, as opposed to a bioreactor on the surface. They indicate that the coal absorbs the CO_2 [byproduct] that's generated, and you get a pure stream of methane. They're out raising money now and trying to commercialize this.

Soheil Saraji: Another thing that I wanted to talk about is a problem with the blockchain that is called the oracle problem. When you deal with digital assets, the blockchain is a perfect matching technology. It is built to work with digital assets; they go hand in hand. But when we connect blockchain to the real world, a physical world, that connection becomes the weak point. It could be manipulated. For example, as you know, Project Canary is a third party that provides monitoring and certification on the field. But when you think about it, there's going to be an incentive misalignment. Who is paying project Canary? Typically the companies that are hiring them certify their operations. What do you think about this problem? Is there any solution that could be used to mitigate this issue? How can we trust the data that is transferred to the blockchain by a party that might have other financial incentives?

Bryan Hassler: Well, certainly, standardization of rules, regulations, and protocol, puts everybody out on a common playing field. And that typically needs to be the kind of government mandates. As much as we don't want government in our business, standardization across states, as opposed to state by state makes things more efficient and more audible. We need to be able to audit what's going on, And I think it brings costs down across the board. We've evaluated a number of different auditing services. We think that project Canary, I mean, we don't know what their proprietary rating is in terms of platinum, gold, or silver, [in terms of] issuance for certificates. We do know that they don't allow for any incidents to be erased that have occurred in the past. You know, we've owned PureWest for almost two years now, but Ultra [previous company before reorganization] may have had a leak five years ago, and it goes against our score [now]. We've had some discussions; we've got procedures in place that take care of the future. Why something that wasn't on our watch [should] be a part of your ratings and evaluations? And I think all the ratings, [for example, for] tech companies, out there have similar type issues. So, commonality would help cure some of those issues.

Soheil Saraji: It was just an example to show that financial misalignment could happen. I am sure PureWest and Project Canary are doing their best as pioneers in establishing RSG. I agree with you. One of the key issues here is the standards. And since the government has not stepped in to develop those standards, then it seems companies have to come up with their own solution.

As PureWest you already experimented with carbon credits/offsets, when do you think is the right time for you to embrace carbon credits or carbon offsets? I mean, as the company strategy to further reduce your carbon footprints. What do you think is not there yet to incentivize you?

Bryan Hassler: Well, we're certainly not moving as rapidly as I'd like to. From the standpoint of monitoring, or having an independent third party validate our production profile. I asked Kelly [Bott] this morning, we have a meeting every two weeks on everything that's ESG. And she said, we're on track by the end of the year, to be fully certified by project Canary. 75 to 80% of our production can easily be monitored because it's pad drilling, and there is sufficient production in place to keep the costs at a reasonable level. So, we are on track there. We are [also] evaluating a number of independent monitors, and you should talk to her about that particular test environment. I asked about solar, because we're looking at solar arrays to reduce our tier two and tier three emissions footprint. It sounds like we've had some stops and starts there, and about ready to respond to Holly [Krutka]'s request for us to fund some of the hydrogen institute [initiatives] that she's looking to put up there. I think when we invest in universities, we want the work that they do with respect to the dollars that we put in to move us towards commercialization versus theoretical study. GM had contacted us on fuel cell vehicles in the field, which I think is probably more viable than electric vehicles, although we'd get two Ford F150 commercial electric vehicles on order. You know, if we have fuel cell vehicles in the field, we need some form of hydrogen generation to refuel the trucks or to fuel the rig fleet that we anticipate having up there. So, these are things we're looking at. We've got a lobbyist in California, working with lawmakers to educate them on responsibly sourced gas and what it does. Citadel sits on our board, and I'll send you a piece from them, kind of showing that responsibly sourced gas with a very low methane intensity and a little bit of RNG give you the same footprint as 100% RNG does.

Interview 3: Jasmine Zhu

Interviewees: Jasmine Zhu, VP of Market Development, Xpansiv
Interviewer: Dr. Soheil Saraji, Associate Professor of Energy and Petroleum Engineering, University of Wyoming
Interview Date: 8 July 2022

Soheil Saraji: Thank you for coming for the interview today. would you introduce yourself and tell us about your background?

Jasmine Zhu: Sure. The resume itself is quite short already. Jasmine Xu, I've been in the energy space since probably 2004, [which] was my first introduction to commodities. I started out in operations within Morgan Stanley's commodities team, and I've spent several years there from 2004 to 2015. And in between that time, I've spent about three years out in their Shanghai-China office helping to develop a location in China that was post financial crisis. So, it's a really interesting time to kind of work on the other side of the world where the demand is all just coming out from this one big country. Obviously, just incorporating some of the existing knowledge that I've honed out in the New York office, and it really had helped to bridge some of that understanding of how commodity is transacted. It was still a very physical-based type of a transaction in Asia and China alone. So, the whole concept of financial instruments, derivatives, and hedging strategies were still somewhat new to them. It was really insightful and a good learning experience for me as well.
Post China assignment I came back to New York and that was in 2013–2014, where the US shale revolution has really kicked off in full speed and there was a ton of new LNG projects that's being developed and looking for long-term off takers, 20-year contracts, take or pay

agreements being signed up, and really all going overseas to Asia to find those buyers. We've seen some success between the Japanese and the Korean buyers because LNG has always been part of their fuel mix. That was very new for Chinese buyers in essence, and they, at that point, have very few receiving terminals. The infrastructure was not there, and they were still very, very dependent on coal.

Yeah, so [I] spent a lot of time and energy and then posts that went to S&P global for three years, covering some of their strategic accounts. And then the whole concept of energy transition has been what all of our clients were looking for. They wanted to understand carbon, they wanted to understand our carbon footprint, greenhouse gas, how do you trade these credits? How do you generate these credits? And then because it's heavily driven by the compliant market, so you are kind of stuck with the fundamental supply and demand concept. And then the whole voluntary market kicked off in 2019–2020, where corporations are making these voluntary sustainability targets and are looking to start getting involved in buying carbon credits to offsets some of their emission footprints. But I think on the flip side of that is they started become more cognizant of their own carbon footprint as well. So, it's not just about buying credits to offset but also trying to reduce what they're doing in terms of their role within the climate awareness.

Soheil Saraji: That is a geat background. As you know, the underlying theme of this book is the energy transition, and how oil and gas industry could adapt. And particularly, we are looking into the carbon economy, blockchain solutions, and things of that nature. What do you think of the global energy transition trend that has been going on for a while? How will it affect the energy industry as a whole and the oil and gas industry as a subset of it?

Jasmine Zhu: Yeah, absolutely. I think the U.S. is probably not leading this effort. Most of these initial efforts has come out mostly from the European side. They've had a carbon market in place for many years, and us too, but it just wasn't enough liquidity and just wasn't volatile enough for people to want to put too much resources in that market. But we've seen a lot of the big oil companies now shifting some of their assets and their portfolio into cleaner energy, whether it be developing hydrogen or coming up with low carbon fuel solutions, or just acquiring smaller companies who are on the verge of making better technology come into place. You've seen Chevron and BP all taking leads in generating renewable natural gas as part of their portfolio. They're acquiring farmland and they're acquiring landfill. So, all these new assets that historically was now part of what you think a traditional energy company would care about, is now a driving division within that organization. I think that's one for some of the bread-and-butter energy companies and then you see, trading shops like the Makarios and the Trafigura and some of the more nimble and more strategic traders in the commodity space. They're doing a lot more innovative type of work, where it's not just about investment, it's also about capturing and optimizing what they currently have. And for them, we're starting to see a lot more innovation coming out in terms of working together with other trading shops, and coming out with plans to how to really be part of that supply chain integration.

Soheil Saraji: I agree with you on one of the apparent strategies for the oil and gas industry, particularly for mega corporations with the resources to try to extend or expand their portfolio to renewables. But also, there are some smaller companies or even medium-sized companies that might have different strategies. From your perspective, what do you think fossil fuels industry will look like in 10 years or so?

Jasmine Zhu: If you were to ask me that last year, I probably would have said the trend is that fossil fuels will be phased out sooner than we would anticipate because we're seeing a lot of new projects. That's purely green-driven, like when projects are out here in New Jersey, and New York, and some of the northeastern states are basically saying we're going to have very aggressive Net Zero plans in place. And now, with everything that's going on with Europe, Ukraine, and Russia. I think this really has put a stop to some of those realities, which have shifted a little bit, and you see a bit of the supply shortage towards Germany, France, and other European countries that may have triggered demand that we in the United States haven't considered or priced in from 10–15-year type of a forecast. Now we're saying we should drill more and produce more liquefied natural gas so that we could send and help our allies overseas. So, the existence of natural gas suddenly becomes a necessity again. I think as much as anyone wants to think 15 years out, we'd like to be in a much greener place. It's hard to say that for now, when there's such a shortage in this market, and people have to really build in certain hours where they cannot have electricity just so that they can manage the grid and the cost of that generated power.

Soheil Saraji: I agree. I think this is very idealistic to think that the oil and gas industry would fade out very soon. I don't know if that's ever possible because of energy security, being a very important topic. To begin with, a hasty or haphazard transition to renewables—which is why we aren't ready for storage—is a major issue right now. There are certain movements within the oil and gas right now that are trying to bring low-carbon fuel like responsibly sourced gas to the market. What do you think? Is there a place for these kinds of products in the market right now or in the future?

Jasmine Zhu: Yeah, I think responsibly sourced gas is a quick fix, a low lift type of effort that I think everyone should play a role in. It's like buying a candy because you know is bad for you, but this is a lower sugar candy, so it has less sugar in that candy bar. It's a pretty simplified way of just thinking about it because you're not changing any behavior or practices per se, in a dramatic effort that's being put towards that reduction, but at the same time, it's really about transparency. It's about putting data out there so that people can now start analyzing data, interpreting that data, and making decisions on the back of that data. The concept of responsibly resourced gas or certified gas is really just to drive that transparency but allow those transparencies to create a market where potentially there could be a premium that can be captured. You're still trying to incentivize producers or give credit to those who have better practices in place from a monetary standpoint. And whether that be your buyers looking to pay you a little bit more for a lower emission MMBtu of gas, or your lenders willing to give you a cheaper rate because you are one of the cleaner producers in the gas segment.

Soheil Saraji: Now that you're talking about markets and low-carbon fuels, I'd like to hear your thoughts. What do you think would be the main force driving this kind of market? Do you think it would be mostly driven by customers? or would it be more of a regulation push?

Jasmine Zhu: I personally still think it's fundamentally driven by supply and demand. So, if you think about natural gas in itself, it's a purely demand driven market, at this point. We've got 96 BCF of gas coming out every day. There's no shortage of that certification being passed down, if producers are willing to do that. If you were to look at renewable natural gas, which is driven by landfills or dairy farms, then you have a much more limited supply. And the demand side is much higher because people are more than willing to pay you multiples of what a regular fossil fuel MMBtu of gas would cost you to help neutralize or clean up their portfolio a little bit more using biomethane. And partially from that one because it's just a cleaner fuel to make from renewable products. But also because there's a market for it, like the LCFS [(Low Carbon Fuel Standards)] market where there's some form of regulatory support, that helped drive the cost in the forms of credit that you could attract in ranges $15 to $50, depending on which market and when you're looking per MMBtu of gas, versus like just $3 plus or minus for natural gas. If later on there are some form of a regulatory mandate or support that comes to say responsibly sourced gas could be considered in the same bucket as a renewable natural gas, then we there'll be a whole new ballgame in terms of where this gas is coming from, who's producing, and how much we're willing to pay for it.

Soheil Saraji: This is a good discussion. I'd like to take a step back, perhaps toward a high-level discussion, and bring up the concept of carbon economy, which has been circulating around. The idea is to create a global carbon economy in which carbon has value and either adds value to a product or removes value from the product, and the whole economy will be impacted by the carbon intensity of different products. That could be something futuristic. Do you think that would be a possibility? Would the carbon economy become something real in the future, in a global sense?

Jasmine Zhu: Yeah, that's the common denominator at the end of the day, When you think about what is your climate exposure, what your carbon exposure, it become less relevant in terms of some of the other VOCs [(Volatile organic Coonhounds)] or greenhouse gas that are out there. Everyone wants to simplify it into what is the cost of carbon, CO2e, and that's where you're seeing the market kind of come into place where they're just trading carbon offsets, and they're putting different values on the different types of carbon projects, whether if it's nature base, or red plus or if its technology driven. So, you're allowing the market to kind of create your own buckets of premium versus discounts and what's worth more and what isn't.

Soheil Saraji: So, you're basically supporting the fact that we have different kinds of pricing depending on our confidence on the underlying technology.

Jasmine Zhu: Yeah, that's what the market is behaving as does the market ultimately want to get to a point where it is regardless of what's behind the scenes, if it's just a simplified product? I do think it makes everyone aligned a little bit more you, kind of bring that uniformity into this one place. And then that's something that companies like Xpansiv is doing to really help simplify some of those transactions and put these projects together to provide the liquidity so that it becomes a lot more seamless and easier to transact instrument.

Soheil Saraji: At this point in time, I don't see how we can come up with a unified carbon value because, as you said, it's not only that we remove one ton of carbon, but the certainty of the process I think is a big deal. For example, the certainty is much lower when using agriculture-based CO_2 removal than when simply injecting it underground for storage.

Jasmine Zhu: Yeah. And for that reason, I think there still needs to be more rigor that that gets apply to how we identified the credit in itself, and then that's what we rely on. Certain certifying bodies like the Varras and the ACRs of the world to remain agnostic and just be purely focused on quantifying those types of projects. And I think the whole concept of quantification becomes very blurry as well, depending on who you talk to. You have academics like yourselves and you have technology developers, you have environmentalists, you have market driven stakeholders, you have investors. So, everyone have different types of motivation and backgrounds in terms of how they think quantifying something is meaningful and then you sit and you wait for the regulator's or some type of trade association to come out and help organize everybody. But there's different agendas being spoken at these panels and everyone kind of find it to be difficult to align themselves. And that's the problem; why some of these things like the responsibly sourced gas isn't moving as quickly as it should because of the uncertainty around which horse do you bet on when it comes down to the certification of your production? Am I using the latest technology or am I using the dial up version? and what does that cost mean, and do I get any ROI on the back of that? Am I going to get really compensated for being the best or can I get by just by doing what I need to do?

Soheil Saraji: Basically, you are saying that some sort of standardization for both data and certification is required. Currently, there is not any accepted uniform carbon accounting. There is also some incentive misalignment because usually the verifiers are paid by the project owners.

Jasmine Zhu: Yeah. The market has been stressing the whole concept of an independent third-party certifier. I do think that's critical. The market could handle a handful of standards, just like we have three rating agencies. It is like Googles and Apples of the world, they do similar things, they have different methodologies, and the market can handle that. It is just we need to eliminate a lot of that noise that's coming out from the sidelines. And any type of direction that the governing agencies could help would be also very impactful because now we're not just thinking domestic implications, we want to make sure that what we measure here and how we measure those emissions, also pass the smell test when we send those LNG cargoes to Europe. And they recognize how we measure and they recognize our practices, so that there is that uniformity from an international standpoint as well.

Soheil Saraji: This is a really good discussion. Maybe it's time for us to learn a little bit about your company. What is Xpansiv doing in this market? Are there any new or innovative solutions that you are providing?

Jasmine Zhu: Yeah, sure. Xpansiv is the global platform for environmental commodities. And what that is, is basically a marketplace right where we help facilitate buyers and sellers around environmental attributes. We cover several markets, one being carbon, and another being water. Then there's also the RECs market, which is the renewable energy credits, i.e., solar, wind, and differentiated fuels, which is the responsibly sourced gas, certified gas. And what it is really just helping to bring transparency to the market. There's price discovery that gets shown on a daily basis. And it's currently the world's largest carbon offset spot exchange. Last year, we traded about 120 million tons of carbon offsets; the year before that, I think, was about 34 million. So, from 2020 to 202, it went from 34 to over 120 million of carbon being transacted. Really, in a sense of, what that visual looks like, is the market shifting towards not just awareness of carbon but willingness to buy as much as they can because it is one of those things that they want to start utilizing it towards, and applying it towards their carbon reduction targets. Currently, because it's such a liquid market, it's an easy solution. You buy some carbon offsets and then you get to make the environmental claim.

Soheil Saraji: I was just looking at your website. There were a few different types of carbon commodities were listed. I was wondering if you could help clarify what those words are: GEO, N-GEO, and C-GEO? What is the difference between them?

Jasmine Zhu: The GEOs are global emission offsets, which are basically backed by different types of carbon projects that are technology driven. The N-GEOs are nature-based global emission offsets, so those projects are more around forestation and whatnot. And the C-GEOs are a new contract that was launched this year. And it's also technology-based, but it's more catered towards carbon capture and some of the other more current technologies that are being implemented.

Soheil Saraji: Okay. In terms of carbon geo- sequestration or sequestering carbon dioxide under the ground, you said that C-GEOs would be the right type of standard. Right?

Jasmine Zhu: I think you have guys like ACR and Varra who're starting to come out with different types of certification that cater to different types of projects, specifically more about CCUS and even things like upstream emission reduction. That's, I think, one of the compliant markets out in Europe, the UERs, so it's still new. I think when people are now coming very creative, in terms of how they could or how they see what they're doing, could generate credits. And we talk to clients that have looked at working around abandoned wells, for example, where there's leakage coming out on a daily basis, and one could quantify that and capture that. But what does that yield for them in terms of credits, and who's willing to pay for that, and where does that go? So, it really comes down to, again, I think the certifying bodies to make that distinction and put that type of project into some meaningful credit and package it in a meaningful way. For us, Xpansiv, because we're agnostic, and we are just taking the output from all these certifying bodies or giving credits to [projects in areas that are] pretty receptive to whatever passes the sort of fires, and let the market kind of speak.

Soheil Saraji: So, as long as you have trust in the certifier, have good track records, and are trustworthy, then it should be easy to add the new type of projects, right?

Jasmine Zhu: I mean, because you assume that they're the experts in that field, so they would be able to kind of make that comparison between project A and project B to some reasonable extent.

Soheil Saraji: There is another commodity item on this list, which is called digital fuels. Can you tell us what digital fuel is?

Jasmine Zhu: Digital fuels are basically digitalizing the actual commodity in itself. The first product that we kicked off in digital fuels was DNG (Digital

Natural Gas). Think of it like a birth certificate that comes with each MMBtu of gas. It's still the gas; it is still being produced the same exact way. But now we're just putting data and attaching that data that has those environmental attributes into an immutable digital file. So, each molecule, comes with a birth certificate that's digitally captured, that gives you the environmental attributes like carbon, methane, VOCs, time it was produced, which facility it was produced [from], where it was produced, and things like that, just basic, preliminary primary data. That one right.

Soheil Saraji: How does it work? Is it similar to RECs? Based on my understanding of RECs, when the renewable energy is generated, a certificate is minted to represent it, but as soon as that renewable energy enters the network, the actual energy and the certificate have their own life.

Jasmine Zhu: Yes. It's very similar to how the RECs work it'd be much more difficult to physically track the actual molecule with that certificate once it's produced. This certificate basically gives the producer right to those digital asset, so whatever the methane intensity is for that one molecule of gas, it is the property rights of the producer, also the scope 1 for that producer of that molecule down to the ability to utilize these data to help with their carbon accounting, as far as what is the producer of scope 1¿ what is the buyers scope 3? because you're buying that molecule that has this file of environmental attributes, which then gets transfer to the buyer, as part of their scope three, so that at the end of the day they have a better measurement around what is that carbon footprint of those gas they just bought.

Soheil Saraji: I understand carbon offsets and carbon credits; they have value because the companies are moving towards their net-zero goals, but who would buy these digital fuels?

Jasmine Zhu: That's a really good question, basically, what the responses toward the gas market are that is being created? It's paying a premium for the data or certificate that shows that the gas that you're buying is cleaner than the gas that has no data. The EPA puts out a methane emission intensity number every year, and the most recent one from 2020, I believe was 0.437 percent. So. if you historically have never measured what it is that you know that came with your natural gas, you could presume that your footprint is the national average (0.347) for methane emissions. But if you start buying, now, gas that has a certificate that tells you that each gas has a molecule or you buy those environmental attributes that basically pairs up with your physical molecule that says, you're buying gas with the 0.08 or 0.05 type of methane intensity, then you could voluntarily make those reduction claims because now you've got data that backs up to the actual commodity that you're purchasing versus just buying the commodities with no transparency of what it is. The benefit of it is really, again, providing that fugitive methane number to help with the scope

3 [emissions] as a buyer. And let's solve the ability to offset the fossil fuel that you're still buying at the end of the day. When you combust that MMBtu of gas, you still have a carbon footprint, which will still require you to buy a carbon offset if you're lucky to be net zero. In addition to digital fuels, the digital natural gas, we're applying that concept to all of our commodities. The second product that was recently launched is digital crude oil, where with third-party certifiers or emission monitoring technology, it is capturing the carbon intensity within one barrel of crude. That certificate is then being traded either with the barrel or separately to refiners or whoever that's interested in buying cleaner barrel of oil.

You think commodity is just a physical movement of something, but now you add one more layer on top of that, which is the data for each one of these commodities. Then you create this whole new market. Like I said, you got the gas, you got the oil, you could do the same thing for hydrogen, sustainable aviation fuel. You could do that for a bushel of wheat, a ton of cotton; I don't know how cotton's traded but basically everything that you are measuring, but not just the methane intensity or carbon intensity, you're measuring all of those environmental attributes like water, how much water went into growing this Almond? When people make these responsibly sourced claims, where's the data behind it? Well, now you got the access to those data, which is all being captured from production, from inception, and is being stored within a digital fuels registry. So, one really could say, today I care about the methane intensity of something, and some of these other environmental attributes like water, and SO_X and NO_X are less important. Fine, maybe I'll pay a little bit more for that methane intensity. Six months later, there's a water shortage, I want to buy the commodity that takes up the least water, and I'll pay your premium for that. So, you get to really build out a lot more markets, because now you've got 20 attributes to work with, rather than just this one MMBtu of gas which is traded for bucks that Henry Hub. Now, the data becomes a lot more richer and the technologies are being developed to kind of put further confidence around that data that's being captured.

Soheil Saraji: Right. Excellent discussion. I really like how futuristic it is. So far as I can tell, the digital data resides with a central party, maybe with the validator. Is this correct?

Jasmine Zhu: The data belongs to the producer or the operator of the facility. There's a handful of data aggregators and data refiners out there that basically work directly with the producers; either they have sensors around the facility, or they have cameras, or they have flyovers of drones, or they send someone out there with a handheld sensor to measure things. Those are all data points that could be captured. And then you've got your meters that are grounded onto the facility. The IoT could pull those data in, then you'd do a mass balance calculation using bottom-up or top-down [approach] with all the different types of instruments that are sitting around and collecting those data points to come up with a reasonable monthly average or daily average or minute by minute. For us, again, we're agnostic to the type of technology one decides to use as long as it checks off certain types of rigor, checkboxes. Is it continuous? Does it have X amount of equipment that kind of help provide XYZ types preliminary data points? and then, is it independently being measured?

Soheil Saraji: Perhaps now is a good time for us to switch our focus to the concept of blockchain, Do you think that it can play a role in any of the things discussed today? Do you think it is a beneficial technology?

Jasmine Zhu: Yeah, I think anything that could provide transparency, it's a move forward. And I think the market should just get comfortable with that, like, this is the direction that we're going, even if you are a national responsible producer, perhaps now it's time to start working towards getting cleaner. And then that type of mandates coming from your investors, coming from SEC, it's coming from your customers. So, I think that the better we are at providing granular data, it will serve the market as a whole, and it'll serve better for our climate agenda as a whole. And as far as, the blockchain concept. Yes. I mean, obviously, we want to figure out a way where we could actually measure the entire supply chain from start to finish, from the cradle to the grave, and everything that sits in between. The market is adapting to that; You have LNG guys that are looking to lead in some of these efforts, like Cheniere Energy Inc. They started tagging their cargoes where they're working with a bunch of different [entities like] academia, technology companies, and producers at the production site. And now they're working with the midstream guys to capture what the pipeline emission looks like when they flow through. And then at the end, in their liquefaction center, [they can think] maybe we should switch out different types of equipment, maybe we should electrify, we shouldn't use diesel for certain things as input. Having that type of data will help them make better, cleaner decisions so that when they sell their LNG or put their LNG on a vessel going to Japan, Korea, Europe, or China, they'll have that data available. And it'll be really interesting to see, then, if certain countries are betting different premiums for that.

If there's ever a carbon penalty, then you'll change the game. Like, oh, my tariff on my tax coming into China is x per carbon, CO2-e, versus my carbon tax if I go into Germany is y, then I might not want to sell it to this place versus the next place.

Soheil Saraji: This is coming. European Union is trying to create a scheme that anything that crosses the EU border must have a known carbon footprint. And then, for example, Russian fuel might not be interesting anymore if you consider the carbon taxes that are going to pay on top of the fuel price.

Jasmine Zhu: Yeah, but you know, Russian fuel data is also something that maybe factor based rather than actual measurement, which is why I think US producers also need to lean in more around direct measurement and actual measurement and not be factor reliant, as much as we can. I don't think anyone can make that switch overnight but we should lean towards direct measurement. And I think the EPA hears that from a lot of the industry stakeholders. They're holding these type of coalition meetings somewhat regularly, that has access to new administration when it comes down to getting the feedbacks from the likes of Project Canary, MiQ, and CleanConnect with all these names or people that's been working in this responsibly sourced methane capturing technologies space. So, they've been giving really like real-time supportive feedback around direct measures versus emission based.

Soheil Saraji: Are you familiar with or aware of any specific projects that are attempting to implement blockchain for different aspects of ESG trading or climate finance?

Jasmine Zhu: I don't know if there's any publicly announced efforts because it's so new I don't think people want to go and over-advertise it, just in case it's not the right move. The concept of blockchain again, we've seen in the latest NASEB (North American Energy Standards Board) publication, where NASEB is the agreement that all the physical gas traders have with each other or gas traders or buyers. So, it's pretty much a standard language that basically guarantees the delivery of a physical MMBtu of gas. There are the trade association that anyone who's in the physical space looks up to in terms of guarantees and quantifying and some of the standard verbiage that goes into documents and contracts. So, NASEB, came out recently about two–three weeks ago, that they want to start incorporating the idea of the digital ledger and smart contracts because of the recent growth that they've seen in certified gas. So, you've got your gas, which is business as usual, and now you've got your certified gas, which is now the certificate. Because it's starting to really pick up momentum NASEB is surveying the market to help draft this language around the smart contract for the certificate.

Then, Kinder Morgan Tennessee gas pipeline, just like three days ago, the July 4 weekend, got FERC approved to offer our two different pools, one they call producer certify gas, which is responsibly sourced gas, and the other is regular gas. So, they're offering two pools that the shippers basically say hey, I've got certificates I'm going to offer my gas to this producer certify gas pool and buyers but then pick which pool they want to buy from, and then they will still use the same kind of nomination and purchasing type of infrastructure as this, but now all the certificates that should come with the gas, the buyer and the sellers will bilaterally agree on, and then I would hope it will sit in a registry like ours like Xpansiv to settle up. It's definitely capturing the attention of trade organizations and regulators when it comes down to the RSG space. Purely because you got these producers who's going to be 100% certified, and there are these big producers like Chesapeake and EQT, who, at some point, are not going to be able to just offer you regular gas and they want to be either distinguished from a monetary standpoint or incentivize or, align themselves differently than everyone else who isn't certified so then you got midstream guys like the pipelines who said okay, you know what, we'll put you guys in a different pool.

Soheil Saraji: It's an interesting approach that you can create markets that are completely separate from one another, or not completely separate, but at least provide options for different types of demand. We're nearing the top of the hour, so I'm wondering if there's anything I haven't asked you that you think would be good to talk about, or if you have any comments or questions.

Jasmine Zhu: I think it's such a new market still, that there's probably a lot more [can happen in] like six months from now if we had this conversation. I think there'll be a lot more changes coming in. And I say that because we're working on a lot of different things where we're trying to prove out the concept of the entire supply chain, we'd like to get to a point where we could share what the entire carbon [footprint], what the mission profile looks like, from production down to this utility, and how we got there. I think Cheniere will be able to start sending off a couple of cargoes that has tags on it, and I think these are all probably six months from now kind of conversation, but a lot of this will start to come into play. And then the mark is going to have a lot more certificates flowing around; right now, it is still a very small segmented type of market and most of the demand is coming out of the Northeast. I would like to think six months from now, the industrials have come into play, the steel or the transportation guys. Lego announced that they're going to build a sustainable factory in Virginia, making Lego blocks, my kids love Lego blocks. So, I've been trying to find someone to introduce me to that project manager person in Virginia and help them really think through the entire front to back of best strategizing for low carbon products on the back of that building, whether it be solar panels, or using cleaner fuel or electrification or buying RECS, or carbon offsets. The ability to help a corporate client like them to really bundle and provide commercial solutions, and things like if you're building a building from scratch and you want to make it net-neutral, what do you do? Solar panels, and electric vehicles could get you credits, charging stations to help you get credits. You take the money that you're getting from these credits, apply it towards buying certified gas, so that is better gas and there's cheaper and then you kind of make this whole plan of how this one project will look like. I think will be really, really exciting. If I ever get a Lego call back, I'll let you know.

Interview 4: John Westerheide

Interviewee:	John Westerheide*, Senior Director of Customer Solutions, Project Canary
Interviewer:	Dr. Soheil Saraji, Associate Professor of Energy and Petroleum Engineering, University of Wyoming
Interview Date:	31 March 2022

***John Westerheide participated in the interview in a personal capacity reflecting on his 15 years of industry experience across multiple employers and multiple segments of the energy value chain. His comments are not attributable to Project Canary or former employers that may be engaged the development and application of blockchain technology within the energy ecosystem.**

Soheil Saraji: Could you give us a little bit of information about your background?

John Westerheide: Yeah, absolutely. So, I have 15 years of experience in the oil and gas industry. I started out of an undergraduate, with an economics degree. So, I do not have an engineering background, but a Marketing and Economics background. I started with General Electric straight out of school to oil and gas. The first six years of my career were spent in what I'll call product strategy, business unit strategy, new market development, new product development, supporting traditional business units inside of GE Oil and Gas, their expansion, and growth. The next six years of my career were spent in technical strategy, innovation, new solution development, etc. This is where I worked with Glenn Murrell at GE global research, working on everything from AI, augmented reality, mixed reality, unmanned vehicles, drones, to chemical processes and separation, CO_2 capture, emissions monitoring, and IoT sensing, etc. So, I have a very broad and diverse background in the innovation space, but that gave me purview inside the company with both GE and Baker Hughes as to technologies that were on the horizon. And then understanding specifically how to adapt those technologies into useful value cases for new solutions and new offerings in the oil field for customers. In October of this year, I started with Project Canary. Canary is a certified B Corp, focused on altering the arc of climate change through delivering independent third-party assessment and certification of best practices for oil and gas development. Think of it as an ESG credit rating score, if you will, on how operators look like. So, we do independent analysis for a trustable certification. And then we follow up that analysis with continuous emissions monitoring, providing a reliable data stream that shows that we just didn't come out to a site and look at it once and then come back a year later. Rather, this is what's going on. So, what we call the measurement economy, data-driven ESG, high fidelity, data-driven ESG metrics that can be used for things like capital markets, and transacting responsibly sourced gas deals, where there are emissions profiles tied to the gas volumes that are being marketed. So that gives you a little bit of background on myself.

Soheil Saraji: Great. I think that blockchain for carbon monitoring is such a fast-moving space that I thought cannot just write my book based only on written literature. So, I decided to reach out to people who are already working or building something in the Blockchain space and are familiar with the energy industry, like yourself. This allows me to get a fresh look at the ongoing for the book, rather than just documenting something that was written two or three years ago, which is most likely not relevant anymore.

John Westerheide: Absolutely. Well, I'm humbled for you to be speaking to me with such an impressive group of authors on this book, so thank you.

Soheil Saraji: I have heard a lot about you from Glen. He told me that John is the person to talk to because he has been looking into Blockchain for many years, especially its applications. So, let's go through some of the questions I have here. I know you have given a talk on sustainability recently. I watched it this morning. How do you think the energy transition will impact the energy industry as a whole and the oil and gas industry as a sector?

John Westerheide: So first of all, I don't think that our energy future and the energy transition is a binary choice between what I'll call traditional fuels, and zero emissions, or what is called Green Energy Fuels. The truth is that energy transition is going to be predicated on different paces at different places. What I mean by that is that each location is on its own energy adoption curve. So, what energy looks like in parts of Africa is very different from what energy looks like in Wyoming or in Oklahoma where I'm sitting. Equally, those different places have different proprietary energy endowments. For example, Costa Rica has a lot of geothermal. Other places may have proprietary endowments for wind and solar and other places may not. So, when people talk about the energy transition, One, it's not bright binary. And two, it's not homogeneous across all geographies. You're always going to be balancing the societal need with the proprietary endowments and economics of energy resources locally, to figure out what that right balance and mix are. In fact, I think it gets a little challenging when certain groups and individuals are only assuming what their world looks like and making policy and regulatory decisions for people whose energy world may look very different. Does that make sense?

And I know that it's a complex answer and a solution. But I think that's the best way to talk about it. Regardless of where you're at, we need to move to lower greenhouse gas intensity and lower carbon intensity sources of energy continuously. And realistically, we've done that, and going from a biomass world to a coal world, to liquid petroleum-driven energy infrastructure, to gas infrastructure, to hydrogen, and to nuclear fuel, etc. What you can see is that whether intentionally or unintentionally, we have always moved toward higher societal outcomes through energy that has lower carbon intensity throughout our history. We need to understand what all those consequences are and what they look like. And that's going to come through measurement. That's going to come through data. That's going to come through transparency to the consumer, and that's really key here. So, we must go from making ideological speculative decisions to data-backed and informed uncompromised decisions on energy. And I don't want to get into this too much, but there definitely are sort of different ideological camps out there on the future of energy. And there needs to be more discussion at the middle ground on continuous improvement and continuous improvement through transparency data and measurement if that makes sense.

Soheil Saraji: I totally agree with you. The thing is that we are not going to sleep tonight and wake up tomorrow and then everything has changed. It's going to take a few years or decades of dedicated work to achieve this. So, we are never going to have 0% emission, but we can reduce it to the barest minimum.

John Westerheide: Our CEO would always say that we don't want to only reward the highest performers, but we want to start rewarding the most improved, right on the field. And what I mean by that is, right now, there's a family in India, let's say, using wood, biomass dung, or whatever to do their home heating, home cooking, etc. And we can switch them to LPG as an example. That's a very material improvement in quality of life for them, while also reducing emissions. Yeah, it may not be solar. Right?

But don't let perfect be the enemy of better. And always try, for continuous improvement, and view this as a long transition rather than a magic silver bullet that solves everything globally for everybody. Right?

Soheil Saraji: Yes, perfect. I think this might be a good time for us to start talking about Responsibly Sourced Gas (RSG). So, what do you think RSG could do in this regard, moving towards lower emissions? I know that you're working on a related project, and you have enough information about this.

John Westerheide: Yes, absolutely. So, it comes down to this: Regardless of the industry that you look at right now, but specifically in B2C industries, this idea of what we consume, being aligned to our personal values, and differentiated along the lines of provenance, sustainability, etc., is important. We saw this in the diamond industry when they started talking about ethical sourcing. We see it in the building materials industry, the food industry, etc. One of my favorite examples is I picked up a water bottle the other day and it told me where it came from, and that the plastic was sustainably sourced or recycled. What is more fungible in this world than two hydrogen molecules attached to an oxygen molecule? But we're differentiating those hydrogen and oxygen molecules based on their provenance, how they're delivered, how they're packaged, etc. Where we haven't seen that yet is in the energy industry. And what I mean by that is, we've seen it between the world of traditional fuels and renewables, or Green Energy Fuels. But we haven't seen it in the energy industry. And what's funny to me about that, is that the energy industry is what makes our world go round. Right?

So, the things that we're most dependent on in this world like the sources of energy: electricity, gasoline, etc., are the things we have the lowest visibility and understanding of their provenance, where it comes from, how it was produced, and the supply chain by which it was delivered to us. In fact, I think that a very large challenge for developed economies is energy literacy. If you must collect your own energy resources daily to cook and heat your home, you will be very linked to your energy supply chain. However, if you grew up in the United States, it's just expected that when you flip the lights on or flip the switch, the light comes on, but you have really no idea where those electrons came from, and the molecules that made up those electrons. So as far as where you're going with your book, I think what blockchain promises to provide is transparency to the world and humankind's greatest and most complex infrastructure, which is our energy infrastructure. It also provides transparency to the consumer about such information as; where did my energy come from? How is it produced? Do I like the value statements of who produced it and how they produced it? and am I willing to make trade-offs on cost, price, reliability, and resiliency associated with those verified attributes of the energy that I'm consuming? I think the biggest thing that blockchain does is lower the administrative cost of creating data-backed transparency in those complex supply chains. To follow the molecule from the reservoir to the burner tip, to the electron, to the wire, to somebody turning on a light bulb using it, to the number of intermediaries or the number of people involved in that transaction for that molecule, how many different assets it hit, who own those assets, etc., is amazingly complex. And before advancements in technology, like blockchain, the administrative cost of that tracking and transparency and collating and curating the data that goes along with it was almost impossible. Although it would be possible, it'd be costly. And I think that both cost and ability have fundamentally changed with the development of new technologies like blockchain and our ability to do that. Does that make sense?

Soheil Saraji: Yeah, that makes sense. And it leads me to the next question. I understand you think the most important application of blockchain in the energy sector is lowering the cost of tracking the energy supply chain from the source all the way to the consumer. And then maybe we could do that on the carbon footprint of the entire energy consumption, Right?

John Westerheide: Right. I think there are a whole lot of opportunities for blockchain in oil and gas traditional energy supply chains. I think as it relates to the energy transition, and driving capital efficient markets, especially around the rise of ESG, investing in that type of deal, the ability to bring low cost, high fidelity understanding of oil and gas supply chains, and being able to differentiate producers, midstream operators, Petrochem plant, utilities, etc., based off of that information from both a consumer perspective, but also from capital markets and investment perspective, is definitely one of the main use cases as it relates to ESG in the energy transition. So hopefully, that makes sense. I don't want to just say that the most important thing that blockchain can do in the oil and gas industry is this, but I do think it's a very valuable use case.

Soheil Saraji: Yeah, I can see that that's at least one of the areas Blockchain itself is changing the energy industry. So maybe we can do more with it in the future, but I believe supply chain tracking to be the most important thing now. What are some applications of blockchain that you see in the energy or oil and gas industry that you think are interesting? It might not be an immediate application, but maybe you think it will be a futuristic application.

John Westerheide: This is a great question. I think there are tons of applications. One that I've always been fascinated with is fractional ownership of minerals and production and being able to open to direct investment from non-accredited investors into oil and gas production. Now, if you think about it, your ability to get in and invest in the development of an oil and gas well, or mineral rights usually takes large capital and many intermediaries. And I think Blockchain's ability to enable fractional ownership, reduce the administrative cost associated with that type of play is very interesting. And it creates a diversity of exposure to assets, when you don't have to go in on long large chunks of minerals or large chunks of wells, you can diversify at a finer point and make up a portfolio that may be more risk tolerant, resilient, etc. if you want to do that. So, I think that's a very interesting application. I also think that there's a lot that we can do on produced water, and understand net water and how water is utilized, etc.

Making sure that in basins or areas where water scarcity is a major concern we can get a better understanding of water use, which again, aligns with that transparency and ESG sort of statistic. Also, I think the ability to relate who was on a well site, what job they performed on the well site, etc. We always say it is the operator's responsibility, and they're the person. But it takes a very complex process and lots of individuals to bring a well into this world, and being able to understand even down to the level of who the field tech on that site at that time was delivering that service, etc. And I think that there may be legal bonding opportunities there for understanding performance, as well as the provenance of that well, and how it was brought into this world. Does that make sense?

Soheil Saraji: Yeah, that makes a lot of sense. I agree with both applications you mentioned.

John Westerheide: And we've seen similar applications, whether it be NFTs (non-fungible tokens) or whatever, and I know it's slightly different. But the idea of, hey, I want to have fractional ownership of an artwork, or I want to have fractional ownership of an expensive car and that type of thing. I think the same sort of principles and applications could be applied to oil and gas investing.

Soheil Saraji: Okay, perfect. So, we have talked about a lot of opportunities available with Blockchain in oil and gas and energy in general. But let's talk a little bit about the challenges. This question will be in two parts. First, what are the challenges of Responsibly Sourced Gas (RSG) and other low carbon intensity fuels that are trying to track their lifecycle in a manner so they can report the emissions? Not in terms of the market but in terms of technicality. Let's address that and then talk about Blockchain.

John Westerheide: Right. So, I was thinking about market development, new solution, and new technology adoption, in terms of carrot and stick, if you will. You have a carrot, which if you're showing great returns on investment, IRR, and multiples, you're reducing costs and driving value, those are easy decisions to make, right? Equally, if it's a regulatory-driven thing, sort of a stock market, if you will, you usually get adoption and quick movement. I think the challenge lies in when you're in between those worlds, right? Where regulatorily there is an emphasis to do that, and even maybe future regulation and challenges for adoption out in the future, but not today. Equally, in capital markets, where some are paying premiums. Some deals require this, but not everything does, right? So, you're sort of in this weird Limbo space where we think something's going to happen, but it hasn't happened yet. So, what is the appropriate timing for making those investments because you want to hit the window where the investment gives you preferred access, and more value upfront, because you're a first mover? But you don't want to make timing where you're the only one doing it, and therefore you're losing money on the investment until the market catches up. So, I think that's maybe where we are, right now with RSG. And I say that from a standpoint of everybody knows where the puck is going. Right? It's just that they're they are skating to the puck at different paces. And there are some that are totally bought in, there are others that are watching and observing. But we're moving out of what I call the pilot phase, where people are just trying to understand what it is like. I think there's a clear understanding as to what it is now. it's a market timing issue based on the carrot and the stick or the incentives that they must make those investments and timing. Does that make sense?

Soheil Saraji: Yeah, I understand that we are in limbo right now. There are some markets, like California for example, that energy companies are looking into as potential markets. I have also heard about an upcoming Methane tax from the federal government. But I don't know the status of it right now. Do you have any updates on that?

John Westerheide: I don't have any updates. I am aware of it, and I am also aware of not only jurisdictions here in the United States, but international jurisdictions that are looking at things similar or maybe even further along. In implementing that, I think there are a couple of challenges, though, if you implement something like that, you really need to move away from inferred standard emissions factors, which is the current standard approach in the industry. And you must get to measure data-driven factors. We wouldn't want the US government inferring what our salary is, and then taxing us accordingly. They need more data than that. So, the thought here is if things do move that way, and there's a likelihood that they will move that way. Fundamentally, how we measure emissions, and how we measure things like methane also need to change for those types of regulation to be implemented. Does that make sense?

Soheil Saraji: Yes, that makes a lot of sense. So, what do you think is the future of these measurements? I heard about a lot of different tools. I learned that there are a variety of sensors being installed in the field, there is drone-based monitoring, and potentially high-resolution satellite imaging in the future. Do you have any knowledge about these different technologies, and which one most interests you?

John Westerheide: Well, first, I think they are all interesting. And I think they'll all be used, not discretely, but in combination with one another. And honestly, another opportunity for blockchain is the ability to document and capture what the satellite said, at the same time as what the continuous monitoring sensor said on the ground, and being able to link that information in an immutable sort of data chain. But what's clear is, if my application is to provide immutable data associated with a natural gas marketing contract for an RSG deal where emissions intensity is written into the structure of that contract, you're going to have continuous monitoring on the ground. You need that high fidelity and low or high intermittency data. Now, if I'm just trying to get an understanding of an aerial basis, where leaks might be occurring, let's say 100 m by 100 m or a kilometer by kilometer, satellites are great for that. But in a place like the Permian, where I may have checkerboarded, lease holdings, and operations, being able to tie that back to a specific action on a specific piece of equipment, for a specific operator is just not realistic. Right? Drone flyovers or just flyovers in general are a great way of improving the fidelity and therefore actionability of operations. But still, they're discrete, right? They're intermittent, etc. So, it's this idea that it's going to take a combination of all three, sort of what I'll call bottoms up continuous monitoring to tops down the satellite, and then getting the data linked on both a spatial as well as its own portal temporarily, to correlate what the bottoms up to tops down actually is, and to build a holistic picture of that. And then, maybe that helps each one of those technologies advance in the future as well. Because you're not just using a single siloed data set, you're using multiple datasets and understanding what's really going on.

Soheil Saraji: Yeah, I undersatnd. There work at different scales and there are different ways to integrate and interpret the data.

John Westerheide: Yeah. But I would say the two vectors that are most important for delivering an application are: what is the fidelity of the information? And then what is the frequency of the information collected?

And determining the application for how the information is going to be used is important for determining fidelity and frequency. Right? If that make sense? If you're looking to do Eldar once a quarter, that's very different from trying to do Cross Border Carbon Adjustments into Europe on LNG. Right. Those are two separate types of things.

Soheil Saraji: Exactly. Thank you for the clarification. So, one thing I see as a challenge for these kinds of measurements, and generally for ESG types of measurement, is a lack of standards. Let's say Project Canary has its own workflow of how to do the measurements, another company comes in with its own different standard. So, what is the industry standard now? I think what this suggests is that the next step for the industry is to come up with a global standard. That when somebody in the US takes a reading, it will have the same value in Europe and the same value everywhere in the world.

John Westerheide: Yeah. That's a great question. And I think standards are going to be jurisdiction-specific. So, I don't see us getting to be a global standard. I don't think that's realistic. Nor do I think it's practical, because as I said in the beginning, at different paces at different places, you may need to understand the need locally to determine what the standard should be. That way, what we're doing here in the US, doesn't dictate what they must do somewhere else that isn't the US, and may not be facing the same challenges. Now what I think the commonality is, regardless of jurisdiction, and location, is the ability to use data to inform the standards, right? I think one of the challenges that we always face is the setting of standards without a true understanding; operational, real-world understanding versus an academic or theoretical understanding. And so, what I hope to see is that when standards are set, those standards are set based on good quality, real-world data, and information that allow us to understand both the pros and cons of the standards that are to be adopted as it relates to where it is in the RSG market right now or looking at ESG profiles associated with hydrocarbons. Yeah, there are independent standards like ours, right? There are regulatory standards, like maybe the state of Colorado or Wyoming, etc. And what I think will occur is there'll be regulatory minimums, and then there's going to be consensus driven by capital markets as to standards, data back standards associated above those regulatory minimums, if that makes sense. So, the idea here being there may not be one comprehensive standard, but there's at least going to be a minimum regulatory standard. And then there might be a variety of standards that play out over time. Another way to think about this is like the credit rating agencies, there's Fitch, there's Moody's, and there's S&P. But there isn't just S&P. Does that make sense? And I see above those regulatory minimum standards, that's what you're going to see also in the ESG space, there probably won't be 100. But there probably will be more than one. Does that make sense?

Soheil Saraji: So, from your discussion so far, I understand that you consider the markets to be the driver of ESG-based changes in the energy sector, rather than the government because usually, government is a little bit behind. I want also to know your opinion about the impact of the market on the ESG (energy consumption side), especially the Responsibly Resource Gas and other products that are trying to achieve low carbon emissions.

John Westerheide: Yeah. So let me say that I don't believe that there's no role for regulatory frameworks and users, but from that standpoint, I believe that markets, capital markets, if given good transparent data are incredibly efficient at allocating capital towards a specific outcome. As sort of economic background, tell me what the incentives are, and I'll tell you what the outcomes are. Right? So, to that point, let's say that tomorrow, there's $100, a metric ton folk, for carbon CO_2, implemented and now capital markets must begin pricing that into their investments based on data. Well, you're immediately going to see you know, a capital shift away from some sectors and move into other sectors. You know, based upon an input Question. So, my personal belief is that the ESG energy transition is going to come about through efficient capital market investments that are tied to specific outcomes that have economic value. Right. And that is going to occur via transparent, high fidelity, and very trusted information to those markets to make those decisions.

And not only that, but the capital markets to some points are also typically more nimble and able to change faster than regulatory cycles. Because of different challenges with the political regulatory cycle. If that makes sense.

Soheil Saraji: Yep, absolutely. So now, let's look at the same question but with respect to blockchain. So what do you think would be the challenges of blockchain technology? I strongly believe that Blockchain will bring a major change in the industry. But what are the challenges, in your opinion, that are likely to arise when trying to apply Blockchain-based solutions?

John Westerheide: So, let me just try and break down those questions into two parts, right? The first part is challenging, I think not unlike what we're seeing in the ESG space, but standards, right? And standards from across the supply chain not within supply chain participants. I think one of the times when blockchain was easily adopted in the beginning, is when it was within a single operator or customer, that they were using that information internally. I think where it becomes a challenge is when you move from, let's say, the production of gas on a pad, to switching over to a gathering and infrastructure, to switching over to a transmission infrastructure, to switching over to an LNG, you know, producer, etc. If they're not using the same standard tech, you know, tech stack to some degree, etc., It ruins the value of the blockchain. Does that make sense? So, I think that is the adaptability for this case.

Supply chain participant need is probably the largest hurdle. At least being able to make sure that the handshakes will match and keep the value and integrity of the blockchain across the value chain. Right. So that's the challenge. Now there was a second part to your question, and I want to make sure that I understand it right. Is it getting more into the technical specifics of blockchain and not the application?

Soheil Saraji: This is an open-ended question. Yes, from the technical specifics of blockchain for the real-world application, or regulations, whatever you think might be a challenge for the blockchain to be adopted as a tool in the industry.

John Westerheide: Right. You know, just from an ESG perspective, I'm very interested in the energy cost and environmental cost of applying blockchain. Right. So, we've talked a lot about the incentives of moving to blockchain: reduced administrative costs, greater integrity of information, you know, the immutability of information, etc. But I'm really interested in what the cost is for blockchain, especially on that energy. To administrate this, we must have a lot of machines running, right? Those machines are going to consume energy. Right? So how do we balance the value versus the cost? And I'm not sure that that's been researched enough at scale. Right, especially if you think about doing something like switching all the natural gas infrastructures to run on something like this. What does that look like? Maybe I'm not as knowledgeable on that part, but that would be something that I would be very interested in and understand.

Soheil Saraji: Absolutely! I think that's one of the biggest challenges of blockchain. And as you said, I just recently saw recent research on the energy consumption of blockchain. They reviewed 10 different blockchains, and they try to characterize their carbon footprint. But yeah, bringing it to the application in energy, as you mentioned, is a completely different story, and I don't think anybody has done this work.

John Westerheide: Yeah, nothing's free, right? Blockchain is not free. So, we got to understand the puts in the takes, the value we're receiving versus the cost we're incurring because we don't want to solve one problem just to create another problem at scale.
What applications are most fascinating to you?

Soheil Saraji: Um, good question. So, one area I am very fascinated about would be carbon emissions and carbon credits. That's the topic that I would like to see more work. And I know there are people who have been working on it, and I believe the first commercial application of blockchain in the energy sector is come likely to be from this space.

John Westerheide: I'm absolutely in total agreement. I can leave you with a bumper sticker, sort of takeaway from our conversation, right? What I would tell you is in the energy transition, the poor decision is to divest rather than differentiate, right? So, we want to differentiate our energy supply based on how it's produced, and what is produced. We don't just want to fully divest entire portions of our new energy infrastructure, right? Because that's not capital efficient and is going to lead to challenges that are being experienced in Germany or are caught when it gets cold. Things like that, right? And the only way that we can differentiate is through data. Right? And blockchain gives us a better ability to store that information and lower administrative costs of differentiating energy associated with it. So, we want to differentiate, not divest. In order to differentiate, we need technology solutions like blockchain.

Interview 5: Kari Hassler

Interviewee: Kari Hassler, Senior Manger - Market Operations, Xcel Energy
Interviewer: Dr. Soheil Saraji, Associate Professor of Energy and Petroleum Engineering, University of Wyoming
Interview Date: 6 July 2022.

**Kari Hassler participated in the interview in a personal capacity reflecting on her professional industry experiences. Her comments are not attributable to Xcel Energy.*

Soheil Saraji: Thank you very much for making some time in the middle of your busy schedule to talk to me. please tell us a little bit about your background. How did you get to where you are at Xcel Energy Company?

Kari Hassler: Yeah. I have a master's degree in applied math. I started working at Boeing as a statistical consultant and stayed there for a few years.

Then, about 25 years ago, I moved to Xcel and started working there as a statistician. For about 10 years, I did that. Then I was put in charge of all eight of our states' renewable energy portfolios, which is like managing the RECs. The RECs were tied to renewable energy production and helped us meet our compliance goals. So, you know, I helped get that going and worked on the tracking systems as they were being made for the RECs and taking care of everything else. I did that for about six years, and then I moved over to commercial operations.

So in there, I managed the dayhit? group, which was responsible for submitting the offers into my MISO [Energy] (Midcontinent Independent System Operator) and SPP (South Power Pool) markets, as well as establishing the commitment plan for Piesco because they are not in a in a regional energy market. Still working on it, but they're not there yet. And I did that for two years and then from there, I moved over to market operations, where my focus is on NSP in the MISO region and coordinating policies that are necessary for maintaining an effective market and bringing on renewable generation, but we also think a lot about the transition from the thermal resources that we have and to more of the intermittent resources. and so we're highly involved in that process with MISO, sometimes it's a difficult sell to them. They just want to jump ahead to the future and think everything's going to be fine. But there is definitely a lot of work that needs to be done for that transition.

Soheil Saraji: You had a very interesting career arc, from being a statistician to a top-level executive, which is a very interesting career trajectory. Given that you have been with Xcel for such a long time and have likely been involved in all aspects of the business, particularly those pertaining to renewable energy, from the perspective of an energy provider. I'd like to begin with the broad topic of energy transition. What do you think the impact of the energy transition will be on the energy sector as a whole and specifically on energy providers like Xcel? What do you see as the biggest challenges or opportunities during this energy transition?

Kari Hassler: Yeah, I think the challenges go back to complacency of the markets and the energy grid as a whole, especially in the Midwest. We've become complacent because there was always a large surplus of capacity. It was typically dispatchable capacity. Even when we first started bringing on the wind, there was still plenty of dispatchable resources to accommodate for that wind. And so I think developing the policies and the tools that are needed to ensure that the ongoing investment whether that be retaining dispatchable resources through their useful life. We've seen independent power producers retire some of their dispatchable resources because they're not economic with the current policies in MISO. I'm most familiar with MISO, so you're going to hear me say that the most, but it's probably happening and or will happen in SPP, too. And the other and the other RTOs. We've already seen it in California. So, one, ensuring that the incentives are there for retaining the dispatchable resources as they're needed, but also encouraging investment in resources that have the flexibility and the availability and the capability to manage the ramps of the intermittent resources. And the other issues that we encounter. I will say that, you know, one of my overarching theories is fuel diversity and resource diversity is critical as we move through this new age of resources. That's great to have renewable resources, but until batteries become more economic, we still need to maintain the diversity of the fleet through the fuel that the resources use as well as the locational diversity as well.

Soheil Saraji: Right. I couldn't agree more with you on this point. I would like to know what percentage of Xcel Energy's energy portfolio consists of renewable resources. Is it a considerable amount, or is it a minor percentage?

Kari Hassler: Considerable. I mean, especially if you include our new nuclear units in there, I'd have to get you the numbers. I don't know the numbers off the top of my head. But you know, across any day we could see 80% of the load being served with renewable and nuclear units.

Soheil Saraji: Wow. That is a very high percentage.

Kari Hassler: Yeah, I mean, you know, in PEISCO we have a lot of wind, in Colorado [for example]. Again, I'd have to look at the records, but the records are pretty amazing at the amount of load that served by our wind resources. You think of a nighttime minimum load, where the wind is blowing very hard. We've become fairly adept at managing those ramps and ensuring that we have, the resources on standby that are needed to manage those. Xcel definitely leads the way in pushing the envelope of intermittent resources, but we also, I think, lead the way in ensuring that we have a reliable system. One of the issues that we encounter in MISO is that it provides 50% capacity accreditation for solar, currently. They're working on this but when you think of solar capability in the middle of the winter in Minnesota, it's nowhere near 50%. And so we've had to make some outside of the box adjustments to say, hey, look, we know that we need gas resources or some kind of dispatchable resources, because solar is not going to be there in the summer. So, we've tried to come at it from that point of view, but we're also you know, promoting a seasonal construct in MISO so that solar, solar and other resources would be accredited differently for different seasons. But then that would also encourage those dispatchable resources to come online.

So, going back to your original question as to what the portfolio is. We have a broad range of resources. I mean, nuclear, coal, gas, hydro, biomass, wind, solar. It's very diverse and we've tried to maintain that diversity with the dispatchability as our call units are nearing the end of their useful life. We've accelerated their retirement because of carbon issues. I think we have a really good resource plan in place to ensure that we bring on dispatchable resources but also meet our carbon targets in the future. Our current plan that was approved by the commission is to allow new gas resources to come online I think 2026–2027 timeline, but they have to be able to be flexible so that they can convert to say a hydrogen or some other renewable type of fuel in the future so that we can meet our carbon targets.

Soheil Saraji: That is great. This will be a good segue to another topic. Hydrogen seems to be getting a lot of attention right now. And it seems that the federal government is looking into developing a few national hubs. What do you think hydrogen's role will be in the future of energy? What do you think about hydrogen and the hydrogen economy as a whole?

Kari Hassler: I think hydrogen will become a critical factor. I'm definitely not an expert on this. But there certainly are improvements that need to be made. Because of the efficiency of creating the hydrogen. It takes a lot of energy to convert and so, we need to definitely make some improvements there. We have to think about location of the hydrogen and location of the resource that will be burning the hydrogen and the liquidity of the markets. I mean, there's so many things to think about to get that off the ground and to be to ensure that it becomes an efficient fuel. I definitely think it will become important because of the carbon targets. But a lot of work needs to be done in that arena.

Soheil Saraji: We talked about the energy transition, carbon footprints, and renewable energy sources. What role do you think the oil and gas industry, or fossil fuels in general, will play in the future of energy? How will it look like in 10 years?

Kari Hassler: I think gas resources have a big role to play in the transition. It's difficult because we are getting a lot of pushback from environmentalists at the time, [who] say no new gas resources. But they're needed for reliability. So that's a difficult role, and that's where we need the regional energy markets. To help us a little more to ensure that we have that dispatch ability, that flexibility. Maybe those resources don't run as frequently as they do now. But certainly, they could be available if they are needed.

Soheil Saraji: Okay. Gas seems to have a lot of future potential in the transition. Also, there is room to use natural gas or coal bed methane as fuel to make hydrogen, like blue hydrogen. So, we talked a bit about the energy transition, how the energy sector would change, and how oil and gas might change in the coming years. And there's "carbon economy". How do you think the carbon economy will change in the future? Do you think it will eventually take off?

Kari Hassler: We really need it to take off. It's going to be difficult to meet the targets without some value of carbon. I mean, when you think of a regional energy market, such as MISO and you think of multiple states within that market, Minnesota [for example] may set one target for their state and North Dakota would set a completely different target for their state because they are coal intensive. They want to keep those units on they want to keep the coal mines open. So how do you ensure that carbon is valued where it needs to be valued without increasing the amount of carbon in another area? For example, if Minnesota says the value of carbon is X per ton, and we include that cost in our offers for our generation. North Dakota is just going to say, well, we don't have a value for carbon. So, our generators aren't going to include that cost. The economics of MISO are going to be what? well, they're going to pick up the lowest priced resource and rent. Well, that doesn't really solve the carbon problem, right?, because we've just increased the carbon in North Dakota, even though we've decreased it in Minnesota. We've got to have some overarching value, I think that is applied everywhere. Otherwise, we're going to see this disconnect between the states. As they go through and approve of value of carbon and it's different in different areas, then it's not going to have much of an impact.

Soheil Saraji: That's true. So, what stops us as a country from getting to that point? what's holding us back from getting there? I mean, do you think it has something to do with technology? Do you think it's a problem with the rules and regulations? Do you think it's a problem with the markets?

Kari Hassler: I think it's probably a regulation problem. I think FERC [Federal Energy Regulation Commission] has been hesitant to create a value for carbon, and they've left it up to the states. In a market-based system, I'm not sure how that's going to work. The federal government certainly has limited the amount of carbon. There's always issues with that. I don't know if you're aware of the with the Casper and the latest Good Neighbor federal implementation plan for Casper and just the ratcheting down of allowances that's for NOx emissions, but similar kind of structure. You know, for us, it'd be best just to put a value on it and just let the market deal with it. Instead of just ratcheting down the allowances in some states but not in others. You get back to the same problem.

Soheil Saraji: Right. So, it seems like you and I are both on the same page. I'm also in favor of letting the market decide. If you set the rules and let the market decide, the market would push carbon emissions down if you put a price and a cap on it. Similar to what California is doing, for instance.

Kari Hassler: I think currently market participants can do that. We can incorporate carbon limitations into our offers. If the state comes up with the value of carbon, we can incorporate it. But again, if you have a market that crosses several states and each state is different, then how do you how do you control that leaping at a carbon from one state to another?

Soheil Saraji: So, do you know if there is a plan or strategy for carbon credits inside the Xcel or among the companies that provide energy? I mean, switching to renewable energy will be one way, but do you have a plan for carbon credits or carbon offsets?

Kari Hassler: We are just starting. We've kind of dabbled in carbon offsets in the past. At one point, we thought carbon credits were going to totally replace renewable energy credits and but they haven't, of course. So, we're just starting development, how are we going to do this? and we have our Piesco operating company in Colorado like I said, we don't participate in a market where we are the balancing authority there. So, that's one set of issues. We also participate in SPP, which is our Texas and New Mexico footprint and then also in MISO. So, we've got a lot of different things going on, and it's probably going to be something that we have to deal with probably first in Piesco and then in SP and then SPS, but we're just starting to kind of think about if we wanted to proceed with the value of carbon and you know, take that up with the regulator's how to go about doing that.

Soheil Saraji: Okay, Have you received any inquiries from your customer side? Is anybody interested in knowing their carbon intensity they're receiving from you?

Kari Hassler: We certainly have. We've seen larger retail customers that are interested in their carbon footprint. Xcel publishes a sustainability report that has our carbon footprint, our carbon impact on it, that we can always point to. I think retail customers are becoming more and more educated in this space. You know, when I was on the renewable side, when they wanted to reduce their carbon they would say, okay, we'll purchase Cimarex to do that. And then we feel like we've met our goal. Well, that's nice, but that doesn't incentivize incremental, renewable resources. So, we are going to go out and have our own purchase power agreement with renewable resources. And now I think we're probably seeing the same thing on the carbon side. I think retail customers are happy with what Xcel has done with the reduction of carbon, but they also want to have an impact and control their own carbon footprint. I think we're starting to see educated customers taking that control themselves instead of relying on the utility provider to do that for them.

Soheil Saraji: Okay. That's a valid point. Because if the customers have interest, then maybe some further regulation would also help to shape the market. One thing that we passed over—and I didn't ask the question at the time because I went in a little bit different direction—was that there are movements in the oil and gas industry to reduce the carbon intensity of fossil fuels. The best example is responsibly sourced gas, which, as you know, a lot of companies are looking into. There are pilots out there. What do you think about RSG from the market's perspective? Do you think there's going to be interest in that kind of product?

Kari Hassler: Definitely. It's definitely a transitional fuel that's needed. And I think we're starting to see a little more interest on it. But it's interesting because Brian [Hassler] and I talked about this because he works on the responsibly sourced gas and the renewable gas. I see a lot of parallels in the development of the RECs tracking systems on the electric side with the tracking systems that are needed for those gas products. And I don't think that structure is there. I know it's kind of a chicken and egg kind of thing. What helped on the electric side was the requirements for renewable portfolios, renewable standards. And then of course, well, if a state has a renewable standard, then there has to be a way tracking it. Perhaps that's an area once that regulation comes out, and it's required, then the tracking systems would follow. It would be nice if the market could encourage the tracking systems to be developed earlier, sooner rather than later or even without the regulation but that typically seems to be the pattern. There's a law put in place; you need to have a way of ensuring that that law is complied with and then tracking systems follow.

Soheil Saraji: Yes, I agree. The lack of standards is also a big the problem right now. A few companies looking into that. As this is the standard for monitoring, let's make it a broad standard that everyone agrees on. I hope there will be a good room for standardization of low-carbon fossil fuels as a key transitional fuel.

Kari Hassler: Yeah, two things that I thought of there. One is this the standards. Absolutely critically important. We saw the same thing on the electric side, we were all kind of floundering around using a multitude of certification systems and one wouldn't talk to the other and it was very fluffy and you never knew if what you were getting was exactly what you were paying for. You thought you're paying for. So absolutely, the standards need to be developed. And the other piece that you said that triggered something in my mind was the 10-year transition, which is correct because that's about the window that we have. And if we get beyond those 10 years, I don't think we're going to need it anymore. Because I think there's going to be maybe 10–15 years, [until] batteries are going to be more efficient and more economic and the technology that is needed is just going to explode here. I think there is maybe a 10–15 year window for this transition. A lot of the policies that we've been pushing it MISO 5–6-7 years ago, they're like, well, maybe we could do this now and I'm like, forget it. It's too late. We don't we don't need it anymore. I think just, you know, thinking of that window and encouraging development of the responsibly sourced gas with the standards and the tracking systems and getting that off the ground sooner rather than later. And knowing as a company that there's opportunities to do that. I think is really critical in this transition period. Otherwise, it'll be too late.

Soheil Saraji: Yes, I agree. So, it's all about the right timing. If we try to switch to renewables too soon and don't have the storage capacity, we'll have the same problem as Germany. I recently heard they are opening one of their recently retired coal-fired plants back online.

Kari Hassler: You know, 10 years from now that coal plant wouldn't even be available, it would have been dismantled, not even available for them to bring it back online. I think that is an interesting piece of the transition and the need for the fuel diversity that we were talking about.

Soheil Saraji: This is a good discussion, but I also want to make sure that I honor your time. So, let's switch a little bit towards a concept that has been a personal interest of mine for a few years: the blockchain technology. I don't know how much you're familiar with the technology, do you see any application of it in the energy sector?

Kari Hassler: Yeah, that's another area that I'm not very familiar with. I kind of have an understanding on the outside looking in. The one aspect, and this really doesn't have to do anything with the oil and gas resources, of blockchain that I thought would be really interesting would be [load management]. Currently, the way the markets are set up is that we have a set amount of load that needs to be served and the generation flexes to serve that load. But in the future, I think it's going to be more common for there to be a fixed amount of generation and the load will flex to use that generation. That, to me is an area where blockchain could be really helpful, [it] is on more of the distribution level aspect instead of the wholesale market. But enabling that flexing of the load to meet the generation or to say, okay, at this price point, you know, you will have your load is firmed. It will definitely be served at this price point. It may be interruptible, [otherwise].

Soheil Saraji: Are you suggesting some kind of market?

Kari Hassler: Yep, similar to a market. I think order 2222 with the dispatchable energy resources, kind of sets that in motion a little bit. We have a long ways to go. But I think that the use of blockchain for that, flipping the flipping the load and the generation in reverse, would be very helpful.

Soheil Saraji: I never thought of it that way. I guess you've answered most of the questions. Is there anything that I didn't ask today that you want to talk about? Is there anything you want to add? I couldn't be happier to hear that.

Kari Hassler: Yeah, I can't really think of anything off the top of my head that you haven't covered. But if I think of something, I'll shoot you a note. And one thing that I always think would be interesting, and I haven't seen that anyone has done it yet but thinking about the transition to renewable resources and thinking about winter storm Yuri in Texas, in particular. I've always thought it would be interesting to calculate the number of batteries or the battery capacity that would have been needed to store the wind energy to meet the requirements for two or three days when that red was all shut down. Everyone says, well, renewables are the answer. Well, let's think about how much that would have cost it would have had to been billions of dollars just based on the capabilities of the batteries in the current day. As far as the need for a transition plan, and the responsibly sourced gas and the renewable gas and the hydrogen, and all those kinds of things. I just think it would be an interesting exercise to calculate what that cost would have been.

Soheil Saraji: It would be a great idea to do that. Or it could be a project for students in one of my classes. This reminds me of hydrogen. In a way, if we turn the sun's energy into electricity, we can't store a lot of it with batteries. What if we use high-voltage electrolysis to turn the sun and wind into hydrogen, store it, and then burn it because hydrogen is easy to burn? Of course, the technology isn't quite there yet. We need to work on it, but, you know, that's another place where hydrogen might be useful. If we can figure out how to store a lot of hydrogen, we might be able to use it as a kind of energy generator or cushion.

Kari Hassler: Yeah, the same can be said for electric vehicles, too. Brian and I have thought several times should we get a battery here we have solar, and we could get a battery. Well, for one thing, our power doesn't go out enough to really require that we need it. But I said for another thing, let's just get an electric vehicle. That's going to be our battery sitting in the garage. And we can use that to capture the excess solar instead of back feeding it to the grid and then we can just use the battery of the vehicle that way. So I think we're going to see a lot of that as well in the future as well.

Soheil Saraji: I agree. And especially with the rise of electric cars, if the number of electric cars reaches millions in the future, it could become a huge energy storage. One challange is that if everyone has an electric car, when will they charge it typically? usually at night. This means that everyone has a high need for energy everywhere at a particular time of the evening that could out extra load on the gride. One solution that has been proposed is to use blockchain for trading energy stored in electrical cars. If you can charge it when nobody needs it, like early in the morning, you might get much better rates for electricity. Then, you may connect your car to the gride to sell it back in the evening, when there is high demand at a higher price.

Kari Hassler: Yeah, in Xcel, especially in PIESCO, we're already doing time of day incentives for electric vehicle charging and discharging with the advanced metering.

Interview 6: Kelly Bott

Interviewee: Kelly Bott, Senior Vice President of ESG, land, and regulatory, Pure West Energy
Interviewer: Dr. Soheil Saraji, Associate Professor of Energy and Petroleum Engineering, University of Wyoming
Interview Date: 1 July 2022.

Soheil Saraji: Kelly, thank you for coming today. Give us a little background on yourself. And what is your educational background? How did you end up in this position that you're in?

Kelly Bott: Sure, I'm the Senior Vice President of ESG, land, and regulatory for pure West energy. I have an electrical engineering degree from the University of Wyoming. I started my career at the Wyoming Department of Environmental Quality and the Air Quality Division. And so, I headed the planning section, which did all of the NEPA coordination. The environmental impact statement for the Jonah field was one of the first projects that came across my desk when I first started. And shortly after that, came the Pinedale Anticline SEIS. So, I've been working on pinedale anticline issues since 2006. That was how I got my foot in the door of the oil and gas industry. And then, in 2012, I transitioned over and started my own consulting company called 307 Consulting, where I coordinated the activities of Ultra Shell and QEP for all of the work that they were doing in the pinedale anticline specific to that record of decision from the BLM. I'm so glad I did it for a few years. And then, in 2014, I came to work for Ultra as the environmental and regulatory manager, and I just kind of worked with that company, even as we transitioned and went through a couple of bankruptcies. In 2021, we rebranded as PureWest Energy. So that's how I ended up where I am today, in charge of all of the ESG efforts.

Soheil Saraji: All right. you're the right person to talk to on the topic of ESG. My first question is about the whole energy transition concept and the energy industry, or the oil and gas industry, within the energy industry. What do you think about it? Especially in terms of the opportunities and challenges that the oil and gas industry will face during the energy transition.

Kelly Bott: Well, at the end of the day, we are a fossil fuel company. And so, you know, we have to find a way to communicate that natural gas is part of the solution toward the energy transition and find a way to present our product as being part of that solution. And so, you know, we've set a high bar for ourselves; part of it is regulatory driven. We're in the Upper Green River Basin, which is an ozone non-attainment area. And that's driven a lot of our improvements, for sure. But we've tried to push that even further. So, if we as an industry want natural gas to be part of the solution, we need to make sure that we're doing everything within our power to make sure that it is produced with as small an environmental footprint as possible. That is certainly true in terms of climate change and energy transition, not only in terms of emissions but also in terms of surface footprint and water impacts, and we must demonstrate that we are doing everything possible. as well as we can.

Soheil Saraji: There has been a history of the oil and gas industry not taking environmental issues seriously. And now maybe, is the time for us to move forward and show that we are serious in this regard. So, let's first answer what are the ESGs? You have already mentioned methane emissions and water.

Kelly Bott: That's an intriguing question. So, someone once asked me, "What is your favorite letter in the ESG acronym?" and it's kind of a similar question. You know, it's because I don't believe that you can do anything in a vacuum. And my response to that question was, "They're all important," because, you know, you can do all this great work on the environmental and social fronts. But if you don't have the governance to show that it's done with integrity, it doesn't mean a whole lot. And so, you know, when it comes to something like this, this notion of transition and what's important, what we're finding is that you have to approach ESG from a system thinking perspective. So, when we're out there doing projects to reduce emissions, especially in an area like the Upper Green River Basin where we have other critical resources like visibility, protected viewsheds, and sage grouse. We have big game and migration corridors.

And so, what we have found over the years is that sometimes when you're doing projects to benefit air quality, like, for example, if we wanted to electrify, it's not straightforward because we have to think about it. So, if we run power lines, we've suddenly created a perch for raptors that pick off sage grouse. So, it's a matter of striking a balance between all of those resources. As a result, you can't do anything in a vacuum; you can't start a project without considering how it will affect all of the other resources. As a result, PureWest must consider not only methane and our overall emissions profile but also how those reductions may impact or benefit some of those other resources. And so, what we find is that when we do that and look at it collectively with a systems thinking kind of approach, we're able to reduce our overall environmental footprint on multiple resources.

Soheil Saraji: That's a great point. I agree with this; you cannot just isolate one thing and try to optimize it because then other things become suboptimal. what are the potential markets for something like responsibly sourced gas? Or, more specifically, what is the commercial or economic incentive for a company like Pure West Energy to follow ESG guidelines?

Kelly Bott: I guess a lot is going on in the space like market development, differentiated gas, and market space. And I guess there are two schools of thought on differentiated gas, including responsibly sourced gas, or anything we're doing to differentiate ourselves; ideally, it would come with some kind of uplift or, you know, payback for us doing that work. The other possibility is that what we might find down the line is that it's a license to operate. You know, companies that are not doing this might get hit with, you know, discounts on their gas or different things like that, or they might not be able to operate at all. So, you know, being responsible and doing this is part of our DNA at PureWest, right? But at the end of the day, it's really about license to operate and finding ways to kind of commoditize that differentiated gas, right?

Soheil Saraji: Yes. So basically, PureWest has certain initiatives in ESG with the hope or forecast that there will be benefits soon, right? There is a concept—I don't know how much you're familiar with it—that has been going around a lot called the carbon economy. This is a big umbrella that includes customers from the transportation supply chain, oil and gas, and everyone else. What do you think about this concept?

Kelly Bott: I think it's gaining a lot of traction, and I think that we are going to see the effects of this "carbon economy" concept, you know, more and more in the oil and gas industry. I think that for us to play a part in that, the first thing that we need to do is have a really solid, defensible accounting of what our carbon footprint is. And that's where, you know, there's a lot of technology coming out of the woodwork around monitoring and quantification of what our CO_2-equivalent emissions look like. In reality, we're very good at calculating them using EPA emission factors and other predefined metrics. But what we need to do is measure, comprehend, be confident, and determine how we fit into that type of carbon economy. And before they were thought of as real contributors to the concept of the "carbon economy,"

Soheil Saraji: Yes, agreed. So, can we get a little more granular with your experiences working with ESG activities, especially concerning responsible source gas? Maybe you could give us a little background on how the projects you're working on works, including involvement from Project Canary and others? And what are the key aspects of the workflow, as well as the data you're capturing?

Kelly Bott: When we started thinking about certifying our gas, there were a couple of things that we were trying to do. It's very helpful to have a third party come in and audit and kind of give a blessing to what we're doing, but also to give some recommendations for improvements. So, as you may know, our methane intensity rate is one of the lowest in the country for 2020, according to the most recent EPA verification, and it is 0.05 percent. If you're familiar with methane intensity rates, and that's again one of the lowest in the country, that means the low-hanging fruit has been picked. As a result, we're constantly looking for ways to reduce that. So, we were debating whether obtaining gas certification from responsibly sourced sources was the best option for us. So, we started looking into the companies that offer this, the first is MiQ, which is a Rocky Mountain Institute-sponsored effort; that's a kind of field-wide certification, and it is very focused on methane. But they don't look at things like water impacts or some of those other things. We looked at equitable origins, and they were very focused on indigenous rights.

So, when we discovered Project Canary, one of the first things we noticed was that they were comprehensive in their certification program. And so, they're looking at everything from our emissions profile and performance to how we protect the land, our reclamation activities, our safety metrics, and how we interact in the community. As a result, it was a complete package. And that's what we were after, because in the Upper Green River Basin, as I mentioned, we've got, you know, protected viewsheds and wildlife, air quality issues, and world-class waters and recreation. And people are very, very interested in making sure that we're doing things right. And so, we started our Project Canary initiative in March last year. We then went through a sort of "phase one." So, we were very concerned about the perception of "greenwashing," and we did not want to be "greenwashing" in any way. So, we used it as we selected our first phase. We kind of used some of our classification metrics, which were more about production and cost flows and things like that and were not environmentally based. And so, we identified our first phase production and we got that done. Last year, it was about 41% of our production. On average, we scored platinum on about 80 percent to 85 percent of that production.

Platinum is the top decile of performers for Project Canary. To get those certifications, we had to provide everything, from every piece of documentation and every planning document that we put together, to Project Canary, who was very interested in knowing if we did things correctly from the start when it came to planning the wellbore, planning the surface, planning for risk, identifying risk, and mitigating risk. They went out and did a field tour to make sure that what they saw in person matched what we said we saw in person. And then they go through and evaluate the risk specific to each site and how we're mitigating that risk. And then they provide this certification to us that gives us our rating and tells us how we're doing right now, along with some suggestions for things that we can do better. And so that was all very, very important to us as we went through the certification process. We have committed to certifying 100% of our production to eliminate any perception of "greenwashing" by the end of this year.

I can tell you that I think we're up to 51% of our production being certified, and we're getting more certifications every day. And it is now 78% platinum and 22% gold. So, we're just plowing through and again plan to certify 100% of our production, which will not only allow us to pursue the differentiated gas market but also give us some ideas for improvement so that we can continue to whittle away at our environmental footprint.

Soheil Saraji: Let's talk about your ESG goals. So, one of the goals that you have is to certify 100% of your wells, right? What are the other ESG goals that your company has?

Kelly Bott: Yes. And we're working hard to get the best ratings, platinum, and gold, on our wells; this is built into our company performance metrics, which is significant. And the reason that we did that is that you know, this trust level certification is so all-encompassing of all things environmental, safety, and community that we felt that was a really good gauge of our overall company performance. So, we've got some metrics for achieving platinum and gold certifications. The other thing that we're doing in conjunction with Project Canary is stationary methane detection. We're running a pilot right now. with Baker Hughes and Project Canary through the trust-building process. They're agnostic when it comes to what kind of sensor we deploy in the field. But they want us to have some kind of sensor out there to detect and monitor our methane emissions.

So, to that end, we've got a couple of pilots going on. And then all of this data kind of feeds into our reporting, our quantification of methane, and all of these kinds of factors. And then we've got another pilot that we're working on with Colorado State University where we're looking at, I believe, seven different technologies on a single pad that we've been doing this summer. So that we can design a field-wide strategy that will be kind of customized to the different types of pads we have out in the field because we've got pads with 50 wells, nine wells, and single wells. And so, what the monitoring strategy looks like for those is going to be different, but it factors into how those projects carry trustworthy certifications.

Soheil Saraji: Great, I didn't know that. I thought project canary is the one responsible for installing the sensors. So, basically you are providing the sensors, collecting the data, and they are auditing your data. Is that right?

Kelly Bott: Yes, so in terms of the TrustWell, certification—they simply want to know: are we monitoring, yes or no? And then we get credit for that. The next step—and frankly, on the monitoring side, it's been a little bit challenging because, as I mentioned, our methane levels are already very low, which means the leaks we're looking for are very small. And we already have a very strict inspection program in place, with optical gas imaging cameras and a fleet of, I believe, five of them. and six inspectors that go out there and survey everything. And we do; more than half of our surveys are voluntary.

We have already increased the frequency of our cameras. What we don't want to do is, you know, if we have a blip on the monitoring data, it could be an allowable emission, it could be something from a pneumatic device, or something else. And what we don't want to do is divert our camera inspection resources from more important sites to track down all those blips. So, we are working right now to analyze the data from the monitors and make sure that we understand the difference between a leak and a pneumatic emission. because again, we are talking about very small emissions. And so, you know, we certainly don't want to be chasing ghosts, as I like to say, and we want to make sure that if we are sending a team out there to repair, we are confident that there is a leak there. So, all hands are on deck right now, working on the analytical side.

Soheil Saraji: Thank you. You have already mentioned a few technologies that you are employing in the field and you are also looking into other technologies that might be useful. I you don't mind, tell us more about other technologies that you are exploring, like drones, satellites, anything like that?

Kelly Bott: We are designing a monitoring strategy that will be kind of multi-level. So, we're looking into the satellites. The capability of the satellites is looking for super emitters. We don't think we have those anywhere unless they're off in a place we haven't found yet. like away from our pads, or, you know, maybe there's a natural seep; we don't know. But that is something that we're pursuing. We're also looking into fixed-wing and drone flyovers; we've already done one, and we had a pilot do one at the southern end of our field just to test the capability. And so that's going to be a part of our strategy. We've got a company that's going to be doing some high-resolution imagery gathering, and they'll be looking for leaks when they do that flyover. So, we're starting that, I believe, this month. And then, of course, we've got the stationary monitoring and the handheld camera. So, it's going to be kind of a multi-level approach to making sure that we understand what's going on out there and that we're responsive to any issues that we find.

Soheil Saraji: So, how do you integrate all of this data? How do you store them anywhere? Do you have any in-house data storage that you use, or are there any third parties involved?

Kelly Bott: That's part of the pilot we're working on; we also have another pilot with a company called Validere Air. And Validere takes multiple data streams. So, they're pulling in everything that we use to build our calculated emission inventory. So, our equipment, our throughputs, and all of our emission factors that's one data stream; they're looking at data from monitors and are also technology agnostic. So, they're looking at multiple different sensor technologies that are coming in. And then they're also doing mass balance calculations. In terms of comparison, what is their mass balance, for example, and what is the emission inventory? Say, what do the stationary monitors say? And how far off are we? And do we need to pay attention to some of this stuff? Based on what they're seeing in the stationary data, they believe we're overreporting on the calculated emissions inventory side. but you're correct. I mean, we're still trying to figure out, like, what do we do? We have all of this data, and it is a lot of data. And how do we distill it down into something useful for somebody like me who just wants to pop it up? And look at it right now; I don't have a lot of time, and I spent a lot of time looking at it. But how do we consolidate all of that in one place? And so, So that's a big focus for us right now.

Soheil Saraji: Okay, this gives us a good segue to move towards regulatory aspects and standards, One of the things I learned from you is that there is not an established method or technique for tracking emissions. In your opinion, what is the role that the regulations or standards could play to just help you in the process?

Kelly Bott: That's a great segue because there's a coalition of folks that are starting to come together in the early stages, where we are working with all of the different monitoring providers and certification bodies including Project Canary, Equitable Origins, and MiQ. When it comes to certified gas, there is a big difference between a field-wide certification like MiQ and a well-by-well certification like Project Canary. But how do you standardize that so that we can all have the same label, which is certified gas? And so, we're working with that coalition called the Differentiated Gas Coordinating Council. And part of that discussion is, "How do we get the regulators involved so that they can kind of weigh in and help us define these standards and potentially adopt them through some regulatory mechanism in the future?" Those discussions are in their early stages, and I believe there is a lot of interest from various regulatory bodies to play a role and kind of bless this, but as with any new market, a lot of questions must be answered before we make any real visible progress.

Soheil Saraji: Exactly. I can emphasize that this is one of the most significant and impactful move from the regulators to resolve the standards and regulations around this topic. It opens up a lot of possibilities for not only your company but for basically any company to start moving in that direction. So, I'd like to bring up blockchain. I know that you are familiar with it. So, what do you think the blockchain can do in the ESG-related activities, or what are the advantages that blockchain can provide for any of the operations that you're working on?

Kelly Bott: For Pure West, I believe that blockchain will become very, very simple because when we talk about differentiated gas, whether it's responsibly sourced gas or some other attributes that we're attaching to it, we need to be able to have a mechanism that assures a buyer that they're buying it that this attribute, per se, hasn't been sold multiple times, that it's real, that it's valid, and that there was a unit of gas produced.

In connection with that, and from what I understand, though I'm no tech expert by any means, the cool thing about blockchain is that we can have a unit of gas that we can sell on the market. Then there are the environmental characteristics, which we may be able to attribute to a completely different person or entity. And so, and it's cool because we can, you know, have a molecule or a unit of gas coming from a platinum pad, and maybe we can attach methane performance to it, or we're one of the first companies in the Rockies to get our freshwater-friendly, verified attribute from Project Canary; that attribute can be attached to that blockchain token, and so, it's really exciting that we'll be able to do that.

Soheil Saraji: Yes, that is wonderful. I agree, there are a lot of other attributed that could be considered, as you mentioned earlier. One needs to consider the entire ecosystem. What do you think would be the challenges of getting Pure Wests from where you are to a place where you are a blockchain-friendly company? What are the barriers in your mind from here to that desired place?

Kelly Bott: I think we're close. We're working with a company to build out a system. I think the biggest challenge is not the technology or the kind of tokenization process. It's the marketplace, and we're getting buyers to understand what this is and how it can help them. I mean, it's kind of like the voluntary and regulatory carbon offset markets, trying to figure out how these tokens can help another company achieve their own sustainability goals.

Soheil Saraji: I've covered most of the questions that I had; there are a few left. We're just getting started at the university with a new course called "Blockchain in Energy." You work in the industry and are a potential employer. What do you want the students to learn in a course like this?

Kelly Bott: You know, one of the things that I like about recent college graduates is that they come up with innovations and ideas of ways that we can use that we haven't thought about. And so, you know, what I'd like the students to bring is, you know, how this technology can benefit us right now based on what we know about it and our field. But, you know, thinking about it from just an innovation perspective, I just think students bring a lot of energy and innovation, so the more you can get them thinking about ways to expand the use of or communicate the use of blockchain, I think, would be helpful.

Soheil Saraji: So, would you say if two students came to you for a job or internship, and one of them has passed the blockchain in energy course, Would you consider this an advantage for them?

Kelly Bott: Yeah, absolutely. In this energy transition thing, I think the first conversation that you and I had about blockchain was really about my introduction to the entire concept, right? So, I'm beginning to see this as a technology that will enable us to both commodify and communicate our environmental attributes. And that's something I hadn't considered, you know, being able to communicate, and I think of it like metadata, right? You have it; you can communicate to a buyer that this is all the great work we've done with this production and using it as a communication tool is cool.

Soheil Saraji: Yes, that is a big factor. Is there anything that I didn't ask that you want to talk about? Or are there any other topics you'd like to discuss?

Kelly Bott: No, I think the only thing that I would mention is that companies like PureWest exist, and I think there are a lot of us out there; I mean, we're looking at not only ways to reduce our environmental footprint but to get more involved in the carbon economy. So, we're looking at, and I know that we're not the only ones looking at, ways that we can develop our carbon offsets and things that we can do to get involved in carbon capture or, you know, really anything. So, when I was talking about the overall system's thinking, I mentioned that we're currently conducting a pilot in the middle of our field to look at reclamation practices and soil. And so, as part of that project, we're looking at soil amendments. We're working with the University of Wyoming on a product called "Ginate," which is a soil amendment but also a potential carbon offset for us. And so, when we talk about the system thinking, I mean, it's real. We're looking at the carbon uptake capacity of soils, we're looking at how we plant, and we're looking at pollinators. I mean, you name it, but at the end of the day, there's always a carbon economy and it is tied to everything we do.

Interview 7: Sriram Srinivasan

Interviewee: Sriram Srinivasan, Chair of Mining & Minerals ESG WG, OriginBX

Interviewer: Dr. Soheil Saraji, Associate Professor of Energy and Petroleum Engineering, University of Wyoming

Interview Date: 14, Sept. 2022

******The interview was conducted after a Discussion Panel on Measuring, Reporting, Verifying, and Tracking using Blockchain at Blockchain in Oil and Gas Conference, Houston, TX, September 14th. The initial part of this conversation was not recorded. However, this is a thought-provoking interview and we decided to include it as is.*

Sriram Srinivasan: So, what I was trying to say is that most transactions today are between buyers and sellers. If you look at someone buying raw materials and then turning those raw materials into finished products, that might take two or three steps. So, there is a source or raw materials, there is likely a trader, a refiner, and a contract manufacturer. In that supply chain, the buyers would like to know where their inventory is. For that you need information to be shared by all of the parties. So IoT signals aren't very useful on their own unless that data is going to be shared or used in another way. To share data you need a system that protects your privacy. This system needs to interoperate with other blockchains since Exxon will soon have its own blockchain. Everyone will likely have a blockchain integration of their own. It's not clear that everyone uses the same blockchain. From what we know about history, that may not be the case. So, you need a way for blockchains to talk to each other. This could mean that large refiners each have their own blockchain, buyers have their own etc. So, there needs to be a way for blockchains to talk to each other. That's where the IBC (Inter Blockchain Protocol) comes in. Different blockchains have thought about it differently, the inter-ledger foundation has spent a lot of time thinking about it, and how all of us will build technology that essentially is interoperable, native. So, if we interoperate, the next thing we need is standards, right? If I'm going to go and pick some information from you and the other person, and they're all in different standards. We have to start writing adapters.

Soheil Saraji: Yes, I understand. However, how does this data get from the field to the blockchain? This could be done by a person manually entering the data, but it is impractical for a long period of time. Eventually, we need to have sensors, correct? This sensor could be completely independent of the blockchain. For example, the information could be entered into a database. Nevertheless, there are a lot of steps involved. The concept is comparable to connecting trusted IoT devices to the blockchain.

As someone responsible for IoT, they will likely have multiple IoT arrays. If some sensors go completed offline, they should be able to replace them without losing data. In a supply chain system, the data should be transferred from one party to another down the chain. There are multiple parties involved and each need to have their own sensors. This way we can follow the product/commodity down the chain with timestamp attached to each transfer. I'm working on a project with the goal of tracking carbon down the chain from capture point, to transport to underground storage. This may involve multiple parties who transfer chain of custody along the process. This is a good application of blockchain.

Sriram Srinivasan: So let me ask you, if there are three entities that are going to join on your system. Why do you need a blockchain?

Soheil Saraji: Good point.

Sriram Srinivasan: What I'm saying is that today, SAAS systems exist and people and their suppliers work on the same system. If I can get people to adopt a system, the question is, why would I want to have them use a blockchain? So you got to think about that right now. Now if the IoT data that I'm producing, I want to share with more than one party, instead of creating separate systems for each party, I have a generic system so then you get into standards. Meaning if I'm going to just work with you, then we can all use a database. But if I am getting data, not just from you to recapture but from somebody else and from a third person, then I need to build systems that get from all three people. If there was a standard and I just adopted to that standard than any new party can just send me the information and I can send information out. So what I'm trying to make is a case for why blockchain, and if you think blockchain, you automatically need standards, otherwise the blockchain becomes not terribly useful.

When you take blockchain and you think standards then you recognize in the world today there are many different blockchains and there are many different companies making independent decisions on which blockchain is the one, so you need interoperability between those. If you think of a petroleum use case, you will first ask the question, why? Why can not they use a database? And if you're using a blockchain, how many parties are there going to be communicating? And what would be the standards? By being in academia, you could go get the data and then formulate your understanding of three or four or five different types of data communication that exists today, what might be a standard? and then that begins. So if you produce a paper on standardized documentation required for direct carbon capture, the market right now needs those because there are so many different bits and pieces. Then, what we who are not part of the oil and gas community, but are in the world of trade working with banks, working with large businesses, can then build adapters to that standard and then financiers can then finance. The banks are using their own blockchains their own systems. And in large enterprises, they're so large, different groups are using different systems. So, how do you bring it on entering this new world where there needs to be the right kinds of incentive, right? So it's protocols, protocols of communication, and standards of data transmission.

Soheil Saraji: Okay. I completely agree. What do you guys do in Origin BX? Is there any room for academia-industry collaboration?

Sriram Srinivasan: So, origin is a sandbox. To the extent that it needs data from industrial participants, and then it needs the ability to formalize that many standards. We've been working with and have relationships with Stanford University. We've also got a relationship with the value chain business school at the University of Michigan Ross, and we recently put a paper together on deep tier finance with the University of Michigan Ross School of Business. We also work with the Georgetown University. We're putting together another paper with an academic group on cash flow analysis within a supply chain and being able to sit back and ask given traditional flows in different industries; if you look at the research, there's a website called spendmatters.com. If you go to that website, you'll notice that they recently published an article about the outstanding number of days for large buyers and what happened between last year and this year with some of the large companies in the US supply chain finance where they increased the number of days. So, I think the average is 111 days. Thus, the buyer buys from the seller using a supply chain program and basically ends up paying 111 days after they make the purchase. I think last year it was really close to 90 days, and it's increasing. Now, what is happening with supply chain issues is that suppliers are not getting paid by the bank on time. However, suppliers are now in a position of strength, and they may not be willing to do that. And so that might increase the cost of goods all across the chain. So those are the kinds of things modeling out cash flow could do. I don't know the movement of cash in the world of petroleum, but if you just look at a high-level view, the killer blockchain application has always been finance. If you look at buyers of petroleum they essentially have lots of letters of credit. And then when they buy oil and the oil is on the ship, maybe title transfers on the ship regularly. So I think that data is important. But the thing that lubricates all of this is finances. From an academic prospective, if there is, I don't know some work done on how is it being financed? Who are the players, what is some transactions? How long does an invoice take to get paid? What is the cost of capital? Where are the stronger companies with stronger balance sheets and the weeker balance sheets? Where is the arbitrage? I think you will find that implementation of blockchain, when you understand that better, suddenly starts to make a lot of sense.

And as a big part of the petroleum world, you have essentially the price of oil that fluctuates, but you've got a lot of other costs across the supply chain. And if there is a way for example, to maybe have the intermediaries be on what they call Cash to Cash basis or positive Cash Conversion Cycle CCC it's going to be interesting. I buy from you. I do something and then I sell it and I get paid and if I can do all of it in a positive cash cycle, I don't need as much working capital. I can increase the capacity of my trading using my blockchain facilities. And I can tap my buyer's, could be a strong buyer, balance sheet to assist with this process. And effectively allow me just to bring my value added without having to be worried about working capital, cost of goods, and things like that, my margin score, then you get paid earlier. They get more visibility, kind of it's like a win win situation. It's a different way of looking at blockchain. So then incentive structures are in place for people to provide that as opposed to saying, Oh, the data is going to help me here and I can do this. I can do accounting. Really, if I was a businessman, I'm making money. So then if you go from that, oh, you use this to make money, it changes the conversation.

Soheil Saraji: I see. So, in your opinion, probably the first application that blockchain could have in energy would be on the financial side. Right?

Sriram Srinivasan: The killer application of the internet was email. When we consider all of the different forms of communication, such as WeChat, WhatsApp, and others, we tend to think of email as if it hasn't changed over the years. In fact, that is absolutely true. I mean, different variations of email have some sort of communication between parties. The killer application of blockchain is finance. You'll see that often applications that become successful will typically be variations of the killer appliation. So you have to think about in the petroleum business, there is data there is process and then there is payment, typically in large companies treasury and procurement are in two silos. What a blockchain does is finally figures out hey, wait a minute, there is a real linkage between those two and in those 30 days and 60 days on those payment cycles, the linkage happens. And many of the deals that happen today happen because of the nature of our intermediaries need to charge they need to charge because they got to hold that inventory for some time. And then there is a risk associated with this transaction and you know, cost of the risk is added to the cost of the goods. So blockchain does not remove us. What it does is it makes the risk transparently known in a suitable way so that the right parties can take the course. So eventually the killer app for the blockchain I think is the is the finance which then brings across the databases.

Soheil Saraji: Well, it was great to get your perspective on that as well. That's very different from everything else that I've heard today because everyone is looking for applications other than financing.

Interview 8: Steve Swanson

Interviewee: Steve Swanson, CEO, North Shore Energy LLC
Interviewer: Dr. Soheil Saraji, Associate Professor of Energy and Petroleum Engineering, University of Wyoming
Interview Date: 8 July 2022

Soheil Saraji: For the purpose of this interview, would you give us a short information about your background and how did you end up at the place that you're at right now?

Steve Swanson: Sure. My name is Steve Swanson. My educational background is a double degree from the University of Michigan in economics and an honors degree in philosophy. And then, I went on to graduate school and got a law degree and an MBA at the University of Denver. I've been in the oil and gas business for over 30 years as an independent. We've run a series of six companies for a private equity group out in New York. This is the sixth, and our assets and focus are primarily on southwest Wyoming almost exclusively. And these are large legacy fields that were the core assets for many of the major US oil companies back in the 1980s and early 90 s, that have moved their way down the food chain as assets in oil and gas often do. And they've ended up with us during the period where we're looking to repurpose them for a lower carbon footprint. So basically, taking legacy conventional oil and gas production and combining that with carbon capture and sequestration, so that we can reduce the carbon intensity and carbon footprint of traditional oil and gas assets.

Soheil Saraji: Great. So, I have a list of questions that I can go through, but I don't want to keep it rigid, so we would go wherever the conversation takes us. You're the right person to talk to because you have a lot of experience in the oil and gas industry, and you are one of the people that recognizes the energy transition happening and that the oil and gas industry has a part to play. So maybe we start with the big picture and then we go down to what are the changes that you think the energy transition will force us as an industry in oil and gas to undergo or what are the opportunities that this transition provides for our industry to play a role in.

Steve Swanson: Well, yeah, we always must look at adversity as an opportunity. Right. And so, you know, we're faced with a little bit of adversity because the politics today are such that renewables are in favor and hydrocarbons are in disfavor. And so, as part of the analysis, both economically and politically, I think we're forced to find a bridge between the two. And that bridge is basically this transition that you're referring to. And so, we're trying to figure out how to maintain our existing hydrocarbon production, but have it be less carbon-intensive. It's almost a disconnect, because at the core, we really are producing hydrocarbons, and they are very efficient. And we've had a good 100-plus years of relying on and refining our experience with hydrocarbons to make them even more efficient. So, we have a whole infrastructure in place that would be hard to just throw out altogether. And so those of us in the industry in hydrocarbon industry are trying to figure out how it is that we can maintain our production of those hydrocarbons but have a lower carbon footprint. Basically, that's the theme and in the hopes that the combination of those two things hydrocarbon production, plus a zero carbon or low carbon footprint, will equate over time with renewables. And so, the way that we've approached that is to look hard and try to deploy carbon capture and sequestration. And the reason we do that is that oil and gas-depleted reservoirs are ideally suited for carbon capture and sequestration. So, if you're looking to make a material impact on taking CO_2 out of the air, and storing it someplace, these reservoirs are very large. The ones that we own are mostly depleted, so they're at low pressure, and easy to sequester CO_2. And so, you know, that's the logical place for an oil and gas producer to look, right? We can do things around the margin like reduce our methane leaks. We can emit less CO_2, we can use waste, heat, and things like that. That will make us a little bit more efficient and a little bit better carbon intensity, but those are things that we spend time doing in our normal course of business. So, I think there's a misconception that oil and gas producers are indifferent to whether there are leaks and environmental damage. That's not true. I think across the board, in my experience, everybody is a steward of you know, the resources and the environment. And it doesn't make sense that we would just be willing to let our valuable products just leak into the air and not get any revenue for it. So, you know, we're sensitive to those types of things around the periphery anyway. But for us to be part of this transition, I think we must do something different and substantive. And I think carbon capture and sequestration fall into that category. So that pathway is the one we've chosen to focus on. It's a combination of our reservoirs being ideally suited for CCS (Carbon Capture ad Storage), but also the fact that we have all the infrastructure in place to do a CCS project. We don't have to start from scratch. In other words, we can repurpose our existing oil and gas assets to incorporate CCS in them and not have to incur huge upfront capital costs to convert them.

Unlike maybe a coal plant that must add an entire carbon capture system before they can accomplish a CCS project, we operate and own large legacy gas processing plants where we can incorporate CCS into the process and sequester the CO_2 directly underground. And so, the logistics for our project are much better than others because we don't have to lay a pipeline to ship CO_2. It's just right there. And so, if there's going to be this transition, and if CCS is part of that transition, I think you have to start with projects that are the least intrusive, and the least logistically challenged. So that's why we've focused heavily on CCS at our existing plants because we think they accomplish both of those purposes.

Soheil Saraji: Right. I agree with you on that point. But I want to learn a bit more about the source of your CO_2 gas processing plant. What is the purpose of the process?

Steve Swanson: So, we produce natural gas, but we sell it into a common carrier pipeline. And so, our job is to produce it. We're an upstream oil and gas producer, so we produce and sell it. We try and cannibalize the least amount possible. But your question is a good one, where's the CO_2 coming from? And so, there are two models for CO_2 sequestration. The first is an aggregation model where you go out and find various emitters of CO_2, like ethanol plants, for example. And then you capture the CO_2 at that point source, and then you ship it via a pipeline that typically you must put in, and you bring it to a common sequestration site. The second model, the one that we ascribe to is a single-point source CO_2 emission that sits on top of the sequestration reservoirs. And we pick that model. For obvious reasons, the logistics challenges are nonexistent because we sit on top of the reservoir so we already have the infrastructure in place. Second, we don't have to deal with multiple third parties to acquire the CO_2. And so those two things not having to lay all that pipeline, which, you know, generates issues with time and money for regulators and private enterprises and multiple ownership groups. So that's a very difficult thing. I refer to it as herding cats. And it's very difficult to accomplish that in a reasonable period. The point source CO_2 generation, which is what we look at, solves those problems. But then you must figure out a way to generate the CO_2. And so, for us, we're looking at putting a different CO_2 process on our plant to produce a different commodity. So, in our case, we're looking at producing either hydrogen ammonia or fertilizer urea, and we don't do that currently, but we can use the natural gas that we produce in combination with an industrial process that generates CO_2 as part of that commodity generating process, whether it's hydrogen ammonia or urea. So, steam methane reforming or auto-thermal reforming are industrial processes that you must use to generate those outputs. And so, when we go through the process of building a project, for those outputs, inherent in that is determining an industrial process that can make those outputs but also generate CO_2, and so that CO_2 we incorporate into our process was captured and sequestered on site. So, we'll be able to produce those outputs hydrogen, ammonia, or urea, with what's virtually a zero-carbon footprint making them blue instead of brown or black. We don't source wind or solar for the power necessary to run those processes. And so, we're not green in that sense. But as we talked about this transition, I think it's critical for anybody who's a realist to see that you're going to have a blue successful project before you have a green successful project.

Soheil Saraji: Yeah, I agree. This is a transition, as the name implies, and during the transition, different things would be important at different stages. Now I understand what you are planning to do in terms of the source of CO_2, so you don't have to ship the CO_2 from somewhere else. So, another bigger picture question is, what do we need to do to make CO_2 sequestration widespread? When I was a Ph.D. student, I researched this topic, and I worked on different aspects of CO_2 storage from EOR (Enhanced Oil Recovery) to storage. To me, it seems that implementing the science and technology involved in CCS is not a problem. Please correct me if I'm wrong, but why has carbon sequestration not become a large industry yet?

Steve Swanson: I agree with you completely. We have the technical ability. We have the willpower. We have the start of some economic incentives. But the big mile, the big roadblock, and the big hurdle to get over are regulatory. And so, as an independent producer, or really anybody who is interested in sequestration, if you're going to use oil and gas, and depleted reservoirs to store the CO_2, you must work with the existing regulatory framework. And so, there are two different types of wells that are permitted for sequestration; a class two well, which is ubiquitous. They're used for enhanced oil recovery in the oil and gas business. And it's generally run through the state's oil and gas commission. That well-understood routine regulatory environment that we're comfortable operating in. The other option for permanent sequestration is through what's called a class six injector Well, that's used for permanent storage. And the class six wells are not well understood. They're not. There are only really two functioning class six wells in the country right now. And the reason why that's the case is there are two hurdles that you have to get over to get a class six permit. The first is a technical analysis that's related to the suitability of the reservoirs and basically all the things that are below the surface, and whether it's suitable for permanent sequestration. And that's been defined generally in several different ways. But you're basically looking at 100 years to make sure that the CO_2 that you inject underground can be stored in a way that it doesn't migrate or leak for a long period of time.

Soheil Saraji: How do you achieve that If I may ask? Is it only by applying models? or do you have sensors on the field? how does that work?

Steve Swanson: Yeah, both. So, you're required to have a static and a dynamic model, to begin with. So, you can track the theoretical migration of the CO_2 plume downhole, but on top of that, you must have an MVR program, which is a long-term monitoring program of that CO_2 plume. So that in real-time or close to real-time you can understand its movements, and then its likelihood of going to places that you don't want it to be, and most of the concern is over. Drinking water aquifers don't want to be contaminated with CO_2. So, as I was saying, there are really two major hurdles. The first one is that that's a hurdle that I think we can get over from a practical point of view. Unfortunately, this technical analysis and suitability of the downhole environment for CO_2 sequestration has been handled by the EPA on a federal level. It is now starting to move to the state level; each state can apply for primacy. Wyoming and North Dakota as we speak, are the only two that have been granted primacy and North Dakota is the only one that's issued class six permit to date. But we, for example, are the first ones to apply for a class six permit in Wyoming. So, we're kind of on the front edge of that and I think we're working our way through the technical review period of that. Wyoming has delegated the regulatory authority for the class six well to the DEQ, Department of Environmental Quality, instead of the oil and gas commission, who regulates the class of wells that I referred to earlier. Okay, so you must stand up an entirely new, regulatory body to deal with this, and then DEQ is an existing body, but they deal mostly with air and water pollution, not CO_2 sequestration. And so over time, hopefully, they'll be experienced enough to review these but it's taking a while for them to get up to speed like the oil and gas commission would.

Soheil Saraji: There is an interest in the state for boosting sequestration activities and I'm sure they are probably working on it but as you mentioned, anything like this will take some time to hire the right people and you know, to put the regulation in place.

Steve Swanson: Yeah. So as far as the technical review goes, I think that that's an easy path to follow. I don't expect it to be dramatic in terms of the time or effort that it takes to get through a technical review. The other hurdle is much larger and potentially fatal to CCS. And that is the financial assurance hurdle. So, there's a two-prong test for these classes. The first is the technical review and I think we can say that that's in hand. The financial assurance piece is not in hand.

It is a serious issue as we speak, and I think the resolution of it will hold the balance and the success or failure of CCS as an industry.

Soheil Saraji: Let me try to expand on it. Is this like insurance that somebody needs to lock some money in, in case anything happens? Is this the money to be used to recover from the disaster or something like that?

Steve Swanson: Yeah, in simple terms, that's exactly right. The regulator wants to make sure that if there is the damage of whatever nature, but mostly environmental damage under that general definition, there are funds available to remediate it. And so, that's no different really than the environmental regulatory environment that we operate in 24/7. So that alone is not the problem. The problem is that when you run your models, and you're looking at permanent sequestration over a long period of time or decades, up to 100 years, and when you start stress testing those models by assuming the worst event possible at the beginning period and extrapolate that out throughout the time of the model. You create an environment where your financial assurance is well beyond the capability of anybody to satisfy. So, for example, hard numbers, the EPA when they granted two class six permits to Archer Daniels Midland back 10–15 years ago, their model generated a financial requirement of $40 million per well. And, in fact, I gave this presentation at a WEA (Wyoming Energy Authority) conference. You may have been there. You know, even ADM, one of the largest publicly listed industrial commodity companies in the world still couldn't satisfy the requirements of the EPA, which had to be modified to accommodate them so they could get approval for that class six permits. The point is that if the states when they received primacy, follow the EPA model for identifying and quantifying financial assurance only major oil companies would be able to qualify. And so, the number of successful permanent storage through class six injector wells will be extremely limited. Right. And our experience so far is not promising. So, as I said, North Dakota has issued one class six permit, and the financial assurance obligation, although not $40 million from EPA is $20 million. And that's what had to be posted in terms of insurance or bonding or surety. Those typical types of things that we use to cover financial obligations into the future had to be met for $20 million as well. And, you know, speaking from our own perspective, as an independent, I can say that that's a chilling effect on our being able to pursue CCS. So, getting back to your original question about why it's not more popular or more ubiquitous in the world, that's the single reason that has kept the industry from sequestering more CO_2.

Soheil Saraji: Right. So, is there any economy of scale with it? If you have multiple wells in one region, can you get the financial assurance to become lower, or is it just fixed per well financial assurance?

Steve Swanson: Well, that chapter has yet to be written because we only have three permits ever issued. Right? So, we don't have a very good sample size for our data analysis at this point. But so far, it appears that financial assurance of at least \$20 in the \$20 to \$40 million range is per well, and there is no benefit from having multiple wells operating.

Soheil Saraji: Right. This is exactly why I connected with you at the WEA conference because we share almost the same perspectives. I always ask this question, why don't we get to the next step in CCS? It seems like there are lots of hurdles that are non-technical that cause the delay.

Steve Swanson: One of the things that we could look at as a resolution to this roadblock is to get out of this mentality that it is frozen. And we must brainstorm about what a possible solution is. And so, one of the things that we've proposed is that we segregate the periods of time and identify risks in each of those periods of time. So, if we have a 100-year total period, let's look at maybe 10-year or five-year slices of that. And determine what the risk is in our model for each of those periods of time, and then try to match the financial requirements to the risk levels. Right. So, you know, that's a practical approach to what seems to be a very difficult and maybe unsolvable problem. And the reason we propose that is not only to break the ice here and try and move forward but also because it reflects that in the initial years of sequestration, you have an extremely low risk of a bad event. In our case, we have mostly depleted very low-pressure reservoirs that have held CO_2 and other gases, nitrogen, and methane, for centuries. And so, for us to fill those bathtubs with CO_2, it's going to take many years before we can start even building pressure, and it's the pressure that drives the bad events. Pressure is going to drive a plume to where you don't want it to be. The pressure is going to drive the CO_2 plume to an aquifer. So, you must parse through what these risks are, and identify them in the years or decades, where they're very small. Our argument is that the risk is so small, and you should have a similar financial assurance package related to that, right? And so, as risks grow over time, and maybe you're in 50 or 60 years, and you're filling up your bathtub and the pressures are higher, and the likelihood of a bad event starts to increase. At that point, it's reasonable to assign a higher financial assurance to that. And so, that's one approach that is very reasonable. But you know, you're dealing in a regulatory environment that is generally not friendly to oil and gas. You're looking at, you know, a political environment that's the same, but also you know, given that we have a precedent with the EPA, and they have models that they have used in the past, we're having to break a mold here to do something that hasn't been accepted. And so, we haven't had this to date. We haven't seen any embracing of a concept like that. And I'm not suggesting this is the only solution to the problem. It's just a practical solution to the problem. But from what we've seen so far, I'm not encouraged.

Soheil Saraji: Right. This seems to be a very big part of the hurdles in carbon sequestration. But let's say there is a solution, and we solve that problem at some point. The next thing that could be a hurdle is the economy, right? Because you are spending more money than in a regular operation to capture your CO_2 and put it on the ground. In your case, it's probably less of a problem because you have one location, and you don't have to deal with transportation and many other issues. Within your own model of operation, how does the economy work? Do you think that the product added value and the fact that your product is low carbon intensity would bring some extra finances to the project?

Steve Swanson: It's possible. We haven't seen that reflected in the market price of these outputs yet. There's a lot of talk about that. But the reality is that blue ammonia is priced the same as black ammonia or brown ammonia or even green ammonia, it doesn't really matter because the same carbon offsets market that's potentially starting to impact that, but we haven't seen it yet. In terms of the cost, that's one of the reasons why we've taken the path we have because we wanted to reduce the operating expenses for sequestering CO_2 as much as possible. So potentially, we could break even with tax credits that are provided through 45Q, right, with the IRS, and we're much closer than anybody else. And so, you know, we're hopeful again, this is a legislative package that, you know, everyone in the industry is hopeful that the 45Q tax credit rates will increase, but if they do, we're fine. We would have covered all our operating costs. And so, your original question about what the economy is doing and whether it needs to be at a high level was not to justify the CO_2 sequestration. I think we can safely say that our model works regardless of the economic activity level. Now, I should point out that if you're not doing permanent sequestration, and you're using the CO_2 for enhanced oil recovery, which is really the only other alternative that is very heavily dependent on the economy and economic activity because the EOR projects won't take place with oil below $60 or $70 (a barrel). Right. It's just recently the oil has gone up over $100, and it justifies those kinds of programs. But given the amount of time we've spent, especially recently at very low levels of crude oil prices, to have a long-term project in place. It's going to take sustained oil prices more than that $60 or $70 (a barrel) level before companies start to plan for these long-term projects. So that hasn't happened yet. Although the economics are there for these EOR projects today, and that is a legitimate sequestration activity. Just not as good as I think. The optics are not as good as permanent sequestration.

Soheil Saraji: Do you also think that you could deploy EOR in your property?

Steve Swanson: Yes. So, we have two sets of properties. One is on the very west-southwest side of the state. It's called the overthrust belt by Evanston and that's where we have our current project, where we got the WEA grant funding for the permanent sequestration. We own another set of properties in the Green River basin or by Rock Springs, and our project there could be a EOR project or permanent sequestration. So yes, we've looked at both.

Soheil Saraji: Well, maybe it would be a good time for me to discuss a project that I'm working on regarding blockchain and how it can play a role in this environment. So, we started looking into carbon credits and offset markets around the world. The European Union has its own scheme. California has a cap and trade scheme. There is even a possibility that the federal government at some point will implement some sort of program in this space. Whether they consider that or not there is a market coming up for carbon offsets/credits. The buyers are typically companies that are pledging to go carbon neutral, like Google, Microsoft, and some oil companies like Occidental Petroleum. So, there are lots of operations you can't use renewable energy for today. It's impossible. We don't have the storage capacity required. Integrating fossil fuel with carbon offsets might be a path forward. There are some projects in the voluntary markets to develop agricultural or plantation-based carbon offsets. Those are very hard to standardize and monitor. In my opinion, one of the biggest potentials in space could be underground carbon storage. Because of all the EPA requirements, we know exactly how much CO_2 goes underground and they're being monitored for years to come. If we find a good way to develop standardization and workflow for measuring, monitoring, and tracking a carbon sequestration project, it could provide the required volume and accuracy for generating trusted carbon offsets and credits.

This is where Blockchain can play a key role as a distributed ledger technology. If we use blockchain-based technology to develop a carbon accounting or carbon monitoring database for a project, like, say, your project, we could connect all the monitoring devices back to the blockchain and know exactly how much net carbon is being stored. The next step is to digitize that stored carbon in the form of carbon tokens and bring them to the market. This can provide extra financial incentives for carbon storage projects that could potentially be competitive (or even more significant) to the 45Q. What do you think about this project?

Steve Swanson: We will volunteer in a heartbeat if they want to use our site and our project to work out the kinks in that. We'd be happy to volunteer.

Soheil Saraji: Perfect. It's very good to know that. We are starting the project, and I would like to just keep in touch with you. Are you doing any sequestration operations at this point?

Steve Swanson: No, we are not. And so, the grant from the WEA that we got is for the feasibility of a CCS project. And so we just closed with them last month and so we're now engaging third parties to help us do that feasibility study. One thing that is different for us, because we have existing operations out there and we are currently an oil and gas operator, is that we can bring a project to deployment much faster than anybody else. Right. And so, we're expecting to have some type of deployment within two years. The timing is good for us to participate in something in terms of the blockchain and its application to accounting for those CO_2 volumes and connected directly to our long-term monitoring programs. So, you can see that we're sequestering certain volumes which are easy to keep track of and that there aren't any leaks or offsets to that.

Soheil Saraji: Exactly. And you know, the final product that I have in mind is something that has minimal interaction with humans. We are working with some IoT devices that we could put in pipes, and the data goes directly on the blockchain. Nobody interacts with the device so nobody is there to tamper with the data. Creating some standards and a use case or rather a pilot test is what I think is missing.

Steve Swanson: I agree. You're doing the right thing and you need a place to deploy it. And so, you should put it out on our project.

Soheil Saraji: Absolutely. That will be great. We'll keep in touch about that. We have gone through most of my questions so far. There are only a few left. If everything goes well, some people suggested that we can create a carbon economy. Meaning that we can treat carbon as a commodity, such that if a product has more carbon in it, it has less value compared to products that have less carbon if we can track it accurately. What do you think about this model?

Steve Swanson: It's not well thought through, because carbon by itself is very valuable. Carbon has been proven to be a very efficient energy source. And I just disagree with the commentary that we ought to focus on the carbon itself. We shouldn't just throw out the baby with the bathwater. I mean, I just don't understand that. The carbon we should use and take advantage of its benefits. And if it has disadvantages, then we ought to deal with those separately. And so, I think the right approach is to deal with the emissions and sequester them. And then effectively have a net zero carbon footprint, but you give the advantages of the carbon.

Soheil Saraji: Okay, I have a question here. I did a quick search, and I found you are a part of Clean Energy Systems. Could you please tell us what that project is and how it's related to the activities you're working on?

Steve Swanson: So, I'm not an owner or part of the management team at clean energy systems. I sit on the board of directors. And their project in California is a conversion of a biomass plant, a shuttered biomass plant, kind of regenerating it so that it can use biomass as the input for the industrial process. In this case, it's oxy-combustion, which has a byproduct of pure CO_2. So, it's a power generation process that then is combined with CCS. And so, it's carbon negative because they're using renewable fuels, generating power, and sequestering the CO_2, so it's very slick. And they've got a couple of projects in California right now. Some big-name partners have signed on at this initial stage. And so, it's very promising. It goes a step beyond what we're doing. Because we're transitioning from conventional hydrocarbon production. They're starting with a biomass plant and then doing CCS on top of that.

Soheil Saraji: Oh, interesting. I just read a little bit about it. Sounds like something that holds a great future. Is there any question that I didn't ask you that you think I should have on this topic? Are there any comments that you want to share?

Steve Swanson: My only comment is the same one I made at the end of my presentation at the WEA. I think that you ought to hand out class six permits to everyone that knocks on the door. I don't think there should be any limitations at all. You should just give them out. And the reason is that there are hurdles enough within the process that solve any concerns you might have. For example, it costs us $10 million to drill a class six well, because CO_2 is an acid gas, and so you must have special metallurgy. And in these deep wells, it's very expensive. And so, you know, people aren't going to abuse the class six permit if they must go through all this. So, the fact that we're doing an analysis of the technical review for Class six is totally fine. Does anyone object to that in the industry? No one objects to that at all. The single objection is the financial hurdle that's put in place that's artificially creating this logjam of potential permit applications and potentially having a chilling effect that would destroy the industry before it gets started. I think the immediate solution, which has no bad consequences for anybody, is to just hand them out. If you've passed the technical review, have at it, here's the permit. Go ahead and spend your money. Try and make it commercial and if some CO_2 leaks at the end, then you must pay penalties. But that's just how we operate in our normal course of business anyway. So, I think that what is potentially tragic for CCS becoming an industry is a problem that shouldn't exist. That's a false problem. It's a false narrative. There isn't that kind of risk that people are seeing, and to hold back permanent sequestration, for that reason, is one of the definitions of insanity. And so, you know, the other finishing point that I'll make while we're on the talk here, is that four million tonnes of CO_2 have been emitted into the air, just since you and I started talking. And how much of that's been sequestered? None of it. And so, the point is also that while you're focused on the risk of migration of the plume or risk of coming into an aquifer, you know, the fact is that by doing nothing, the risks are much higher.

Closing Thoughts

We have learned a lot working on this book.

Together, we have explored the oil and gas industry's many initiatives for the climate problem, from carbon capture to eliminating methane emissions to sustainable fuels and plastics. We have spoken with people in the industry who are working seriously on these and other efforts to reduce emissions. Finally, we've looked at how, with the help of the blockchain, these initiatives could fundamentally transform the industry. Altogether, this will make the oil and gas industry more sustainable, and in more ways than one.

The transition to a zero-carbon economy creates new opportunities for the industry to get on a more solid financial footing. It could diversify its product mix beyond just commodity oil and gas into low-carbon energy. Industry participants, especially small and mid-sized ones, could differentiate their products based on embedded emissions or carbon intensity. This could, in turn, lead to premium prices and long-term purchase agreements, which would produce stable revenue streams that lock in profits and secure more favorable terms from lenders. Thus, the industry should embrace the energy transition not just to benefit the environment but also its own balance sheets and investors.

This would also mean a transition of the oil and gas industry from one of physical commodities to digital products. The liquids and gasses that get loaded into tankers or flow through pipelines may still be the same, but the digital certificates of origin and carbon intensity will be the ones that determine their value. Low or zero-carbon products with proven origins will command higher prices from customers and get better terms from investors and lenders.

This brings us to the blockchain. The past year (2022) showed us that it's certainly not magic pixie dust on its own. Rather, it's a technology for collaboration without needing a centralized authority, such as a government regulator. This means that it could enable the industry to prove the climate benefits of its new products. It could then help raise capital, track production across multiple jurisdictions, establish self-regulatory regimes, and create new markets.

S. Saraji and S. Chen, *Sustainable Oil and Gas Using Blockchain*, Lecture Notes in Energy 98, https://doi.org/10.1007/978-3-031-30697-6

For example, let's say that we accept that our advanced society needs reliable energy and that rising living standards in developing economies means we must produce more energy, not less. At the same time, we accept that climate change is real and must stop emitting the greenhouse gasses that cause it. So now comes the hard part: What is a real plan that would provide for our energy needs while meeting the world's climate goals?

The oil and gas industry may argue that solar panels and lithium-ion batteries alone could not meet the world's energy demand, so natural gas is needed. But unless it could put forth a *credible* climate strategy, who would believe it?

Next, let's say the oil and gas industry puts forth a proposal for "carbon-neutral natural gas." It will minimize the emissions footprint from producing natural gas, including eliminating methane flaring and venting and then removing the emissions from burning natural gas with carbon capture and storage. If the industry tries the usual "Trust us" or "Trust the government," environmentalists would predictably protest "greenwashing" and "lobbying." Investors, banks, and the general public would probably be skeptical at best of the industry's claims. So instead of the usual, how about the industry say, "Let's figure this out together, based on open standards and equal access to the data."

This is where the blockchain comes in. All the stakeholders, including the oil and gas industry, environmentalists, banks, investors, and government agencies, could jointly define a standard for carbon-neutral natural gas together by answering questions such as "Would the methane leakage in natural gas production and transportation be added as emissions?" "Would CO_2 removed by carbon capture and storage from natural gas combustion be deducted from emissions?" Even controversial questions such as "What carbon credits, if any, could be counted against natural gas emissions, and for how long?"

Once a standard is established, it could be coded on the blockchain and used to certify the carbon-neutral natural gas as it is produced. All parties could have equal access to both the standard and all the data needed to verify its practice. Further, the blockchain could transfer the emissions reduction from buyers, support long-term offtake agreements, and structure financing transactions. All of this could be done at a fraction of the cost of traditional processes.

Wishful thinking? Perhaps. But didn't you just meet, in the course of reading this book, people in the oil and gas industry who are working on real projects to reduce emissions? If they are serious about fixing the climate problem, shouldn't we join them? Or do we need a reminder from Benjamin Franklin, who famously said back in 1776: "*We must all hang together or surely we will hang separately.*"